Beyond the Sacred Forest

New Ecologies for
the Twenty-First Century

SERIES EDITORS
Arturo Escobar, University of North Carolina, Chapel Hill
Dianne Rocheleau, Clark University

Beyond the Sacred Forest

COMPLICATING CONSERVATION IN SOUTHEAST ASIA

EDITED BY

Michael R. Dove,

Percy E. Sajise, and

Amity A. Doolittle

Duke University Press   Durham and London   2011

© 2011 Duke University Press

All rights reserved

Designed by Heather Hensley

Typeset in Minion Pro by Keystone Typesetting, Inc.

Library of Congress Cataloging-in-Publication Data
appear on the last printed page of this book.

CONTENTS

ABOUT THE SERIES

This series addresses two trends: critical conversations in academic fields about nature, sustainability, globalization, and culture—including constructive engagements between the natural, social, and human sciences—and intellectual and political conversations among social movements and other nonacademic knowledge producers about alternative practices and socionatural worlds. Its objective is to establish a synergy between these theoretical and political developments in both academic and nonacademic arenas. This synergy is a sine qua non for new thinking about the real promise of emergent ecologies. The series includes works that envision more lasting and just ways of being-in-place and being-in-networks with a diversity of humans and other living and nonliving beings.

"New Ecologies for the Twenty-First Century" aims to promote a dialogue between those who are transforming the understanding of the relationship between nature and culture. The series revisits existing fields such as environmental history, historical ecology, environmental anthropology, ecological economics, and cultural and political ecology. It addresses emerging tendencies, such as the use of complexity theory, to rethink a range of questions on the nature-culture axis. It also deals with epistemological and ontological concerns, building bridges between the various forms of knowing and ways of being embedded in the multiplicity of practices of social actors worldwide. This series hopes to foster convergences among differently located actors and to provide a forum for authors and readers to widen the fields of theoretical inquiry, professional practice, and social struggles that characterize the current environmental arena.

ACKNOWLEDGMENTS

The publication of this volume was made possible by the support given by a number of people and institutions during its long gestation. Institutions that have lent their support to this publication and/or the research on which it is based include the East-West Center's Program on Environment, the SEAMEO Regional Center for Graduate Study and Research in Agriculture (SEARCA) in Los Baños, Yale University's School of Forestry and Council on Southeast Asian Studies, and the John D. and Catherine T. MacArthur Foundation, which generously funded this project by means of grants to Michael R. Dove and Percy E. Sajise. The support of current and past MacArthur Foundation staff, including Kuswata Kartawinata, Dan Martin, and Judith Rayter, has been invaluable to the project editors and directors.

The editors also acknowledge the permission of the Koninklijk Instituut voor Taal-, Land en Volkenkunde to publish here M. R. Dove's essay on non-Western conceptions of sustainability, an earlier version of which was published as "Living Rubber, Dead Land, and Persisting Systems in Borneo: Indigenous Representations of Sustainability" in 1998 in the journal *Bijdragen* (154 [1]: 20–54); the permission of the journal *Human Ecology* to publish here A. A. Doolittle's essay on native customary rights, an earlier version of which was published in 2001 as "From Village Land to 'Native Reserve': Changes in Property Rights in Sabah, 1950–1996" (29 [1]: 69–99); and the permission of the journal *Agricultural Systems* to publish here Endah Sulistyawati's essay modeling the Kantu' swidden system, an earlier version of which was published with Ian R. Noble and Michael L. Roderick in 2005 as "A Simulation Model to Study Land Use Strategies in Swidden Agriculture Systems" (85 [3]: 271–88). M. R. Dove is grateful to his research assistants Caroline Simmonds, Yuliya Shmidt, and Katie Hawkes for their tireless work on early drafts of the manuscript, and all of the editors are grateful to Heather G. Salome of Metaglyfix for her work on the volume's illustrations. Thanks also are due to the students in the Carol Carpenter/Michael Dove advanced seminar "Social Science of Development and Conservation: Advanced Readings" in spring 2008 in Yale's School of Forestry and Environmental Studies, where the penultimate draft of this volume was presented and discussed.

Finally, the editors are very grateful to two anonymous and extraordinarily thorough reviewers for Duke University Press; to senior editor Valerie Millholland, for patiently nursing our project along to its final conclusion; and to the series editors Arturo Escobar and Dianne Rocheleau, for accepting our collection into the ranks of their illustrious series.

None of the foregoing people or institutions are responsible for matters of fact and opinion published here, which are the sole responsibility of the joint editors and individual authors.

PREFACE

In the early 1990s, the sense that existing approaches to conservation and development in Southeast Asia were not only not working but were making it difficult to even think of alternatives brought together two of Southeast Asia's most influential academic programs on the environment, those of the East-West Center (EWC/ENV) and the Southeast Asian Universities Agroecosystem Network (SUAN).[1] From 1993 through 1997, the John D. and Catherine T. MacArthur Foundation funded an effort by Michael R. Dove from the EWC/ENV and Percy E. Sajise from SUAN to develop a collaborative, regional project titled "The Conditions of Biodiversity Maintenance in Asia: The Policy Linkages Between Environmental Conservation and Sustainable Development." Sajise is a Cornell-trained biologist and Dove a Stanford-trained anthropologist. Amity A. Doolittle, a Yale-trained social ecologist, joined the project at a later date to assist in the analysis and publication of the project's findings. This project, instead of dwelling on the conditions that bring about the loss of biodiversity, focused on the conditions that successfully maintain it; it sought to identify the biosocial circumstances responsible for biodiversity maintenance in Southeast Asia; and it asked how the maintenance of biodiversity relates to the maintenance of society.[2] The findings of this project were published in Dove, Doolittle, and Sajise (2005).

Building on the findings of this project, the principal investigators designed a second project, also funded by the MacArthur Foundation (1996–2004), called "The Institutional Context of Biodiversity Maintenance in Asia: Trans-national, Cross-sectoral, and Inter-disciplinary Approaches." This second project was coordinated by the SEAMEO Regional Center for Graduate Study and Research in Agriculture (SEARCA) in Los Baños (then headed by Sajise), the primary agricultural research and educational institution for the ASEAN nations, and also, again, by the East-West Center's Program on the Environment. The first project had revealed how inadequate existing conceptual tools were to the task of trying to promote genuine change in conservation and development in the region. The aim of this second project was, first, to rethink the translation of environmental concepts between East and West, in particular the basic concepts of "nature" and "culture." Second, the aim was to rethink what is meant by the concept "conservation," including how behavior glossed by this term is defined and bounded; and to look beyond state and institutional discourses of conservation to folk, community, and vernacular discourses. Finally, the aim was to focus on how conservation policy is applied and transformed, especially in dynamic real-world systems.

To achieve its end of problematizing and reconceptualizing conservation thinking in the region, the project constituted three successive cadres of researchers, of varying disciplines and seniority, over three successive years. Each cadre or group was convened in a week-long workshop at SEARCA's headquarters in Los Baños to develop research projects for each of its members. The workshops were designed to help the participants transcend their own routinized ways of approaching conservation problems in the region. Each workshop participant was requested to prepare a brief written proposal for a research project, which was then presented to and discussed by the entire group, then analyzed in one-on-one sessions with the principal investigators, and then revised. This entire process was repeated three times for each participant during the week-long workshops. The aim of the three-fold iteration was to help the participants make explicit the structure of inquiry of their approach to conservation and, by so doing, to better enable this structure to be critiqued and transcended. The research proposals produced by this process were then funded for one year of field study, which was followed by another round of writing, feedback, and revision for a final report. A selection of these reports, further edited for publication, are presented here. All of these reports underwent further review and revision, which encompassed, in the case of the contributors whose first language is not English, extensive work to render them into colloquial English.

The editors believe that the reports stemming from this pedagogical process are distinctive in several ways. First, as a result of the emphasis in the workshops on making theoretical premises explicit, these reports are self-reflexive: their authors are thinking about their conceptual tools while employing them. The emphasis in the workshops on problematizing conservation and development orthodoxy has produced studies that critique the localizing bias of so much conservation practice and take a political ecological view of wider determinants, but not at the expense of empirical grounding. The studies presented here are all ethnographically rich and "thick" with detail. Much of this detail is historical: the contributions to this volume battle the presentist sentiment of conservation orthodoxy by explicitly moving away from synchronic and toward diachronic approaches, with a corollary shift away from assumptions of equilibrium toward assumptions of disequilibrium. Finally, the pedagogical structure of the project has yielded studies quite interdisciplinary in character: though dominated by anthropology, they reach out to a multitude of other disciplines, freely crossing the natural and social sciences. These studies also freely cross the boundaries between regional and latitudinal traditions of scholarship: with multiple contributions from (not merely on) each of the participating countries—Indonesia, Malaysia, the Philippines, and the United States—the resulting volume is a product neither of the global North nor of the global South, but hopefully something that spans and bridges the two, a true "world science" (cf. Restrepo and Escobar 2005).

MICHAEL R. DOVE, PERCY E. SAJISE, AMITY A. DOOLITTLE

August 31, 2009, Yale University School of Forestry and Environmental Studies, New Haven, and Southeast Asian Regional Center for Graduate Study and Research in Agriculture and University of the Philippines, Los Baños

Notes

1. The East-West Center is a public, nonprofit, federally funded U.S. organization that has carried out programs of education and research in the Asia-Pacific region since 1960; and SUAN is an association of University-based research groups in Indonesia, the Philippines, the People's Republic of China, Thailand, and Vietnam.
2. The regional workshops for this project were coordinated by Pajajaran University's Institute of Ecology (IOE) in Indonesia and by the University of the Philippines at Los Baños Institute of Environmental Science and Management (IESAM). IESAM and IOE are two of the foremost institutions in the region carrying out research on the interface between society and environment.

CHANGING WAYS OF THINKING ABOUT THE
RELATIONS BETWEEN SOCIETY AND ENVIRONMENT

The field of conservation and development has become complicated, and strengthened, by a number of recent paradigm shifts in social as well as natural science. In recent years the theory, methods, goals, and even definitions of basic concepts in conservation and development have all been questioned. The application of discrete local measures to restore and maintain ecosystem stability, using unproblematized theories derived from North American and West European academia, now not only appears inadequate for solving environmental problems in less-developed countries but actually seems to contribute to these problems in some cases. There is now widespread recognition of the importance of extralocal determinants, the reality of inherent and adaptive disturbance in ecosystems, and the limitations and historically contingent nature of North American and West European science. The intellectual stance of most scholarship if not practice in conservation and development today could be characterized as postlocal, postequilibrium, poststructural, and post-Western.

We will discuss these four points in the first section of our introduction to this volume, focusing on the general theoretical context of our work. In the second section, we discuss a number of specific themes that are taken up in the chapters to follow and cross-cut the volume. The third and final section consists of a chapter-by-chapter discussion of the major points of each contributor to the volume.

Postlocal

One of the most important ways in which the science of conservation and development has developed in the past generation has involved problematizing the traditional bounding of the local ecological and human community as the unit of study. The widening of the unit of study reflects recognition of the relation of modernity to locality (Appadurai 1996; Hornborg 1996). This recognition helped drive the rise of the field of political ecology, spanning anthropology, geography, and rural sociology. The original, central tenet of this field was that local environmental relations cannot be understood, or managed properly, without situating them properly within a wider socioecological context, and in particular without considering the significance of wider political-economic structures (Blaikie 1985; Blaikie and Brookfield 1987; cf. Forsyth 2003; Peet and Watts 1996; Robbins 2004; Rocheleau, Thomas-Slayter, and Wangari 1996; Zimmerer and Bassett 2003). Political ecologists have criticized the routinized blaming of environmental degradation on local communities (Doolittle 2003; Neumann 1998; Peluso 1993; Ramachandra Guha 1989); they have shifted the problematizing gaze from the livelihood practices of local peoples to the consumption patterns and economic growth of the wider national and even global community (Dove 1993a); they have exposed the wider political, economic, and social forces that influence local decisions regarding resource use; and they have shown how externally driven conservation initiatives often infringe on and violate the political, economic, and human rights of local inhabitants (Zimmerer and Young 1998: 20; see also Doolittle 2007). This is exemplified, for example, by the impact of global trade and intellectual property rights on agroecosystem diversity and on access and rights to biodiversity.

In keeping with the wider contextualization of the local human community, the orthodox concept of the community itself—bounded, timeless, homogeneous—and the validity of environmental analyses resting on this concept have been critiqued (Creed 2006; Mosse 1997; Netting 1990; Wilmsen 1989). This critique has come to encompass analyses of intracommunity differentiation, especially along gender lines (Schroeder 1993, 1999; Rocheleau and Ross 1995; Rocheleau and Edmunds 1997). Work in this field has included not only critiques of the local focus but also analyses of how the local is socially constructed (Bloch 1995; Frake 1996) and strategically deployed. This is exemplified by the concepts of the indigenous community and of indigenous knowledge. Once valorized as counterpoints to the hegemony of the discourse of modernity and Western-style economic development, these concepts have now come under intense scrutiny from scholars (Agrawal 1995; Berkes 1999; Dove 2000). Many studies of indigeneity now focus on the way that communities strategically invoke, articulate, or deploy their identity to their own advantage (Conklin and Graham 1995; Hirtz 2003; Jung 2008; Li 2000; Niezen 2003; Tsing 1999).

Postequilibrium

This widening of the analytic gaze beyond closed, self-regulating communities has been accompanied by a rethinking of the opposition between harmony and equilibrium, on the one hand, and disturbance and disequilibrium, on the other. For much of the twentieth century, and although some scholars argued against this, scientific views of nature were dominated by the climax and steady-state assumptions of scholars like Clements (1916) and later Odum (1953). But in the final quarter of the twentieth century a new paradigm emerged, one sometimes called the "new ecology" (Scoones 1999), which disputes assumptions of equilibrium, takes adaptive disturbance to be important, and emphasizes dynamic, historical, and partly unknowable relations between society and the environment (Fiedler, White, and Lediy 1997; McCabe 2004; Pickett and Ostfeld 1995). The best work in the field now looks at flux simultaneously in natural and human communities (Gillson, Sheridan, and Brockington 2003) and at traditional topics such as indigenous knowledge in circumstances of perturbation (Ellen 2007). As with the post-local scholarship, some of the new work in this field has focused on what it means, given what is now understood about equilibrium, when equilibrium values are invoked as desirable and perceived disequilibrium is deprecated.

The postequilibrium paradigm shift has led to a radically altered view of the roles of contradiction and conflict in conservation and development research. There is a new awareness of diversity and of the potential for diverging and even conflicting views in local and also in policy communities.[1] A pioneering study by M. Thompson, Warburton, and Hatley (1986) demonstrated that the search for common ground among the multiple stakeholders involved in most conservation and development projects is ill considered; differing and even conflicting viewpoints are to be expected (see also Vasan 2006). Equally to be expected are differences between policy and practice. Scholars like Mosse (2004) argue that instead of trying to bring policy and practice into line with one another, we must recognize that the two have different purposes and can never agree. Similarly, Robbins et al. (2005) argue that rule breaking in conservation (e.g., illegal practices in protected areas) needs to be seen not as anomalous but as an integral part of the system, even contributing to conservation goals in some cases. According to postequilibrium thinking, even conflict, far from signifying a breakdown in planning, may be a normal mechanism by which environmental and resource relations are managed. The contradicitions that these forces create in landscapes are being systematically studied (e.g., Robbins and Fraser 2003).

Planning and policy failures themselves have come in for rethinking. Studies of apparent conservation and development failure have made an about-face from asking why policy interventions fall short to asking what is achieved in spite or even because of chronic failure (Escobar 1995; Ferguson 1990; see also Doolittle 2006). This interest in contradiction and failure does not represent a turn away from the need for empirical research but just the reverse: it represents a call for greater rigor in the scientific grounding of conservation and development policy through extending the analytic gaze to encompass phenomena previously exempted from consideration.

According to this more dynamic view of conservation and development, socioecological behavior is recursive in nature. Following Giddens (1984: 2), "Human social activities . . . are not brought into being by social actors but continually recreated by them by the very means whereby they express themselves *as* actors. In and through their activities agents reproduce the conditions that make these activities possible." To the agency that Giddens sees in human society must be added the agency of the environment, which changes, often in unexpected ways, in response to human activity. These views of human and nonhuman agency have contributed to a developing belief that there are no final or fixed endpoints in conservation and development management and intervention. Hence the emergent interest in "adaptive management" (Folke, Hahn, et al. 2005; Gunderson, Holling, and Light 1995).

There is a new appreciation for a historical perspective in conservation and development that flies in the face of the rigorously presentist sentiment that has long dominated the environmental movement, driven as it has been by an ongoing discourse of crisis. This appreciation is manifested in the rise over the past several decades of the subfields of historical ecology (Balée 1994; Crumley 1994), environmental history (Cronon 1983, 1991; Worster 1979), and the history of conservation itself (Spence 1999), in particular its colonial-era origins (Grove 1995).

Poststructural/Postmodern

Related to the shift to postequilibrium thinking has been a shift to a post-structural, postmodern view of nature, culture, and, indeed, the presumed distinction between the two. This refers to new attempts to objectify, rather than unconsciously employ, the various "totalizing discourses" of modernity. One example is the discourse of natural disaster. Supported by the shift to nonequilibrium thinking, and by a waning tendency to dichotomize culture and nature, today's scholars of disaster both normalize the concept of environmental perturbation and seek the explanation of "disastrous" impact in political-economic institutions (Hewitt 1983; see also Freudenburg et al. 2009). Some scholars, like Rocheleau, Steinberg, and Benjamin (1995: 1238), further argue that the social construction of crisis by conservation and development practitioners materially feeds into crisis conditions (another example of recursive dynamics): "The representation of crisis is, however, inseparable from the construction of actual crisis conditions." A similar evolution has taken place with respect to our understanding of underdevelopment, poverty, and famine (Frank 1991; Sen 1981; Yapa 1993). Especially relevant here is the work done on the colonial and postcolonial global deployment of conservation discourses (Brosius 1999a, 1999b; Hughes 2006; Neumann 1998; Schroeder 1999; Sundberg 2006). This includes the famous case of Chapin's (2004) criticism of the actual practices versus rhetoric of the largest international environmental NGOs with respect to the rights of indigenous peoples. The argument made in this and related cases is that the discourse of conservation disguises other, more political ends and that it may contain within itself subtle mechanisms of state discipline.

The concepts of nature and culture that underpin these discourses have also come under scrutiny. For example, there has been an extensive debate in the literature, and quite prominently within environmental anthropology, of the perceived archetypal naturalness of tropical forests.[2] Drawing on this and other cases, an extensive literature has developed on the social construction of nature (Braun 2002; Castree 2005; Fedick 2003; Whatmore 2001; Williams 1980). The political implications of this constructivist stance for conservation science and politics have been extensively discussed as well (Botkin 1990; Cronon 1996; Soulé and Lease 1995). Escobar (1996, 1999) suggests that the ideology of "naturalism" has historically played an important role in human environmental relations but that it may no longer be tenable.

Post-Western

As suggested by the debate over constructivist interpretations of nature, the relationship of scholarship to the field of conservation and development has come into question. The problem, some scholars posit, is that practice is "dumbed down" theory. For example, Agrawal and Redford (2006) suggest that many of the problems of implementing conservation and development projects are due to the fact that the projects employ concepts of biodiversity and poverty, for example, that are far more simplistic than those developed in the academic literatures on these respective topics. Others interpret this difference as a developmental time lag, perhaps inevitable, between theory and practice. Notwithstanding the academic shift in thinking over the past generation to a post-equilibrium paradigm, managers of protected areas, for example, still tend to manage with an implicit, and sometimes explicit, goal of maintaining a static equilibrium (Wallington, Hobbs, and Moore 2005). A few scholars, like Mosse (2004), Mathews (2005), and Robbins et al. (2005), argue that the discontinuities between theory or policy, on the one hand, and practice, on the other hand, are inevitable and in some sense adaptive and need to be re-interpreted in this light. Finally, some scholars argue that the linkage of theory to practice in this field is problematic because it inescapably raises the risk of unintended and undesired real-world consequences, stemming, for example, from the study of environmental movements (Brosius 1999a), indigeneity (Gupta 1998; Kuper 2003; Li 2000), and even global warming (Latour 2004).

There is an East-West or North-South dimension to most of these cases of discontinuity between theory and practice, as in the overly facile application of conservation and development concepts developed in Western Europe and North America to societies and environments in less developed countries. The most important example involves the concept of conservation itself, which is often deployed in unintentionally privileging ways. The formidable but greatly understudied challenges of cross-cultural translation of this and related concepts have generated an ever-growing literature (Keller 2008; Lye 2004; Persoon and Perez 2008; Sodikoff 2009; Velásquez Runk 2009; West 2005; cf. Lansing 2006). Hill and Coomes (2004) have even critiqued the application of nonequilibrium models to indigenous systems of natural resource management in non-Western settings.[3] Kirsch (2006, 2008), Lowe (2006), and West (2006), among others, have taken such critique a step further by using their analyses of non-Western systems of environmental relations to interrogate the Western concepts involved, a "repatriating" stance that has its antecedents in the work of Rappaport (1979) and others a generation earlier (cf. Marcus and Fischer 1986). Such efforts contribute to the emergence of what Restrepo and Escobar (2005: 100) call a "world anthropology," which is not an indigenous, native, Southern, or peripheral anthropology crippled by asymmetries, but a plural and more equal landscape of "other anthropologies" and "anthropology otherwise."

The Sacred Forest

These recent intellectual developments in conservation and development can be illustrated by the case of the sacred forest. There has been a vigorous debate as to whether natural resource management in some traditional societies constitutes "conservation." For some analysts the litmus test is intention: traditional practices that conserve resources without *intending* to do so are not judged to be true acts of conservation (Stearman 1994; cf. Krech 1999). One institution that seems to meet this test of self-conscious conservation is the so-called sacred grove or sacred forest observed in many traditional societies around the world (e.g., Castro 1990; Fowler 2003; J. Freeman 1999; Gomez-Pompa, Flores, and Fernandez 1990; Lebbie and Guries 1995; Sheridan and Nyamwera 2008; Sponsel 2005; Wadley and Colfer 2004). Some analysts claim evidence of lower deforestation rates and the promotion of biodiversity conservation in sacred forests (Sponsel 2005), and although some observers call management based on sacred forests "invisible" (Colding and Folke 2001), these apparent consequences of religious doctrine have been interpreted by others as reflecting a purposive and effective commitment to conservation. This image of non-Western peoples using spiritual doctrine to protect dwindling natural resources, often from Western-inspired or Western-driven development, has proven attractive to Western audiences (e.g., Chipko women tying themselves to trees in India or the Penan hunter-gatherers setting up blockades on logging roads in Borneo). Such practices seem to hold up a critical mirror to the secular character of resource-degrading life in industrialized countries.

Anthropological interest in sacred trees and forests reached an early peak in Frazer's *The Golden Bough* (1890/1951). It received passing support from the development of (now abandoned) neofunctional analyses of ritual in the 1960s (Harris 1966; Rappaport 1968). But in the past two decades, there has been an efflorescence of studies of sacred forests—typically islands of relatively more mature and diverse vegetation. Examples include the *tembawang* (community forests) of Borneo (Padoch 1995; Peluso 1996) and the consecration of trees in Thailand (Darlington 2003). This can be attributed to the waxing of interest in community-based conservation and the burgeoning interest in the involvement of global religious communities in conservation (Tucker 2003; Tucker and Grim 1994), something otherwise called "spiritual ecology" (Sponsel 2001a, 2001b).

It is worth examining the concept of the sacred forest not simply as something that describes a certain type of rural space but also as something that *produces* a certain type of rural space (Lefebvre 1991). The concept of sacred forests is most often used today not by anthropologists or other scholars but by community and environmental activists, and this usage is typically quite normative and uncritical in character. The lesser enthusiasm among most academics for the concept of sacred forests (e.g., Langdon 2003;

Vayda 2009) results from a number of perceived problems with it. First, the concept often ignores cultural ecological history. In practice, the sacralization of fragments of forest often, and of individual trees almost always, occurs as part of a wider process of deforestation and landscape transformation (Bloch 1995; Dove 2003). To characterize the partial end products of this historical process as evidence of spiritually driven conservation artificially abstracts what is happening on a small part of the landscape at one point in time from what has happened on the landscape as a whole over a long period of time. It also reflects what is often a non-native view of what remains. In Java and elsewhere in Southeast Asia, for example, sacred trees and groves in and around villages actually symbolize the forest that the village ancestors cut down: whereas conservationists are interested in what *remains* (the sacred grove), the villagers focus on what has been *lost* (the historic forest).

Environmentalist interest in sacred forests also tends to ignore or erase the history of *labor*—referring here specifically to productive, economic activity.[4] Indeed, for conservationists sacred forests are places literally defined by the absence of profane, productive activity. In fact, in many sacred forests in Southeast Asia, hunting, gathering, and even the extraction of timber and other materials for construction is permitted. The relevant distinction in such cases is not between sacred and profane uses but between everyday needs (e.g., gathering fuel for cooking), which might be sanctioned, and singular ones (e.g., felling timber to build a house), which might be permitted. The tendency to instead interpret sacred forest in terms of the sacred/profane dichotomy is related to a misleading belief that still dominates much conservation thinking: namely, that a forest sufficiently intact to be worth conserving must by definition be free of any history of economic exploitation.[5]

Much of the environmentalist thinking concerning sacred forests reproduces a Cartesian divide between nature and culture. Religion and the sacred are of course cultural, but the concept of the sacred forest excludes culture in the sense of the mundane, everyday business of society. By focusing our conservation attention on the nature that remains in tiny enclaves of sacred forest, our attention is directed away from the majority of lands on which people live, such as grasslands and savannas (Dove 2008). The focus on sacred forests represents an example of what Buttel (1992: 19) calls "forest fundamentalism" (cf. Stott 1991). Sacred forests alone offer a very limited basis for conservation (but see Hackkenberg 2008): the sacred character of such areas can be fragile, they are tiny, they are rarely expandable, and the focus on them diverts attention from the much larger secular proportion of the landscape.[6] Cronon (1996) critiques conservation focused on protected areas for analogous reasons: he does not believe that environments that are not fully integrated into society can be conserved.

Excessive attention to sacred forests leads to neglect of the dynamics of the wider natural landscape, of which it forms just a small part. It also leads to neglect of the wider political landscape. Indeed, since the popular conception of the sacred forest is of a place where profane matters like politics are absent, admirers tend to see such places as less involved in the wider political economy than is often the case. However, this dichotomization between the conserved environment and the world of politics is belied by everything that the field of political ecology has discovered over the past quarter century about the inability to understand environmental relations without placing them within wider relations of power and politics. An understanding of the linkages between politics and the sacred, and the implications of this for a more nuanced understanding of sacred forests, is reflected in the work of people like Sheridan (2008), who indeed regards such forests not as places exempt from politics but just the opposite, as "places of power."

Sacred forests have been seen by some observers as places where the environment is protected because nothing changes, because change is held at bay by religion. But this view derives from the now discredited equilibrium-based view of the environment. Postequilibrium-based models render problematic the idea that a landscape must be insulated from change to be conserved. These new models are actually in greater accord with traditional cosmologies in Southeast Asia, which are based on continuous cycles of creation and destruction. Sacred forests can be as dynamic as any other part of the rural landscape (Sheridan 2008). They are part of history, and their sacred character is a product of history. Even when their sacredness is holding steadfast, this does not mean that they are not playing a part in wider, ongoing processes of social change.

The tenacity of the conception of sacred forests as isolated and static in the face of this contrary evidence raises the questions: Why is this vision so attractive and so powerful? What does it achieve? Most obviously, this vision of sacred forests manufactures difference, difference between the metropolitan cultures of environmentalists and the marginal cultures of those having sacred groves. That it valorizes this difference, as a desirable alternative to the secular protected areas of Western conservation principles, is well intentioned, but this fact also obscures its deleterious consequences. The ambivalence and contradiction is well illustrated by the work of Bartlett (2008: 97, 98, 105n7), one of the most astute early observers of the environment and society in South and Southeast Asia. He presents a highly sympathetic portrait of sacred forests in the region:

[There are] little reserves left here and there by aborigines, which except for the advance of "civilization," would have sufficed as sources of seed for the restoration of some species of the original flora. . . . A most regrettable feature of missionary activity among these tribes [in India] is that converts were encouraged by their teachers to desecrate sacred places by cutting wood and other depredations. . . . The same author said that, of the sacred groves of the Munda [of Chota Nagpur], only a few trees were left and that some *sarna* [sacred groves] had been pre-empted for Christian mission stations. Thus the traditional conservation was negated by the advance of civilization!

Thus Bartlett presents the sacred forests as victims in the contest between East and West, and his sympathies clearly lie with the former. But consider what his account leaves out: he sees the sacred forests as premodern and does not begin to ask what role, if any, the forces of modernization might have played in their establishment. He sees them as a valuable but inherently weak and defenseless cultural institution, without agency or the power to resist Western inroads. He demonizes the Western Christian missionaries but says nothing about the role of the wider and much more powerful colonial project of which they were part. He depicts the fall of the sacred forests as a curiously narrow, localized battle.

Bartlett is typical in seeing sacred forests as something fundamentally alien to Western traditions of conservation, which is clearly its attraction but also its vulnerability. The idea of the sacred forest is obviously Orientalist in Said's (1978) sense, and carries the same peril. The present volume seeks to go beyond this idea of the sacred forest in the sense of trying to complicate the easy differentiation of modern and traditional, moving beyond the forest to agriculture and other land uses, leaving behind orthodox notions of the sacred, discarding outdated ideas of environmental harmony and stasis, and rejecting views of the environment that seek to avoid or escape politics.

Cross-cutting Themes of the Book

The foregoing critique of the orthodox thinking about nature and culture in Southeast Asia constitutes the intellectual setting for the contributions to this volume, as reflected in several cross-cutting themes that run through the chapters: the complexity of local communities and natural resource management; the insights afforded by historical perspectives on natural resource use; the ideals versus the reality of planning and relations of power; post-structural insights into the importance of knowledge and discourse; and an ongoing grappling with the concepts of nature, culture, and the role of science.[7]

THE COMPLEXITY OF THE LOCAL AND
NATURAL RESOURCE MANAGEMENT

The research project on which this volume is based was explicitly designed to move beyond a focus on the community level, although the contributors uniformly begin their analyses at this level, in recognition of the value of grounding analyses of nature and culture in community dynamics. Some contributors actually begin their analyses at the even finer level of the household and its dynamics, in particular its domestic cycle. Much of the contemporary interest in this cycle owes its origin to the republication in 1966 of Chayanov's classic work on the subject, which illuminated for social scientists the implications for farm work of the changes through time in the balance between producers and consumers within the household (cf. Durrenberger 1984; Sahlins 1971). Sulistyawati in particular draws on Chayanov's theories to examine, in computer simulations, the implications of the domestic cycle for resource-use patterns in West Kalimantan, Indonesia.[8] Her essay contains an in-depth analysis of the ability of Dayak households with more producers (especially males) to clear more primary forest for swiddens and thereby establish title to more land. The greater the household's landholdings, the longer the periods of forest fallow that it can afford after each instance of swidden-making, which is a critical determinant of the sustainability of the swidden cycle (but see Mertz et al. 2008). This essay demonstrates the importance to sustainable resource use of something as fundamental but largely ignored as the life cycle of the household.[9]

One of the most important dimensions of the household's pattern of resource use is diversification, which is linked to both economic and ecological stability. Sulistyawati's models demonstrate the advantages when a single household makes different types of swiddens, meaning in different ecotypes (primary forest, secondary forest, and grass or swampland). Her models also demonstrate the advantages to the individual household of cultivating groves of rubber trees (*Hevea brasiliensis*) in addition to rice swiddens (cf. Dove 1993c), simultaneously producing for the market as well as for the household's own subsistence (Gudeman 1986; Mayer 2002). Of most interest, Sulistyawati's modeling exercises demonstrate how rubber production can, in the context of increasing population pressure on forest resources, contribute to the long-term sustainability of the swidden system. Dove and Kathirithamby-Wells also examine in their contributions the combination of rubber and swidden or paddy cultivation. Both argue that the cultivation of rubber by smallholder rice farmers is not only sounder in ecological terms than its cultivation on estates but that it is also more efficient economically. The threat posed by this efficiency to the state-supported estate sector has resulted in a century of state efforts to suppress the smallholder sector in Indonesia and Malaysia, in spite of which the smallholders have flourished.[10]

Diversity among as well as within households is another critical determinant of resource-use patterns. All the contributors to this volume recognize the need to move beyond older visions of communities as internally undifferentiated. A central accomplishment of Sulistyawati's modeling is to illustrate what the long-term economic and ecological consequences are of initial differences in household composition. Most notably, her models show that population growth over time drives up land pressure for most but not all households, so that sustainable swidden agriculture remains a possibility for only a few households. Equally central in Djalins's essay on South Sumatra is the distinction between *marga* clan leaders and commoners, with the former employing conservation discourses to defend traditional social hierarchies and the latter employing the same discourses to promote individual rights to forest resources. Doolittle similarly notes in her essay that when her study village in Sabah, East Malaysia, established a native forest reserve in the 1950s, some households wanted individual rather than communal forest rights; today some village households need communal property resources but some do not. And Winarto draws a sharp distinction in her essay between the practices of Javanese rice farmers who have studied and adopted integrated pest management and those who have not.

There are important resource-related differences not only among households but among communities, often following ethnic lines. Much of Lye's chapter is devoted to an analysis of the implications for environmental conservation and indigenous rights of the contrasting ways that Malays and Batek view the forest in peninsular Malaysia. In her essay Harwell examines an analogous difference in resource-use paradigms between Malays and Iban (Dayak) in West Kalimantan. The Malays live downriver and are more involved in fishing than swiddening, whereas the Iban live upriver and have the reverse emphasis. The difference in both ecotype and resource-use regime is reflected in differing cultural perceptions of resources and property, as well as in outsiders' opinions regarding ethnic rights and practices.

The contributors to this volume all appreciate the value of local, traditional resource-use regimes. Djalins's essay on the management of *damar* gardens in Krui, South Sumatra, describes a unique example of the local cultivation of resin-yielding dipterocarps, the reproductive biology of which long eluded even the experts (Michon et al. 2000). Kathirithamby-Wells and Dove both laud the virtues of smallholder versus estate cultivation of rubber. Winarto concludes her analysis of Indonesia's integrated pest management program by characterizing it as an effort to recapture some of the decentralized, nondeterministic character of Javanese agriculture extinguished by the government's program of central planning and rationalization (cf. Lansing 1991). Studies such as these present further evidence that the developmental search for sustainable resource use in places like Southeast Asia is disingenuous. Given such widely documented examples of existing, traditional, sustainable systems of resource use, this spurious search actually reflects not a research challenge but a political struggle over the redistribution of power and resources.

A prominent element in current interest in the local and global environment is the character of non-Western peoples' knowledge of the environment. A Rousseauian idea of "primitive environmentalism" has been central to a number of important environmental campaigns and discourses of conservation over the past generation, in both developed and less-developed countries. Whereas the contributors to this volume all acknowledge the value of local systems of resource use, the elucidation of this value is not their main focus, as would have been the case a few years ago. The possibility that such systems have value, as distinct from models from the Western, industrialized countries, is no longer as contested an issue as it was. As a result, scholars today are devoting less of their time and space to documenting the value of local resource-use systems and more to analyzing why these values are still routinely ignored and what the consequences of this are for both local people and the environment. For example, Kathirithamby-Wells and Dove both examine the reasons for Southeast Asia's history of state suppression of smallholder rubber cultivation in favor of estate production and the negative implications that this policy has had over time for the region's environment. Winarto looks at the state erasure and then partial revival of a role for farmer knowledge and decision-making in Java over the past generation, prompted by the ecological crises of green revolution–based rice agriculture. Such studies demonstrate that the sustainability of local resource-use systems depends not simply on their own economic and ecological viability but also on the existence of conceptual space for them in overarching state systems of knowledge (cf. Li 2000). For this reason, among others, the contributors to this volume all emphasize that there exist no strictly local solutions to sustainable resource management.

A number of the contributors to this volume explain current natural resource-management regimes by tracing their historical development. For example, Winarto interprets contemporary government agricultural policy in Indonesia in the context of more than half a century of intervention in rice production by the postcolonial state. Sulistyawati's chapter models history for a Dayak swidden community, formally tracing the ties between the agricultural present and a different agricultural past. Some of the contributors look at the special histories of resource claims and disputes. Doolittle interprets current conflicts over native reserves in Sabah in light of half a century of land-use contests in the area. Harwell examines the increasing importance of archaic resource claims (e.g., to bee trees) in major conflicts today over human settlement and resource use in West Kalimantan. Most of the contributors maintain that we need to understand colonial resource regimes to understand postcolonial ones, although none are so naive as to say that the latter simply "apply" the former. Thus Lye begins her analysis of Taman Negara National Park in Malaysia with an account of its creation by colonial British officers; and Duhaylungsod argues that we need to understand the history of colonial Spanish and U.S. land policies in the Philippines to understand contemporary land and resource conflicts and the politics of indigeneity.

The objective in presenting these historical perspectives is, in part, to denaturalize present resource-management regimes. Thus Djalins reinterprets the deployment of the *adat* (customary law) argument in current resource-management debates in South Sumatra in light of a century of shifting use of this argument by both central authorities and local communities. Dove shows that the relatively invisible contemporary system of smallholder rubber production represents a successful historical adaptation to the tropical forest environment, repressive governments, and volatile global markets. Conversely, Kathirithamby-Wells shows that the prevailing history of estate development in Malaysia obscures the state and private-sector coercion that made this development possible. As a result of this coerced development of the estate sector, she argues, a historic opportunity to reconcile economic development and environmental conservation, through an alternative emphasis on smallholder development, was lost.

This project of denaturalization shows that the management systems we see today are neither necessary nor inevitable but are the products of particular historical processes. Thus Harwell's historical analysis shows that the resource regimes of the Kapuas Lakes region are far from timeless, and Duhaylungsod similarly notes that the current land uses of the T'Boli in Mindanao cannot in any sense be considered traditional. Sulistyawati's modeling shows how the interaction of social and ecological variables have led to the evolution over time of the particular combination of land uses that can be observed today among the Kantu'.

Conceptions of natural resource-management systems are equally the product of historic processes. Harwell shows that *adat* territories are in large part remnants of colonial Dutch policies toward Malay sultanates and Iban raiding parties, intended to sedentarize and keep the two groups separate. Today some Iban and Malay elites employ these histories as a narrative of ethnic entitlement to secure access to economically valuable timber and fish resources in the face of conservationists' efforts to secure protected area status. Duhaylungsod similarly argues that the indigeneity of the T'Boli and their resource-management systems was historically constructed in response to resource conflicts with outsiders (cf. Li 2000; Benjamin 2000). And Djalins argues that the *adat* argument being used today to defend local resource rights in Krui is but the latest invocation of a local tradition in a century-old contest for power between community and state in this region.

A historical perspective on discourses of natural resource use shows how contingent they are on the politics of the times. Djalins' analysis shows that whereas the *adat* discourse was used to disempower local communities in colonial times, today it has the opposite effect and so *adat* is being revived by community activists. Similarly, Doolittle shows that whereas the establishment of native reserves was unequivocally beneficial for tribal communities in 1950, today this is less clearly the case. The lesson is that regimes of natural resource management, together with their conceptual underpinnings, are constantly evolving.

The contributors to this volume argue not only that the systems for managing natural resources are constantly changing but also that this change is nonunilinear in character. There are surprises (cf. Holling, Taylor, and Thompson 1991). For example, Kathirithamby-Wells points out that the expansion of rice cultivation in peninsular Malaysia has actually declined since the 1970s, in conjunction with the development of the nation's industrial sector, which upsets simplistic neo-Malthusian projections of population growth and land pressure in the countryside. Doolittle observes a similar development underway in rural Sabah, as her informants tell her that they do not expect their children to be farmers. All the contributors employ a nonequilibrium paradigm in their explanations of environmental relations.[11] This is explicitly the case in Winarto's analysis, wherein she examines the impact on resource governance of massive, pest-driven perturbations in Indonesia's rice cultivation.

A dynamic, postequilibrium vision of natural resource management has important implications for governance. There is an inherent tension in the fact that, as Doolittle notes, natural resource-management regimes are dynamic, whereas regimes of governance tend to be static. As Dove writes, there are—perhaps by definition—no static solutions to sustainability: they must constantly evolve.[12] This contradiction is reflected in Lye's analysis of Malaysia's Taman Negara National Park: whereas the environment of the park clearly reflects the country's changing political-ecological relations over time, park governance is focused on preserving images of the past, including a premodern image of the aboriginal Batek.

IDEALS OF PLANNING AND RELATIONS OF POWER

A number of the contributors to this volume engage with one of the most problematic themes of modern conservation policy making, namely, the tendency to see local problems and central solutions. There is a robust literature on the politics of this asymmetry, much of it coming out of the field of political ecology. Less studied are the implications for sustainable environmental relations of the deterministic, centralized planning of natural resource use. The potential consequences of such planning are reflected in the ecological disaster that ensued when the Indonesian government sought to rationalize the country's rice cultivation within a top-down, rigidly enforced green revolution model. The system of integrated pest management described here by Winarto, which was a response to this disaster, was intended to devolve some of the state's authority back to the individual farmer. But tenacious institutional pressures favoring centralized control have resulted in a continual gap between the ideal and the reality of this program, which is one of the main findings of Winarto's study.

Malaysia's Federal Land Development Authority (FELDA) program for smallholder commodity production is the ideological equivalent to Indonesia's green revolution rice cultivation; both represent the epitome of high-modern state efforts to control agricultural production (cf. Scott 1998). Kathirithamby-Wells argues in her essay that, in practice, FELDA has undercut farmers' economic and technological self-reliance and magnified the environmental problems of commodity production. This is only the latest episode in a century of state support in Malaysia for the most environmentally destructive system of export crop production, namely, large-scale, heavily capitalized, centrally run estates. Kathirithamby-Wells argues that the Malaysian government had a historic opportunity to wed ecological to economic concerns by introducing cloned seedlings to smallholdings instead of, or in addition to, estates, but it failed to take it.

The failures of central planning reflect, in part, the existence of multiple and conflicting agendas among the parties to the planning process (cf. Edmunds and Wollenberg 2001). The contributors to this volume recognize a diversity of interests in local communities, but they also recognize a less-studied source of diversity in the apparatuses of government. In some cases this diversity is reflected in a gap between government policy as centrally formulated and locally applied. Winarto attributes the discontinuties between pest management policy and practice on Java to parties in the government and the private sector who continue to oppose limitations on pesticide use out of economic self-interest. Doolittle writes that villagers in Sabah are subjected to conflicting development discourses from different parts of the state and national government. Djalins similarly writes of conflicts in Krui between the agendas of local and central government and also between different government departments, such as those responsible for food crops versus estate crops.

The contributors to this volume are unwilling to regard such conflicts and contradictions merely as epiphenomena and rather see them as inevitable and integral sociopolitical dynamics.[13] For example, Harwell describes the multiple, overlapping, and conflicting tenurial regimes for natural resources in the Kapuas Lakes region. She argues that these conflicting regimes reflect the social reality on the ground and can indeed coexist.[14] The social complexity of this region is thus belied by orthodox developmental efforts to "clarify" the situation and produce a simplified tenurial regime without internal contradiction. This shift from seeing contradiction and conflict as anomalous to seeing them as integral to society reflects, in part, the shift in science from equilibrium- to nonequilibrium-based views of society and the environment.

In keeping with this shift, some of the contributors to this volume argue that too much state coordination can actually be a bad thing (cf. Holling 1995). Most dramatically, Winarto points out that it was the Indonesian government's successful effort to coordinate the country's system of rice production that actually destabilized it: nation-wide coordination created an ecological niche for so-called superpests that had never before existed. There may also be less obvious but equally threatening political consequences to the coordination of development strategies between national and local actors. Thus Harwell argues that there are costs as well as benefits when local communities adopt state mapping technologies in defense of community control of resources (Sletto 2009). And both Djalins and Doolittle suggest that the legal recognition of community-based resource management regimes may or may not be empowering for the communities involved (cf. Nelson and Agrawal 2008).

A corollary thesis is that uncoordinated state activity, stemming from the breakdown or failure of state policies, can sometimes be advantageous for both environment and society. A number of contributors discuss the heretical notion that the failure of high-modern centralized planning may actually provide an auspicious space for the growth of local, sustainable regimes of resource management. Thus the inability or unwillingness of colonial and postcolonial governments in Indonesia to embrace smallholder cash-crop production has kept this sector, as Dove shows in his essay, technologically backward but also—and in consequence—independent and thus amazingly resilient. The very lack of state integration of this sector has contributed to its success and sustainability. Lye similarly argues that although the dehumanizing of the Batek by the Malaysian state has had ill consequences, the resulting lack of an official gaze has also meant that there has been relatively little state interference in Batek lives.

This heretical view of planning has radical implications for our understanding of what constitutes good policy. Lye documents the contradictions in Malaysian government policy toward the Batek—alternately regarding their perceived closeness to nature as an indication that they should be developed or that they are undevelopable—but then concludes that since these very contradictions give the Batek greater space in which to pursue their own lives, the government policy, like the Batek, should simply be left alone. On the other hand, government policies that lack any such internal contradictions and that have consequently been more effective in determining and indeed overdetermining the relations between people and natural resources cannot be left alone. In such cases, as in the governmental rationalization of Indonesian rice cultivation under the green revolution, a process of "de-determinization" is called for. Winarto argues that this is, in effect, what the integrated pest management (IPM) program represents: an effort to recapture some of the less deterministic, less centralized traditional agricultural practices that previously safeguarded the country's agricultural systems.

This turn away from determinism implies a greater acceptance of contradiction and conflict as the natural order of things. This means a subtle but far-reaching change in the orientation of management interventions. It entails, as Harwell points out, a shift in emphasis from resolving disputes to managing them.

Central to most of the contributions to this volume is the idea that knowledge and values are imbricated in power relations and vice versa. Thus Lye argues that the historical vicissitudes of the Batek in Taman Negara National Park reflect not just a political contest but conflicting and embedded values. The state treatment of people and resources within and without the park can only be understood, she writes, in terms of the historical construction of a division between commerce and the "wild." Since such constructs are often naturalized and thereby exempted from scrutiny, a number of contributors, like Lye, endeavor to denaturalize them. For example, by revealing the flaws in the colonial and postcolonial Malaysian state's vision of the rice peasant, its disdain for smallholder rubber producers, and its unequivocal support for estate production, Kathirithamby-Wells demonstrates that the historic rise of the estate sector was not the natural or inevitable development that state and industry accounts have made it seem. On the other side of the coin, Dove describes the constellation of smallholder practices and values that made them such a threat to the estates, including smallholders' beliefs about the need to keep subsistence production separate from (and thus protected from the volatility of) market production.

Such differences in value systems have material consequences and are contested. As Berry (1988: 66) writes, "struggles over meaning are as much a part of the process of resource allocation as are struggles over surplus or the labour process." These struggles are a focus of a number of the essays in this volume. Thus both Djalins and Winarto discuss state deprecation of local systems of knowledge of resource use. This deprecation is all the more powerful when it is couched within the apparently neutral language of modern, Western science. Thus Winarto argues that the green revolution in Indonesian agriculture was not simply a technological revolution but also an epistemological one (cf. Yapa 1993) that gravely diminished the status and scope of a millennia-old heritage of farmer knowledge. As the manifest risks and costs of such technological revolutions have become apparent during the past several decades, the deprecation of traditional farmer practices and knowledge has been reversed in some development circles, as reflected in the rise of a paradigm of community-based resource management. In theory this represents a dramatic shift away from state-driven models, which offers greater potential to meet local livelihood goals by strengthening local society and valorizing local production rationality (Escobar 1998: 75), but it has met with mixed success (Agrawal and Gibson 2001; Brosius, Tsing, and Zerner 2005; Nelson and Agrawal 2008). Both Djalins and Winarto argue that state efforts to recognize and support local systems of resource use may still wind up, perversely, undermining them.

In short, efforts to support local knowledge and devolve authority from the central state back to local communities may have the opposite effect, by means of a process of governmentality. The example presented by Winarto is especially clear. The Indonesian government initially subjected the farmers to classic techniques of high-modern surveillance and discipline in order to implement its nationwide green revolution program, but when this program was on the verge of failure, it resorted to the governmentalizing program of integrated pest management. The two efforts are distinguished by the fact that the green revolution was imposed from without, whereas integrated pest management was voluntarily adopted and internalized. In the first case state experts thus instructed the farmers, while in the second case the farmers were trained as experts themselves, so that they could instruct one another.[15] This analysis of integrated pest management is counterintuitive, because it seems so benign in seemingly shifting some agency from the state to the peasant. The effort to grant freedom and agency is not opposed to governmentality, however; it is a tactic within it. Some scholars, however, are expressing second thoughts about a totalizing vision of governmentality as completely inimical to local freedom and welfare (e.g., Agrawal 2005; Appadurai 2001).

Local communities are not without voice or agency in this struggle over opposing systems of meaning. Winarto states that Indonesia's integrated pest management program has increased not just the participants' confidence in their own farming decisions but also their willingness to question the efficacy of government agricultural programs and extension assistance. Harwell writes that the Iban of West Kalimantan found themselves left out of mapping efforts by the conservation project to codify village territories. As a result, some Iban understandings of territory were invisible on the conservationists' maps, which consequently proved ineffective in settling territorial disputes. Faced with intractable disputes that emerged with rising timber markets, the Iban turned to the traditional dispute-settlement mechanism of cockfighting. For the Iban, the purportedly neutral Western technology of mapping proved to be more politically inflected than the randomizing and ritually sanctioned practice of cockfighting. These essays show how things that are taken as givens—like maps, *adat*, even local history—are actually resources that can be interpreted and deployed to different ends by different parties and are therefore contested (cf. Davidson and Henley 2007). They also show that state relations with local communities are recursive, as discussed earlier: the essays by Doolittle, Djalins, and Harwell all show that state ideas about local communities are appropriated and transformed by these same communities, which in turn affects subsequent state relations with them.

The construction, deployment, and circulation of politically powerful discourses of resource use and conservation is the focus of a number of the essays in this volume. The familiar state discourse of centrally mandated development and modernization is discussed and critiqued in the chapters by Lye, Kathirithamby-Wells, Doolittle, and Winarto. Other chapters focus on powerful discourses employed by local communities. Thus Duhaylung-sod analyzes the strategic use of the discourse of indigeneity by activists and local peoples in Mindanao. Harwell looks at the adoption and use by the Iban of West Kalimantan of discourses of conservation ethics, ethnic resource management, and local sovereignty. And Djalins examines the local appropriation of discourses of biodiversity conservation, indigeneity, and traditional *adat* in West Lampung, South Sumatra. All these discourses are metropolitan in origin, originally hailing from national and international political centers. In part for this reason the local appropriation and deployment of these discourses is often empowering, although it can also prove problematic (cf. Li 2010).

The successful use of such discourses entails some degree of commitment to their terms, and this can entail costs as well as benefits. Thus Lye argues that the Malaysian state discourse of "natural man" that the Batek have bought into is both positive and negative for them: on the one hand, it effectively exempts them from state-coerced modernization, but on the other hand, it marginalizes them from the economic progress enjoyed by the wider society. Djalins suggests that what works locally versus extralocally can never be the same, that we cannot expect the same discourse to serve two such disparate audiences, that there is an inherent conflict here.[16] Because of this conflict, Djalins worries about the unanticipated long-term consequences, versus short-term strategic benefits, of appropriating extralocal discourses of society and the environment. Driven by the same worry, Duhaylungsod suggests that we do not know enough about the long-term effects of commitment to an extralocal discourse like that of indigeneity.

NATURE, CULTURE, AND SCIENCE

A final theme of this volume is the social construction of a divide between nature and culture. The extraordinary tenacity of this divide in contemporary conservation policy is reflected in the backlash against integrated conservation and development projects (ICDPs) by conservationists, who argue that human society is fundamentally incompatible with conservation goals

(Peres and Terborgh 1995; Brandon, Redford, and Sanderson 1998; Terborgh 1999).[17] The continued impact of this divide is attested to by the evidence presented in a number of the essays. For example, Lye suggests that we can only understand the historic development of Malaysia's Taman Negara National Park in terms of a constructed boundary between wild and tame, nature and commerce, and Batek and Malay, and the problematizing of anything—whether people or wildlife—that crosses it. This same dichotomy is reflected in other essays as the tendency of central governments to perceive less intensive systems of natural resource management as part of nature versus culture (cf. Williams 1980). Thus Djalins argues that successive external governments have failed to recognize or credit the investment of human labor in the *damar* forests of Krui, instead seeing these as natural forests. Dove sees a similar history in the smallholder rubber groves of West Kalimantan: both colonial and postcolonial governments have interpreted their ecologically beneficial low-intensity management as either poor management or a complete lack of management (i.e., a complete lack of culture) and have dealt with them accordingly.[18]

The tenacity of the nature-culture dichotomy is reflected in the way in which it is reimposed when events lead to its temporary breakdown. Thus Kathirithamby-Wells and Dove describe how the spontaneous development of smallholder rubber cultivation in Southeast Asia presented a historic opportunity to combine economic and ecological priorities, which was squandered by the steadfast preference of the colonial and postcolonial states for more intensive, capitalized, industrialized estates. Lye describes how the Malaysian state responded to the presence of the Batek in the area gazetted for the Taman Negara—which problematized the state's construction of the park as a natural versus cultural space—by exaggerating the Batek's wildness, their natural rather than cultural dimension. Uncomfortable at the prospect of conflating perceived natural and cultural elements in the park, the state simply reconstructed the culture there—the Batek—as nature.

The contributors to this volume problematize the nature-culture divide by crossing, straddling, or eliding it in their analyses, by acknowledging the role of humans and their labor in the productions of the landscape, and by recognizing that many non-Western societies, instead of seeing nature and culture as two opposing forces, perceive a continuity between the biophysical, supernatural, and human worlds (Escobar 1998: 61). Thus Kathirithamby-Wells compares forest, smallholder, and estate ecologies; Djalins analyzes the Krui villagers' technologies for dipterocarp reproduction; Harwell links natural and socioeconomic flux in the Kapuas Lakes; and Winarto analyzes folk interpretations of pest ecology. This problematizing of the nature-culture divide is also reflected in the interdisciplinary crossing that underpins many of the chapters (cf. Dove 2001), including the study of pest ecology by an anthropologist (Winarto), swidden economics by an ecologist (Sulistyawati), and agriculture by both an anthropologist (Dove) and a historian (Kathirithamby-Wells).

The principal challenge to environmental studies today may not be the divide between nature and culture or between natural science and social science, however, but rather the divide between science and its object. A hoary ambivalence in the academy about confronting the real world has greatly lessened during the past generation, but it has not disappeared, even in such seemingly engaged fields as environment and conservation. Indeed, Whitten, Holmes, and MacKinnon (2001: 3), in an effort to explain the general failure of global conservation efforts during the previous two decades (cf. Speth 2004), ask if the research and development work in this field is merely a "displacement activity," meaning behavior intended to give conservation biologists in academia the illusion of engagement while avoiding its political challenges. Whitten, Holmes, and MacKinnon (2001) see such behavior, thus, as one of the subtle, self-deceptive practices by which science limits its scope.

Many of the contributions to this volume address the question of the scope of scientific inquiry. This includes a recognition that the gap between policy and practice is not just a problem of implementing conservation and development programs but a systemic phenomenon worthy of attention in its own right; that many of the analytic concepts used by scholars and policy makers—like tradition, indigeneity, and modernity—actually have contingent meanings; and, finally, that the use of such concepts may have unanticipated, secondary consequences. Underpinning all these points is the recognition of the need for greater reflexivity, of the need to problematize conservation research and intervention themselves.

Whereas social scientists are sometimes more familiar with the idea of such reflexivity than natural scientists, it is not a question of social science versus natural science per se. It might be more accurate to say that it is a question of scientific practice. Current interest in the practice of science dates from such diverse works as Lenoir (1997), Latour (1988), and Pickering (1995). Running through current contributions to this field is an interest in describing the quotidian dimensions of scientific practice as opposed to its idealized principles. This opposition maps onto the differences discussed earlier between conservation policy and practice, something of interest to all of the contributors to this volume. A refusal to elide this discontinuity is reflected in a corollary refusal to force closure (e.g., Dove, Duhaylungsod,

Lye, Winarto). A number of the essays are analytically messier as a result, because the reality—sustainability, indigeneity, the nature/culture divide, the intentions of policy makers—is itself messy.

The Organization of the Book

The contributions to the book are divided into three main sections: "The Boundary between Natural and Social Reproduction," "Community Rights Discourses through Time," and "Reconstructing and Representing Indigenous Environmental Knowledge."

THE BOUNDARY BETWEEN NATURAL AND SOCIAL REPRODUCTION

In the first section of the book, Lye Tuck-Po, Jeyamalar Kathirithamby-Wells, and Michael R. Dove ask how and why particular resource-management regimes have emerged in particular localities. All three authors argue that local natural resource management cannot be understood without taking a historical approach, looking at social relations, and looking beyond the local system to wider political-economic structures. A wider gaze shows the prevalence of hybrid management institutions and practices, the importance of not just material practices but their symbolic dimensions, and the centrality of contests over meaning (like naturalness and efficiency).

In chapter 1, Lye, an anthropologist, examines the social history of Taman Negara National Park in Malaysia and the livelihood practices of the Batek hunter-gatherers who live in it. Lye demonstrates that the park administration is committed to an ideal of wilderness, represented by a human-free forest, which is enshrined in national legal structures and originally derives from North American concepts of protected areas. As a result, officials simply ignore the history of human modification of the park landscape by the Batek, but they have a harder time ignoring the Batek themselves. Lye argues that cognitive ambivalence about the relationship between people, forest, and wildlife in Malaysia dominate park policy.

As a result, the Batek have a complex, hybrid status: they are simultaneously regarded as wild and domesticated. They straddle not only an ecological boundary between nature and culture but also a bureaucratic one between state institutions focused on preservation and those focused on modernization. Because the state gaze toward them is so complicated and conflicted, the Batek are able to escape it to some extent, while they continue to benefit from the fact that the protected-area status of the park also protects their hunting-gathering territory. Their continued presence in the national park reflects, therefore, some limitations to the reach and effectiveness of state governance.

In this essay Lye moves beyond simple political ecological types of analysis. Her story of the Batek is not about political contest but about conflicting and embedded values. She also problematizes, complicates, and expands the concept of a protected area. She shows that a protected area like Malaysia's Taman Negara National Park is dedicated to preserving ideas, images, and memories, as much as ecology.

In chapter 2, the historian Kathirithamby-Wells analyzes the history of colonial and postcolonial policy in Malaysia toward the highly profitable but environmentally disruptive rubber estate economy. The estates were a late-coming, alien intrusion on a historic landscape that was a composite of swidden and irrigated cultivation of rice, smallholder cultivation of rubber and other cash crops, and fallowed land lying under secondary forest. Smallholders were able to integrate rubber into their swidden and irrigated systems of production with a modest investment of capital and labor and by developing flexible regimes of tapping and marketing. This composite agricultural system provided a buffer against the characteristic boom-and-bust cycle of rubber and demonstrates that environmental and economic imperatives need not be antithetical to one another. Indeed, smallholder rubber production played a significant role in Malaysia's entrance into the twentieth-century global economy.

Kathirithamby-Wells shows that the colonial perception of this traditional, composite agricultural landscape proved hugely important. Its cash-cropping component inspired the European estate planters with a vision of what they could accomplish in the way of market production (cf. van Beusekom [2002] on Africa). But they abstracted this one component from its composite setting. The composite mixture itself—of swiddens, irrigated fields, smallholder cash cropping, and forest fallow—was offensive to the colonial landscape aesthetic and the political economy that underlay it. Colonial observers saw the market-oriented component as too risky for native cultivators, as bad for their political-economic stability, as intrinsically "nonnative." They saw the remaining nonmarket component of the landscape as the true native landscape, belonging to a pure rice peasantry; but this was an imagined tradition. This misapprehension and reconfiguring of Malaya's agricultural and ethnic landscape sowed the seeds for more than a century of social and environmental degradation involving monocropped estates and impoverished coolies.

Postcolonial Malaysia has carried on the colonial legacy of promoting estate agriculture and monocultures. Its megaplantations have initiated a new cycle of forest loss and environmental degradation and become the prime agent of modern forest loss. But Kathirithamby-Wells's history of monocropped estates shows their contingency, not their inevitability; it shows that other development paths could have been taken but were not. In particular, she says, there was a lost opportunity to combine high-producing clones with environmentally sensitive smallholdings, which would have resulted in a modern rural Malaysia with a very different agricultural ecology and peasant economy.

In chapter 3, the anthropologist Dove explores the implications of the assertion among Kantu' tribesmen in West Kalimantan that when they plant rubber in their swiddens or swidden fallows the land becomes *tanah mati* (dead land). Dove argues that the Kantu' characterization of rubber lands as dead has to do with a pervasive ideology of exchange involving both society and environment. Rubber is not part of this system, and accordingly it "kills" exchange and thus the land. Exchange cannot, however, be equated with sustainability, nor can we conclude that rubber cultivation is destructive, for in many respects it is just the opposite of this. An exegesis of Kantu' beliefs and practices involving rubber and dead land reveals the inadequacy of the global conservation discourse for discussing sustainable environmental relations cross culturally.

Of most importance, sustainability in the Kantu' system has to do not with a fixed state but with cyclic relations. Whereas the concept of sustainability in global conservation discourse generally implies a state of conservation and unchanging preservation, the Kantu' concept focuses on a process of active natural resource management and change. In addition, the Kantu' concept focuses at least as much on social relations as it does on material entities like forest cover. The effort to understand their concept of sustainability is complicated by the fact that a composite system of resource management is involved, one with both local and extralocal orientations, and sustainability is a product of the way the two systems interrelate. Interacting with the global economy, but to a large extent on their own terms, is key to sustainable resource management among the Kantu'.

Dove's study shows that there are limits to the ability of the nation-state and of global capitalism to discipline marginal peoples like the Kantu'. His study shows how cultural values, in this case of the morality of exchange, are invoked to resist such disciplining. The Kantu' discourse of killing the land identifies practices that fall on the other side of a boundary drawn between two different transactional orders, one subsistence oriented and one market oriented. The cultural articulation of this boundary inhibits the governmentalizing consequences of the involvement in state-dominated commodity production. This study has implications for current interest in bio-fuel development and carbon sequestration in tropical countries because it shows that there are alternatives to estate schemes and that integrating local and global values and modes of production is hugely complicated.

In the second section of the book, Upik Djalins, Amity A. Doolittle, Emily E. Harwell, and Levita Duhaylungsod explore colonial, native, and NGO discourses used to justify claims to land and control over natural resources. The authors all identify multiple and often competing resource-related discourses in the communities they studied (e.g., indigeneity, conservation, development). Nor are the social relations and practices involved fixed; they are dynamic, evolving, recursive. One source of this dynamism is a paradox of local-extralocal relations: affirmation by the central state of local values and institutions seems inevitably to undermine them. Even benevolent intervention, in short, is problematic.

In chapter 4, the rural sociologist Djalins examines the role of *adat* in struggles over forest resources in Krui, Sumatra. She focuses on forest gardens of *damar* (*Shorea javanica*), which have a long history of being managed by the local communities for resin (Mary and Michon 1987; Michon et al. 2000). When local peoples faced the possibility of losing control over these lands to the Indonesian state, NGOs stepped in and invoked the history of native customary rights to the forest, in part to help garner international support for the local resource claims. This strategy succeeded in winning greater state recognition for local people's resource rights. But the attendant revival of customary law gave new powers to traditional community leaders and, in consequence, alienated the common farmers who had worked hardest to manage and conserve the *damar* forest. This analysis shows that neither the state nor the local community is monolithic. *Marga* (clan) leaders and commoners have different interests; and so do central and local governments, as well as the government offices in charge of smallholdings versus estates.

One of the most important findings of Djalins's study is that *adat* is a dynamic institution. Its political loading, for example, changed radically between the colonial era and the present. Whereas in the 1920s the colonial Dutch government sought to revive traditional *adat* to suppress the rural population, in the 1990s NGOs are reviving it to empower this same population. The connotations of change itself are mutable and subject to contestation. Whereas today many state representatives charge that change deprives *adat* of its authority, and even some community leaders argue that more political weight should be given to tradition than to innovation, many native rights advocates, both within and outside the community, maintain that the dynamism of *adat* is an asset, because it shows that *adat* is a resource for adaptation. The set of actors that is able to invoke the authority of *adat* changes according to whether *adat* has a conservative or radical connotation at a given time and place. *Adat* is thus not institutionally given; it is changing, contested, and manipulated. And this is a recursive process: what was a radical discourse of *adat* became conservative when traditional clan leaders took control of it.

Djalins' study is one of several in this volume that identify a paradox in local-extralocal relations. There are material consequences in representing the local community to suit external expectations, and these may be the opposite of those intended by external actors. The case of Krui shows that it is difficult to win state recognition of local community authority without undermining that authority in the process. There often, although not always (see the preceding discussion of Dove's essay), is a pernicious conflict between what works politically within the community and what works politically beyond it.

Doolittle, a social ecologist, explores the interactions between state law and customary law in the formation of contemporary property regimes in a village in Sabah, Malaysia, in chapter 5. This village was one of the few villages in Sabah that rejected the colonial policy of titling land to individual property owners. Instead, village leaders successfully negotiated with colonial officials for their village lands to be designated as a "native reserve," meaning corporately held village property. While the native reserve served to protect village access to jointly held property in the colonial period, however, it has become less advantageous in the postcolonial political economy.

As in Djalins's study, we see here another case showing that state intervention, no matter how well intentioned, can be inherently problematic. Even the colonial state's recognition of the native reserve was partly disempowering, because it increased state control over local land use. The impact of this was not the same for all members of the local community, however. In another similarity to chapter 4, Doolittle shows that the local community is not monolithic. In the 1950s as well as today, different villagers pursue different land-use strategies and, consequently, have different tenurial requirements. This is due, in part, to the impact of multiple, competing village-level discourses of, for example, development and modernization, indigenous knowledge and initiative, and community-based natural resource management. Doolittle's study challenges the arguments that colonial governments constituted a monolithic force imposing laws on an unresisting native population, that the community is always an appropriate unit for natural resource management, and that reinvigorating customary law always lies in the interests of native peoples.

In chapter 6, Harwell, another social ecologist, analyzes the history of resource claims in the Danau Sentarum Wildlife Reserve (DSWR) in West Kalimantan, Indonesia. She examines the factors—environment, markets, and social-political relations—that have led to changing conceptions of aquatic and terrestrial resources in the reserve, how this has affected concepts of territory, and how this ultimately affects the social and physical landscape. Harwell shows that resource claims to territory are complex and dynamic and do not translate readily into unambiguous and static boundaries. However, she finds that colonial administrators, protected-area managers, and community advocates have all interpreted territorial contradictions as historical inaccuracies—boundaries in need of "tidying up"—rather than as accurate reflections of ongoing claims and counterclaims. Instead of attempting to enforce discrete and stable boundaries between contesting parties, an alternative is to adopt a "dispute-management" not "dispute-resolution" approach to boundary making, one that assumes difference as opposed to consensus and normalizes disputes as opposed to harmony. Harwell argues that such an approach is likely to be more fruitful than trying to resolve all boundary disputes into permanent lines on a map.

Harwell's study has important implications for our understanding of community. Whereas Djalins and Doolittle discuss intracommunity differentiation, Harwell discusses differentiation between communities. In her study, the local community is comprised of Malays and Iban, who have completely different interests and visions. To give just one example, whereas the Iban want to distance themselves as much as possible from the state, the Malays want to be as close to it as possible. To this ethnic-based differentiation between communities, the state has added a layer of bureaucratic differentiation. The modern Indonesian state's conceptual construction of the *desa* as the only officially recognized community unit imposed a completely new, dubious, and disruptive vision on existing Iban and Malay villages. Finally and of greatest importance, Harwell's study questions equilibrium-based approaches to community dynamics. Government calls to "clarify" the tenurial relations of places like the DSWR belie their actual complexity, assume a monistic, homogeneous reality, and further assume that a single correct view of tenure is attainable. Instead, Harwell shows that the uncertainty of supposed facts is the reality, contradictions in rights is not an error but the truth, conflicting tenurial regimes coexist, and contest and negotiation are ongoing in communities that are not in equilibrium. She shows that the region's history itself, like *adat*, is a resource, one to be deployed and contested.

In chapter 7, Duhaylungsod, an anthropologist, examines the land struggles of the T'boli in the Philippines. She shows how the concepts of indigeneity and indigenous environmental relations are invoked and manipulated as communities reconstruct their social and ecological realities in response to new conditions. Following the trajectory of many ethnic minorities in Southeast Asia, the T'boli have utilized the discourse and politics of indigeneity to reassert their cultural and historical rights of ownership to ancestral lands. However, this struggle has pushed them into closer engagement with state agencies. This, in turn, has led to their inclusion in new green development projects and a gradual move away from traditional swidden practices and toward more sedentary and market-oriented agriculture. Although the T'boli's resource-management system remains the basis of their newly articulated indigenous identity, it is itself being transformed. Increasingly, the T'boli are exploring social and livelihood alternatives and creating new possibilities for themselves under the discourse of sustainable development.

Popular conceptions notwithstanding, indigenous identity and sustainable environmental relations do not necessarily mean the same thing. As seen in the other chapters in this section, there are multiple, often competing discourses in the countryside. Duhaylungsod presents another example of how indigenous identity arises from a history of conflict over land tenure and natural resources. External recognition of this identity by the state is a double-edged sword, however. Most obviously, it may simply result in corporate land grabs being replaced with state ones; less obviously, even benignly intended state recognition of indigenous identity may undermine it. Finally, indigeneity is not a fixed end state. It is a recursive process of definition, much like the *adat* analyzed by Djalins.

RECONSTRUCTING AND REPRESENTING INDIGENOUS
ENVIRONMENTAL KNOWLEDGE

In the third and final section of the book, Endah Sulistyawati and Yunita T. Winarto examine the wisdom of traditional practices of resource management, the challenge of representing and reconstructing this wisdom and these practices, and the problem of accounting for recursive feedback processes over time.

In chapter 8, Sulistyawati, a biologist, examines the historical demography of resource use in a tribal community in Kalimantan. Building on Dove's original 1974–76 study of the Kantu' of West Kalimantan, Sulistyawati develops a formal computer model to analyze the dynamics over time of this group's composite system of agriculture. Her model follows the life history of specific people and households, thus showing how the dynamics of the domestic cycle affect the agricultural system. Her analysis shows that most households are able to achieve high levels of rice sufficiency in this system of production, although the number of self-sufficient households declines if population pressure intensifies, in part due to decreasing access to productive land. Her model also shows that pressure for land from population growth is not felt uniformly among all households; less land becomes available for some households but not others. Most important, her model demonstrates that the negative impact of increasing land pressure is not unvarying; households can mitigate it by maintaining a flexible, composite agricultural strategy. By continually shifting the balance of labor and land use between swidden cultivation and rubber tapping according to relative economic efficiency at different moments in time, the community can maintain economic and ecological stability. These findings contradict the still accepted wisdom in agricultural development circles of encouraging swidden cultivators to shift entirely to the intensive, short-fallow cultivation of market crops. Population pressure, land scarcity, and market perturbations all make this a less sustainable strategy than the traditional, portfolio-based production strategy of the Kantu'.

As other contributors have done, Sulistyawati shows the importance of social differentiation; there are no generic households here. Not only are all households dissimilar but even within a single household key variables fluctuate over time as it moves through its domestic cycle. These short-term cyclic changes within households contribute to long-term, noncyclic changes in the community and the environment. Sulistyawati's model shows how the community becomes differentiated over time, and she links this differentiation to transformations in the environment—thus linking social and environmental changes. By linking quotidian practices to the overarching, socioecological system, by modeling the means by which the former produces the latter, she offers new insights into the relationship of individual actors to overarching social structures.

A tribal system of agriculture is an uncommon subject for a formal exercise in computer modeling. This is not the "thick description" (in the sense of Geertz 1973) that has dominated studies of such subject matter for at least a generation;[19] it hearkens back to an earlier tradition of formal ecosystem-oriented modeling that reached a peak in the 1960s (cf. Lansing 2006). Sulistyawati's model eliminates some complexity, as does any formal model, but it thereby acquires considerable power—explanatory, imaginatory, and political. Its analytical power lies in the way it helps us reimagine the past, present, and future. Its political power lies in the authority that it potentially can give to policy statements sympathetic to swidden cultivation and smallholder rubber tapping.

In chapter 9, Winarto, an anthropologist, examines the social and ecological implications of central versus local governance of agricultural strategies in West Java. She begins with the loss of local agro-ecological knowledge that occurred when, in the course of the green revolution of the 1960s, Javanese rice farmers were directed to replace their traditional rice varieties with new high-yielding ones that required chemical and technical inputs of which they had no knowledge. The consequences of this agricultural "de-skilling" (see Stone 2007) were dramatically demonstrated when unprecedented infestations of rice hoppers ravaged the new green revolution fields of Java in the 1970s. This calamity prompted a dramatic about-face in government agricultural policy, as reflected in development of the Integrated Pest Management (IPM) program in the 1980s. Based on a new philosophy of the decentralization of agricultural management, the IPM program was designed to give farmers greater opportunity to analyze their own pest problems and develop their own responses. This process of reconstruction of local farmer knowledge and initiative inaugurated by the IPM program is the subject of Winarto's analysis.

The story does not end with the inauguration of the much-lauded IPM program, however, and herein lies the greatest strength of Winarto's analysis. IPM is an effort to turn back the clock, to recapture the decentralized and nondeterministic character of traditional agriculture in which individual farmers make independent analyses and decisions in response to changing local conditions. But Winarto raises problems with this vision at a number of levels. First, there are prevarication, reversals, and resistance to IPM by farmers and bureaucrats alike. The government itself is not monolithic in this regard: for example, there are internal conflicts between different sectors regarding the wisdom of committing state resources to the IPM program as opposed to pesticide subsidies.

Of most importance, however, Winarto raises questions about the fundamental logic of reinventing agrarian tradition. The history of the green revolution itself has made the effort to return to a pre–green revolution condition problematic. For one thing, many farmers have by now internalized the philosophy and values of the green revolution and are suspicious of attempts to wean them off of its chemical foundation; farmers cling as stubbornly to these practices as they earlier did to their pre–green revolution ones. But more tellingly, the traditional concept of the independent farmer-operator that the IPM seeks to revive has been inevitably, if unconsciously, affected by the implicit epistemology of the green revolution. Thus, in the IPM program farmers are being reimagined as farmer experts, farmer-researchers, and so on, roles and statuses that simply did not exist prior to the green revolution. This reimagination thus represents a conceptual shift. Modern concepts are clearly being used to recapture the premodern (Hirtz 2003); they are being used to return not to traditional local knowledge and practices but to a modern construction of local knowledge and practices.

Taken together the essays in this volume provide powerful examples of why social, political, historical, and economic factors are central to the success or failure of conservation initiatives. Natural resource managers and policy makers who accept and work with these factors are likely to enjoy greater and more enduring success than those who simply seek to remove the influence and impact of humans from the landscape. This is no mean task and requires, as many of the essays suggest, the ability to manage contradictions, to relinquish many orthodox ideas of what conservation looks like, and to practice continuously adaptive management techniques. It requires practitioners who are deeply reflexive and able to focus less on achieving short-term goals and more on a long-term engagement in the complex relationship between people and nature.

Notes

1. See the critique by Leach, Mearns, and Scoones (1999) of the premise of a "consensual community" in community-based natural resource management (see also Berry 2004).
2. For example, see Balée (1993, 1994), Denevan (1992), Posey (1985), and Raffles (2002) on the Amazon; Fedick (2003) and Gómez-Pompa and Kaus (1992) on Central America; Fairhead and Leach (1996) on Africa; and Ellen (1999), Neidel (2006), and Puri (2005) on the Indonesian archipelago.
3. See Kartikasari (2008) for local interpretations of national concepts of conservation in Sulawesi.
4. See Peck 2006 on recent efforts to build bridges between environmental and labor histories.
5. Recent research, like Neidel's (2006) in Kerinci National Park in Sumatra, demonstrates that even landscapes deemed worthy of protected-area status have often been the site of long human occupation and landscape modification.
6. See the study by Jarvis, Padoch, and Cooper (2007) of biodiversity in agro-ecosystems for an example of expanding conservation interest in unorthodox spaces.
7. For another example of rethinking of conservation paradigms in the region, see Sodhi et al. (2008).

8. Sulistyawati's chapter demonstrates the power but also limits of formal modeling since she is obliged to assume an artificially closed community—with no labor coming in or going out.

9. See Walker and Homma (1996) and more recently Perz et al. (2006) for similar analyses from the Amazon.

10. Predictions during the past two decades of the imminent demise of smallholder rubber production in the region have been soundly trumped by the recent sky-rocketing price of rubber, stimulated by the demand from China's booming economy.

11. During the past generation appreciation has grown of the importance of distur-bance to the integrity of ecosystem functions (Botkin 1990; Fedick 1996, 2003; Pickett, Parker, and Fiedler 1992; Pickett and White 1985; Scoones 1999; Zimmerer 2000). Research has demonstrated the shortcomings of the earlier view of eco-systems as characterized by stability, balance, and equilibrium.

12. The "active adaptive management" approach to resource use and conservation is an effort to match dynamism in governance with dynamism in the environment (Agrawal 2000; Holling 1978).

13. See Mosse's (2004) argument that contradictions between policy and practice are not merely inevitable but necessary.

14. See Davidson and Henley (2007) on the role that conflict plays in local dispute resolution.

15. One obstacle to implementing integrated pest management has been that the green revolution logic has been internalized by farmers themselves, which some-times leads them to doubt the wisdom of IPM.

16. See the discussion in chapter 3 of the necessary discontinuity between the eco-nomic morality of the household and that of the market. See also the observation in chapter 5 that villagers in Sabah do not always agree with local activists' visions of indigeneity and community-based resource management.

17. This backlash is perhaps also reflected in the growing divergence of interests between the major international environmental NGOs, on the one hand, and indigenous rights groups, on the other hand (Chapin 2004). See also Sodikoff's (2009) analysis of ICDPS from a Marxian perspective.

18. The establishment of extractive reserves in the Brazilian Amazon represents one of the most important recent efforts to grant official recognition to the rights of the local community to manage resources in conservation areas (Schwartzman, Moreira, and Nepstad 2000; Schwartzman, Nepstad, and Moreira 2000).

19. There have been minor exceptions, such as the literature on so-called optimal-foraging theory (see Dove and Carpenter 2007: 34–37).

I

THE BOUNDARY BETWEEN NATURAL AND SOCIAL REPRODUCTION

THE WILD AND THE TAME IN PROTECTED-AREAS
MANAGEMENT IN PENINSULAR MALAYSIA

This essay's title is an explicit acknowledgment of ideological issues in protected-areas management. The ethnographic focus is a specific national park, the 4,343 square kilometers of Taman Negara in peninsular Malaysia.[1] My fieldwork has uncovered a range of problems, primarily in the relationship between managers (officers and rangers of the Department of Wildlife and National Parks; hereafter DWNP) and the resident community (the forest-dwelling hunting-and-gathering Batek of Pahang state).[2] In earlier writings (Lye 1997, 2002, 2005), I regarded park administration as an imposition on the culture of the forest people. Here, with the people in the background, my concern is more with the DWNP and the broader context of managing wildlife and habitat.

A key characteristic of the Taman Negara administration is its ambivalence regarding the place of people in the park. The people are recognized as belonging there and having a visible presence, but there is no administrative procedure for dealing with the implications of that presence. I will argue that the administrative ambivalence reflects a deeper cognitive unease with the relationships among people, the forest, and wildlife. Of most concern to me is how and why certain places, peoples, and times are defined as wild or tame and how such definitions are woven through the administration of the national park. The most familiar definition that comes to mind, as legitimized in scientific conservation, draws on Euro-American conceptions of the "wilderness," which often deny that landscapes have different use values for different peoples and therefore are social rather than natural spaces (Cronon 1996).[3] Though the tension between scientific and local conservation practices exists in peninsular Malaysia (hereafter the Peninsula), of more interest are the local variations on the concept of "wild."

One issue that will become clear (see Lye 2002) is the artificiality of boundaries between folk/local and scientific/global. For the "local" in this case is both Batek, the people in the park, and Malay, the people running the park and the country. Protected-areas management may appear to be global in its managerial and scientific orientation, but it is filtered through embedded Malay understandings of forest-people relations. And, in a mutually reciprocal process, Malay understandings are filtered through what can be observed of the forest peoples' behavior and environmental relations. As such, the ideological context is extremely complex, and the problem for this essay will be to trace how different symbolic representations come together and are enshrined in official policy.

Protected-areas establishment (and similar resource-management projects) has come in for much criticism, as well as for proposals for improvement. Ongoing issues include the rights of resident populations in protected areas (Colchester and Erni 1999), the failure of parks to improve local livelihoods (Ghimire and Pimbert 1997) and/or to protect biodiversity (Brandon, Redford, and Sanderson 1998), and doubts about the effectiveness of co-management in widening access to resources and decision making (Brosius, Tsing, and Zerner 2005; Colfer, Wadley, and Venkateswarlu 1999; Leach and Fairhead 2001). Concurrently, other critiques (e.g., those of the "cultural construction of nature" approach) look critically at the relations of power and ideas that support resource-management regimes; these demonstrate that the ideas are historically contingent, even in the place of origin, socially specific, and often elitist, and that they may be based on an erroneous understanding of culture-nature relations (Dwyer 1996; Neumann 1998).

My approach in this essay draws from these various strands of discussion. While acknowledging the need to integrate conservation with social-political interests, I maintain that these are only surface problems. There are underlying ideological issues that ultimately shape how managers' objectives are defined, interpreted, and realized in practice. Why reality is constructed one way rather than another is a function of contentions *between* systems of ideas and values that change over time. Because these ideas are deeply embedded, they will not be dislodged with bureaucratic reform. I am not, however, arguing for the opposite position, a paradigm shift. Rather, the question is how ideology—expressed here as symbolic representations of the forest, people, and wildlife—intersects with the management of a national park, and what are the political consequences of viewing the world in a certain way.

The Founding of the Park

The DWNP and Taman Negara originated in the 1930s, long before the current environmental crisis. As an inheritance from the colonial era, it is a product of pre–Second World War concerns rather than today's. If Malaysians wish to congratulate themselves, they can cite Taman Negara as an example of an environmental achievement. But in view of the park's colonial origins, we must ask how the authorities today perceive the park and its people and how colonial images of nature continue to shape the administration.

My point of departure is to investigate what problems and assumptions brought the DWNP and Taman Negara into existence.[4] The relevant context was the agricultural expansion and forest clearance activities in the decades leading up to the Second World War (H. Burkill 1971: 207). Consonant with the establishment of early forestry departments, tentative efforts were made —from the 1880s onward—to set land aside as timber reserves. Related to the largely unplanned growth of the commercial sector were planters' (especially rubber growers') perceptions of animals: to them, "wild life was all right in its place, but if it caused damage to crops, it should be ruthlessly controlled" (W. Stevens 1968: 11). Alarm over the impact of agriculture on animals, especially the larger fauna like elephants and tigers, led to early attempts at conservation. The first game laws—isolated interventions at the state level without wide applicability—emerged around this time, with the first game reserve (Chior in Perak State) not created until 1902 (Hubback 1932: 2:15–16). H. Burkill (1971) claimed that wildlife protection was the culmination of a rising swell of public opinion, but the evidence is not conclusive. For example, as judged by evidence collected in the *Wild Life Commission Report* (Hubback 1932, see below), opinions about conservation measures were sharply divided across communities, with restrictions on commerce and livelihood activities (e.g., hunting and trading restrictions, and land excision for wildlife sanctuaries) generating much suspicion and heated debate.

Into this scenario moved a rather unusual individual, T. R. (Theodore) Hubback. He was, in H. S. Barlow's estimation, "the first genuine conservationist in the modern sense of the term" (2000: 21).[5] An engineer by training, he was a well-known big game hunter turned gamekeeper who first came to Malaya to join the Selangor state government: "He bounced round the civil service, never staying for long in any one job" (H. Barlow 2000: 20). He seems to have been something of a misfit and eventually settled in Kuala Lipis, Pahang (the former state capital and original headquarters of the national park). In 1922 he was confirmed as the honorary (i.e., unsalaried) game warden of Pahang, and in 1930 he was appointed by the governor of the

Straits Settlements to head the Wild Life Commission of Malaya. We have no information on why the commission was formed, but it was at least partially an attempt to address the impact of animals on agricultural expansion, rather than a move toward conservation as such. Specifically, the commission was tasked to examine the existing game laws, investigate reports of wildlife damage to agriculture, suggest methods to deal with the problem, and, finally, to inquire how regulations to preserve wildlife might be organized (Hubback 1932: 1:1–10). Hubback's appointment was controversial, especially to the powerful Rubber Growers' Association (H. Barlow 2000: 22; see, e.g., Hubback 1932: 3:276, 293), which blamed conservation for an increase in wildlife depredations on crops and doubted that Hubback, a noted conservationist, would be sympathetic to their interests.

The three-volume *Report* was perhaps Hubback's (1932) most lasting contribution to posterity. A fascinating document, it contains evidence taken from over 550 witnesses from various socioeconomic sectors and all social strata, presented in transcript or tabular form (on methodology, see Hubback 1932: 1:12–13, 23–24). In light of today's concerns with representation (see, e.g., Brosius 1997a; Leach and Fairhead 2001), we should note how the goal of broad inclusivity was belied by actual practice. As Hubback phrased it: "Evidence was taken from anyone who wished to come forward, from High Government Officials to *Sakai*" (Hubback 1932: 1:12). Yet "Sakai" (i.e., Orang Asli) voices remain largely absent from the resulting document.[6] Though the aboriginals were spoken about *by others* (both Malays and colonials),[7] particularly to consider whether their subsistence practices caused damage to wildlife, and though a document was submitted on behalf of the Sakai of Pahang East (Hubback 1932: 1:378–79), the *Report* shows little evidence of consultation with them.[8] Despite the strategic territorial importance of the Batek (termed "Pangan"), they appear only as Hubback's informants on trails and geography (e.g., Hubback 1932: 2:138). One problem was, perhaps, the unfamiliarity of the consultative process. More significantly, suspicion toward the work of the commission evoked another problem, an unwillingness to speak openly and the misrepresentation of facts and opinions (Hubback 1932: 1:12–13). It is also possible that aboriginal communities were just too remote to be aware of the proceedings and that the commission had neither the resources nor the time to approach them directly. Whatever the reason, the ideology of wildlife conservation had at the outset selective origins. Wherever attention was paid to the aboriginals, their activities and concerns were viewed through external frames of reference, especially that of species conservation.

These limitations aside, the *Report* was indeed a watershed in advancing the first systematic program of wildlife management for the Peninsula. The recommendations, as summarized by Barlow (2000: 22), were: standardize the game laws, have one central agency to administer the laws, establish "inviolable sanctuaries," enforce closed seasons for the hunting of animals and birds, and conserve river fish. With some refinements and adjustments, these recommendations form essentially the core of the program that exists today and the raison d'être of the DWNP.

Hubback stressed the need to bring local regulations in line with "those basic principles of conservation that are now generally accepted throughout the civilized world" (Hubback 1932: 2:7). One of these, a reflection of his admiration for the American model (Gordon Means, personal communication), was to complement the existing system of sanctuaries with a national park, "where the larger animals can breed without disturbance and the smaller fauna can be sure of freedom from the continual harassing that has been their lot" (Hubback 1932: 2:139). He first proposed the Tahan Game Reserve as the national park location in 1927.[9] By 1930, as shown in the commission letters (e.g., Hubback 1932: 1:3), the idea had gained sanction, and Hubback was tasked to investigate whether an expanded area around the game reserve would be suitable for the purpose. It is difficult to uncover an ecological justification for that location above any other in the Peninsula. The most important consideration may have been land availability: the Tahan Game Reserve had the advantage of habitat diversity, a sizable tract of unbroken forest, and low population density: "None of this land is occupied by any sort of permanent settlement, although there are a few wild tribes [the Batek or "Pangan Tanah" in his terminology] which wander through its jungles" (Hubback 1932: 2:133). In H. M. Burkill's words, the area "was too remote to fall under the avaricious eyes of commerce so that its land was not coveted" (1971: 206–7). Finally, in 1938 and 1939, the state enactments that established King George V National Park (the original name of Taman Negara) were passed, and Hubback became the first chief game warden.[10] His story ends here, for he disappeared during the war, just a few years later.[11]

To return to the main themes of this essay, the national park formed the centerpiece of the proposed network of "inviolable sanctuaries." The idea of these sanctuaries was a brilliant solution that advanced something new while *essentially leaving the commercial establishment intact.* It was a radical initiative for the times, but it was also pro-establishment. In view of the planters' perceptions of agricultural pests, the sanctuary idea left planters free to dispose of pests (with restrictions) while, in a metaphorical sense, "shepherding" remaining wildlife into designated sanctuaries (various versions of this position are enunciated throughout the *Report*). Setting land aside for animal protection was politically marginal to the needs of and for capital. Sixty years later, the same problems and assumptions remain.

Some animals would be sacrificed and some protected through systematic management. Certainly an idea for wildlife protection was born. There was now an official perception of what the *place* of animals was and how to keep them there. And, of course, there was also a new perception of the forest and how it could be used. Parts of the forest would be taken out of the hands of commerce and would come to be defined by the presence of animals and the laws maintaining them there; no wonder that Hubback represented such a threat to the rubber growers.

He was also a threat to local Malay villagers. At least one settlement he encountered on the fringe of the national park area was "occupied by persons who do not like the restrictions placed on them by rules and regulations and who find that they can live easier and with less effort in such [remote] places" (Hubback 1932: 2:138). In drawing up the park boundaries, Hubback appears to have taken pains to leave such settlements out of the proposed area, but he did not provide for their larger extractive zones.[12] Communities that were accustomed to exploiting forest resources with impunity were now locked out of a huge swathes of forestland and saw their activities criminalized under the new conservation regulations (drafted in, for example, Hubback 1932: 2:153–57). It is not hard to imagine the responses to such restrictions. Kuala Tahan villagers were, indeed, expressing their resentment to me in the early 1990s, and poaching and encroachment remain a big problem in Taman Negara.[13] But therein lies an important duality, for the "wild tribes," the ancestors of today's Batek, were allowed to remain in the park. Officially, "the Taman Negara Enactments make no reference to the aboriginal population that naturally inhabit the Park" (DWNP 1987: 24). The unofficial picture, as gleaned from the *Report*, is more complex.

Hubback was undoubtedly sympathetic to the aborigines: he was also (informally) the protector of aborigines for the Pahang River (Gordon Means, personal communication). He used the medium of the Wild Life Commission inquiry to investigate whether the protection of aboriginal interests could be accommodated with the protection of wildlife: "Would the Game Reserves supply a suitable location for these aborigines?" The question was posed to Fetherstonhaugh, Pahang's assistant game warden, who replied: "Yes, most suitable. . . . It should be the duty of the Game Warden to see that the aboriginals have a place where they may evolve on natural lines; to supply them with medical assistance; to ensure that their work is justly remunerated; and to assist them to form more settled habits and generally to improve their mode of life, free from fear of exploitation" (Hubback 1932: 1:317). This was the most explicit response reproduced in the *Report*, but it was a minority position. Nevertheless, Theodore Hubback and Gerald Hawkins, the commission's assessor, did try to submit a case for it. Hawkins summarized thus:

> The aborigines, Sakai, Jakun and Negrito should have some protection and might be settled in a portion of a large game reserve where they could slowly evolve instead of falling victims to a civilisation for which many centuries of seclusion have made them for the present unsuited. . . . The aborigines should have, not only special reservations, but also an area large enough to permit them the satisfaction of their nomadic habits. Such areas can only be found in Wild Life Sanctuaries or National Parks, the administrator of which should have as part of his duties the promotion of the gradual evolution of these Wild People, as well as the work of protecting them from exploitation by other races." (Hubback 1932: 1:28–29)

Although neo-evolutionist and paternalistic in the style of the times (Nicholas 2000: 78–80), these proposals were remarkably foresighted in anticipating today's concerns with the exclusion of indigenous communities from protected areas.

Hubback reasoned that "where [the aborigines] were domiciled in areas which are or may be sanctuaries for wild life there they should be encouraged to live their primitive lives as part of the natural resources of that area. But they will have to conform to the laws for conservation which are necessary for such areas and their education on these and other lines must be part of the duties of the Commissioner for Wild Life and his staff" (Hubback 1932: 2:49). For unknown reasons, these proposals did not appear in the final enactments, but they help explain why the Batek and other aborigines were not evicted from wildlife conservation areas. Formally, administrators could not secure land reserves for the aborigines, nor could they establish statutory provisions for them in the wildlife reserves. But they achieved the same effect by simply not resettling them. Today the Batek seem unaware that Taman Negara, the bureaucratic entity, dates from the 1930s and not from the recent boom in tourism, thus suggesting an absence of efforts to contact them and explain what was happening. Ironically, the situation as left by Hubback (which arguably might have changed had the war not intervened) helped render the Batek institutionally invisible and remains largely unresolved today.

As the foregoing sketch indicates, it is difficult to understand Taman

Negara and the operating procedures of the DWNP today without going back to the founding events. My research has traced many developments to the persona and orientation of one man in particular, Hubback. A man of his times and a visionary, what he did or could not do have effects that reverberate to this day. On the matter of wildlife conservation, as mentioned, the problems he confronted have persisted (with, of course, changes in the ethnic composition of actors and interest groups). Wildlife conservation at its inception trotted timidly at the heels of commerce and, many would argue, remains peripheral to government.

How Wild the Forest

As filtered through Hubback and his interest in game conservation, Taman Negara was founded on American ideals of wilderness protection. But, as Cronon's history (among others) shows, the prototypical "wilderness" was fundamentally an "illusion" wrought of the expulsion of Native American inhabitants from their homelands (1996: 79; see also Botkin 1990; Neumann 1998). The American National Parks Service provides little guidance on the question of *people* in the forest and how to respond to their changing needs and claims.

At heart is a contest between different conceptions of people-forest or culture-nature relations. In the American (or, rather, the Yellowstone) model, "nature" was a deliberate political creation that denied the human history of the park. In Taman Negara, however, only *some* locals, the Malays, became personae non gratae. There was (and is) room for incorporating indigenes into park management. This is a curious inversion of the widespread tendency in Malaysia to valorize Malays (the ethnic and political majority) as "in" and indigenes as "out there." At least when Taman Negara was founded, the Malay farmers were positioned in the *out* category, distanced from and personifying threats to nature, while the non-Malay hunters were deemed part of and permitted to remain *in* nature.

Another contention is between the goals of conservation and those of commerce. The colonial enterprise was committed to profit maximization, and this meant that "empty jungles" had to be populated and lands opened up for commercial agriculture and other activities like mining (Cant 1973; Lim Teck Ghee 1976). Forests would be replaced, and if conservationists and foresters were to achieve their objectives, they, too, would have to make a claim for the land. Though this seems like competition between incompatible interests, I argue that, whether colonials favored exploitation or management and conservation, at heart there was congruence in their respective perceptions of the forest, which is revealed in their construction of indigenous economies and identities.

In early writings, it is clear that while colonials were looking ahead to the commercial possibilities of lands they prospected in, they did not overlook the presence of indigenous peoples (Clifford 1897; Skinner 1878). But the latter were not a political force and the official policy one of "benign neglect" (which lasted until the early years of the 1948–60 Emergency). As we saw above, administrators sensitive to the people's needs for land security faced an uphill struggle. If the forest was "empty land" waiting for exploitation and clearance, how could its original inhabitants be perceived as other than nonentities? To create something new, whether wildlife sanctuaries or timber reserves, "emptiness" would have to be enshrined as part of the new construction of forest. It is not surprising that the Taman Negara enactments made no provisions for the Batek. It would have been very odd otherwise.

The assimilation of indigenous communities into conservation areas has a parallel with the national park situation in Africa: there the colonials' "Edenic vision of the landscape was capable of accommodating an African presence, because incorporated in the Eden myth is the myth of the noble savage" (Neumann 1998: 18). In the Malayan "conquest narrative," the people and their distinct ways of life were incorporated into colonial aesthetics, expressed most markedly through romantic idioms. For example, Frank Swettenham, who personally felt oppressed by the "impenetrable gloom" of the forest he traveled through, approvingly described the indigenes he encountered (Senoi shifting cultivators) as "clever gardeners" (quoted in Aiken 1973: 145–46).[14] In his *Malay Sketches* (1895), he makes an explicit connection between indigenes and wildlife, inviting the reader to consider "the primeval forest, the home of wild beasts and Sakai people, aboriginal tribes almost as shy and untamed as the elephant, the bison and the rhinoceros, with which they share the forests of the interior" (quoted in Nicholas 2000: 70).

The general perception was later elaborated by Skeat (Skeat and Blagden 1966: 11), whose description here shows a budding appreciation of something other than brute existence, namely, knowledge, skill, and technology: "The catholicity of their tastes necessitates at once a most thorough and accurate knowledge of the habits of the varied denizens of the jungle, and a considerable amount of ingenuity and mechanical skill in the contrivance of traps, pitfalls, springes, and nooses for securing their quarry, and this knowledge, skill, and ingenuity the wild races certainly possess in a very marked degree." As Rambo (1982: 282–83), not usually the most romantic of observers, sums up: "Although admittedly involving a great amount of subjectivity, it is fair to say that most ethnologists who have direct experience with the Orang Asli have formed highly positive impressions of their overall quality of life. Indeed, the Orang Asli appear to come as close as any group in the world to representing the classical anthropological ideal of the 'noble savage' living in symbiotic harmony with nature."

Aboriginals exemplified a Rousseauian idyll, one that admittedly would not last but while it lasted should not be interfered with.[15] Administratively, associating the people with nature was another way of saying that they were not vital, and this may be one reason why they were admitted into conservation projects like Taman Negara (after all, if Hubback's proposal to include them in the wildlife protection enactments could be overruled, then so could the practice of allowing them to remain where they were). Perceptually, they were largely just "out there." Their cognitive, political, and economic concerns might have intrinsic interest but could not be fitted into administrative policy. Part of the irony, of course, was that winds of change were augured by every colonial exploration and encounter. As we have seen, there was an irremediable tension between preservation and production that also shows up in perceptions of wildlife. The protection instinct would be met concretely through Hubback's proposal of "inviolable sanctuaries" that could protect both people and animals.

People, Bureaucracies, Policies, Tensions

Taman Negara today is an interesting administrative, ethnological, and biological phenomenon. In the enabling enactments, its functions are centrally "the propagation, protection, and preservation of the indigenous fauna and flora of Malaya and of the preservation of objects and places of aesthetic, historical or scientific interest" (quoted in DWNP 1987: 3). This biological diversity is "to be kept in trust for all mankind, with controlled access for scientific research, education and tourism" ("Taman Negara" 1971: 133). As I have argued (Lye 2002), this official articulation of function privileges the role of *observer* (outsider) rather than that of *participant* (insider) for people.

Not only is the Batek role in maintaining the forest not recognized in the founding laws but there also continues to be no official legitimization of the Batek-DWNP relationship. In the *Taman Negara Master Plan* (DWNP 1987), developed fifty years after the park's founding, the policy position on the Batek covers just two pages (in a 137-page document). We can see a glimmer of tension between how the park is defined and how the DWNP practically runs it in the statement that the agency is "accepting the fact that a certain number of Orang Asli have always lived within its borders and will continue to do so in the future" (DWNP 1987: 23). How, then, are the people incorporated into the broader ideology of wildlife management?

Administratively, the Batek are treated as an enigma. As the DWNP expresses it in the *Master Plan*, it has no "jurisdiction" (1987: 24) over the Batek. Orang Asli administration comes under a separate agency, the Jabatan Hal-Ehwal Orang Asli (JHEOA; Department of Orang Asli Affairs) of the Ministry of Rural Development.[16] While, on the one hand, the Batek's territory is managed by the DWNP, on the other, the Batek's territorial rights and social well-being are the responsibility of the JHEOA. And if the DWNP has no jurisdiction over the Batek, the JHEOA has no jurisdiction in Taman Negara. A fence demarcates responsibilities between these agencies: what comes between them—the Batek—is a "buffer concern" that confounds accepted practice. In any case, the JHEOA is unlikely to advocate the kind of policy for the Batek that would be acceptable in a national park.

For the DWNP—echoing Hubback's views—the Batek's activities should not jeopardize Taman Negara's goal of protecting the natural environment. Following provisions laid down in the Wildlife Protection Act of 1972 (especially part 5, section 52), which permits subsistence hunting by Orang Asli peoples, the DWNP "supposes that this mode of life—traditional hunting and forest-produce collecting for domestic use—does not have any long-term harmful effects on the ecosystem" (van der Schot 1986: 15). As long as the Batek keep to this idealized set of hunting-and-gathering activities, they will remain within this frame. Missing from the DWNP's image of the "traditional" life is the long history of forest-product collection, both for consumption and trade exchange, that has long formed a part of regional history (Dunn 1982) and is the Batek's chief source of income today.

There is a sharp conflict between the DWNP's definition of acceptable activities and the assimilationist policy of the JHEOA. The JHEOA's intervention in Orang Asli lives often begins by shaping their mode of food procurement, namely, by promoting agriculture (critiqued in Zahid 1990). Their policy, reminiscent of antique neo-evolutionism, is to encourage sedentarization and swidden farming for hunter-gatherers and permanent farming for swiddeners. The Ministry of the Interior's "Statement of Policy Regarding the Long Term Administration of the Aboriginal Peoples in the Federation of Malaya" (JHEOA 1961: 18) states "no encouragement should be given to the perpetuation of the present nomadic way of life."[17] There is no place in this schema for a way of life based on hunting and gathering as both production and value system.

The Batek's presence reveals clearly the cracks between these bureaucratic agencies. While the JHEOA does not support hunting and gathering, to the DWNP hunting and gathering is ideal for the national park since it is "closest to nature." While the JHEOA demands that the Batek change as a condition of joining the broader society, the DWNP demands that they *not* change as a condition of remaining where they are. The DWNP does not make a case for "the preservation of the ethnicity of the Batek in Taman Negara" (DWNP 1987: 24) or indeed show any inclination to challenge JHEOA policies. Its approach seems to suggest that as long as the Batek are here, they might as well be, or should be, hunter-gatherers.

The center of DWNP's official vision—note the wording in the founding enactments, cited earlier—is apparently focused on a preservationist image of the wilderness, in which "nature undisturbed achieved a constancy that was desirable and was disrupted in an undesirable way only by human actions" (Botkin 1990: 156). As van der Schot (1986: 16) continues, "The DWNP is of the opinion that a mode of life which departs from the supposed pattern might well be harmful to the natural ecosystem [and] has taken some restrictive measures with respect to the aborigines." The most controversial of these "restrictive measures" is the banning of commercial collecting within park borders, which the Batek resent.[18] In effect, the DWNP's policies continue to promote the wild, human-free image of the forest enshrined in the founding laws, rather than the material history of long-term modification.

Controversy over the establishment of Kuala Yong, a Batek settlement inside park borders near the headquarters (reviewed in Lye 1997: chap. 11), also stems from this issue. Kuala Yong is no more than a built-over forest camp, and the planting activities there (mainly of cassava, maize, bananas and other fruits) are another expression of a hunter-gatherer strategy of resource enrichment. The "gardens" there are in some sense a site for food storage that is, given the Batek's dislike of agriculture, unlikely to expand further spatially and technologically, and there is never a permanent population living on site. But while to the likes of agriculture advocates like the JHEOA, Kuala Yong shows arrested development, to managers long accustomed to associating the park with woodlands and to contrasting forest with croplands, these planting and building activities constitute an act of illegal trespass. The managers' arguments might be more convincing if they were also seen to practice minimal impact administration. But ever since the rise of tourism development, such is not the case.

Tourism, as envisioned by Hubback, was a justification for the conservation project (to assure critics that the park could have a social role) and never constituted the main purpose of Taman Negara: "It is abundantly clear that this [recreational development] and any other use of the Park are to be subject to its primary function, that of preserving in perpetuity its unique flora and fauna and other features" (DWNP 1987: 46). Until the early 1970s, even scientists were not freely permitted access to the park. This exclusivity could not have been sustained, however. My calculations from 1980–96 visitor statistics show an average annual increase of 18 percent for both Malaysian and non-Malaysian tourists (statistics released by the DWNP). The park, that is, has been gaining an increasingly public profile.[19] Furthermore, the DWNP was under political pressure to commoditize the park in line with the government's decision to switch to tourism as a foreign exchange earner. Pressure to privatize the recreational facilities in the park, mentioned in the *Master Plan*, was also materializing.

In the mid-1980s, concern for tourism's impact spread among environmentalists (especially the Malaysian Nature Society and the World Wildlife Fund for Nature, Malaysia), who urged the agency to adopt a long-term management plan. These organizations were also, to some extent, forcing the DWNP's hand by producing, or collaborating in, planning proposals (DWNP et al. 1986; WWFM 1986). This was the context for the *Taman Negara Master Plan*. In line with Hubback's original vision, this document continues to link Taman Negara to the American National Park System (NPS) model. And, overriding internal doubts over the place of tourism, the *Master Plan* shows the DWNP making certain concessions: "Taman Negara is the heritage of all Malaysians, none of whom should be denied the reasonable opportunity to visit their National Park" (DWNP 1987: 11).

In the early 1990s, the franchise for park accommodations was privatized; the Taman Negara Resort was opened. The resort's arrival ushered in a new era of tourism. A parallel development was the building of a new road linking the park headquarters to the district capital of Jerantut (finally completed around the turn of the twenty-first century). Before, the final phase of any visit necessitated a three-hour boat journey up the Tembeling River; with road access, it is possible to drive to the park from the capital city, Kuala Lumpur, in just four hours (as timed by myself).[20] Where Taman Negara had once primarily attracted backpackers and outdoor enthusiasts, it could now service an increasingly wealthy clientele and time-sensitive "package tourists," who also needed more efficient service and faster routes in and out. The local Malay villagers benefited from the tourism, for they monopolize the labor supply. As symbolized by the growth of the built environment, which has degraded riverine and wildlife ecology (WWFM 1986; Yong 1992), everybody in the vicinity—including the DWNP—has been changing. But the Batek, to retain their right to remain in the park, are not supposed to change.

The Wild and the Tame

Thus far we have seen how representations of the park, as revealed in administrative policies, are founded on colonial constructions of the wild. Both in the colonial period and today, there is a tendency to associate forest peoples with wildlife and to privilege the imagery of "timelessness." However, as I pointed out at the outset, we need to also recognize the strength of Malay perceptions in environmental protection. Selected bits of Malay as well as colonial perceptions show up in different kinds of policy interventions and enactments. In Taman Negara, we find these tensions and ambiguities exemplified in both the Batek's persona and the park. The continuity of Taman Negara in some sense preserves these older Malay images of the forest and people. The colonials did not, ideologically speaking, "sow their seeds" on completely infertile land. I have elsewhere concluded (Lye 2000, 2002) that the wilderness concept was a legacy of the colonial experience that was appropriated and indigenized. Here I will suggest that the colonial experience offered a definition of wild places that did not replace so much as reorganize and relocate older environmental ideas.

To begin with forest-people relations, George Maxwell (1982: 7–8) writes: "The Malays always consider themselves as intruders when they enter the forest, and never forget their awe of and reverence for it. They seldom go into the forest alone. . . . While it is true that the forest lies almost at their doors, they never forget not merely that no man knows its extent, but that it actually is without bound or limit." As Maxwell correctly suggests, the Malay classification system (common enough among farmers) places the forest "out there" beyond the world of people. The idea of "intruding" on the forest is critical, suggesting a deep respect for the forest's powers and acknowledgment that it is a whole other world in which one might get irretrievably lost.

This ontological separation of the village from the forest was manifested in the supernatural realm: "The forest in the Malay belief system was not simply a collection of huge trees. It was inhabited by good as well as evil spirits; it was feared and awed [sic]; it was considered to be a source of potent power" (Kato 1991: 150). According to Endicott (1970: 64), who pulls together the key colonial writings on traditional (pre-Islamic) Malay religion and was inspired by Mary Douglas's symbolic approach (see below), "Many times the spirits represent the fears into which the Malay divides his experience of hostile nature . . . on the whole the forces of nature . . . have negative values for the Malay." George Maxwell (1982: 7–9) reiterates: "To the Malays the great enveloping forest is full of supernatural powers. . . . They [various kinds of spirits] are known as *Hantu Hutan*—the Spirits of the Forest—and are as real to the Malays and as much dreaded as the tigers and other wild animals of the forest." Because of this fear of losing the self in the forest and being enveloped by these destructive spirits, "human contact with the forest, whether in the form of felling trees or collecting forest products, always required a ritual mediation" (Kato 1991: 150).

As for "ritual mediation," George Maxwell (1982: 8–9) continues:

> Men, such as rattan-cutters and gutta-hunters, whose vocations take them into the forest, repeat a short charm to avert the wrath or displeasure of these spirits; and the farther they go from home the more careful they are to make use of due ceremony and incantation. . . . Similarly, when a party of Malays sets out to drive deer, the commonest of all game, they may go no more than a few hundred yards away from the village; but none the less the leader of the party will utter this preface to his prayer to the spirits—

> "Hail! All hail!
> We crave permission to enter on this domain
> And to tie our noose to these trees."

As Maxwell's observation suggests, there is an ambivalent relationship between the Malays and the forest. At heart, despite an intrinsic fear of the forest, there was a capacity to recognize affinities with it and its wildlife. Charms and "incantations"—protective devices that resolved the spirits' "wrath and displeasure"—refer to their doubts about the moral position of the human intruders. Asking permission to enter humbles forest extractors in the face of the forest. As Wessing (1986) has shown, there was also ambivalence toward the wildlife. For example, like the Batek and many Orang Asli communities today, some Malays apparently regarded tiger visitation as a comment on moral conditions in the village, although in other circumstances they might have viewed tigers as pests to be eradicated (see also Boomgaard 2001: 175–76; Hubback 1932: 1). This perception is strikingly different from the colonials'. Recall, for instance, that Swettenham felt alienated from the forest; he saw nothing but gloom. Malays saw more than gloom.

In Malay theories of personhood, boundaries between categories are not impermeable. There is a hierarchy of values that takes the form of "a graded series" (Endicott 1970: 80). The possession of different kinds or degrees of soul-stuff (*semangat*) was traditionally the criterion determining membership in any category. Negritos (Semang)[21]—the ethnolinguistic group in which the Batek are placed—are grouped with spirits and animals. They are "on the boundary between man and non-man . . . regarded with fear and awe mixed with contempt" (Endicott 1970: 80–81). Evans (1937: 32) records similar perceptions: "The Malay reasons that the Negritos, being to all intents and purposes animals, must, therefore, be in league not only with the wild beasts, but with all the supernatural and inhuman powers of the forest that he dreads so much. On occasion, therefore, and individually, he is somewhat chary of offending the Negritos too deeply." Here, reiterating similar themes, is Schebesta (1973: 13) on the subject: "The people living around the dwarfs relate among themselves very extraordinary things about these tribes, but always in whispers, lest a dwarf hear. The story goes that they are not men at all, for they suddenly bob out of the ground in the most unexpected places; that they have glowing eyes, are ignorant of the use of fire, and eat everything raw."

If there was a clear separation between village and forest, "here" and "there," there was also, to express these ideas another way, a bridge crossed and represented by a host of spirits, animals, and Semang peoples. The Semang were ipso facto symbols of anomaly, "matter out of place" (Douglas 1966: 35). They violated established order and were therefore threatening and dangerous. This symbolic classification recalls settlers' perceptions of pests, mentioned earlier, and colonial perceptions of animals. To cite W. E. Stevens again, colonial planters thought that "wild life was all right in its place, but if it caused damage to crops, it should be ruthlessly controlled" (1968: 11). For planters, wildlife had its place in nature, but when it disrupted the domesticated world of agriculture, it needed to be "physically controlled" (Douglas 1966: 35). The fantastical images collected by Schebesta emphasize the anomalous status of the Semang: bobbing out of the ground (between seen and unseen, above and below), glowing eyes (seeing beyond the ken of mere mortals), eating uncooked food (having human and animal appetites), and so on.[22] The "wild" in this respect was not really the forest out there, which seems to have been well classified, but the forest *people*, whose persona contained both human and inhuman qualities. Being mobile and not fixed to physical location, they were polluting and therefore threatening to the world of the village. The Malay dislike of mobility has been described as "irrational" (Carey 1976), but within this larger classificatory order, that dislike makes perfect sense.

Symbolic classifications are reinforced by (and reciprocally give meaning to) material conditions. As Malay settlers expanded into the Peninsula, and different relations of dependency and subjection developed with the indigenes, and as lands and resources were appropriated and controlled, so did symbolic classifications assume political legitimacy. Most prominent are the binary adjectives used generally to describe aborigines: *liar* (wild) and *jinak* (tame). The latter referred to established trading partners, slaves, or former slaves, while the former were those unacculturated groups that were valued (and feared) for their freedom and autonomy (Couillard 1984; Dentan 1997).

The ambiguity and duality in these Malay perceptions persist. *Liar*, especially, has stuck and remains among the insults the Batek bristle at today. They are well aware that the Malay term is associated negatively with animals and "running all over the forest." For example, in one exchange that I observed between a Malay man and some Batek youths, the former argued passionately that the latter had to stop being so *liar*, because they were denying themselves the chance to settle down and "progress." In such encounters, the Batek respond quietly; in privacy, away from visitors, they retort that the Malays do not know how to think, for the Malays are defining through their own classificatory lens what the Batek consider a valid form of cultural behavior.

Irreconcilable Wilds

The foregoing discussion shows that Semang have a well-defined place in Malay environmental ideologies. These older ideas of the wild predate colonialism and have been subdued, but not destroyed, by Euro-American landscape values. Where the latter postulates a world without people as the ideal Eden (with a politically peripheral place for noble savages who do not disrupt the landscape), the former puts the primacy on people, not place. When the American ideal of wild *places* lands in a conservation area like Taman Negara, it joins hands with Malay ideas about wild *peoples*. The mix is the source of the administrative headache of DWNP. The situation in the national park today—where the Batek's needs and resource practices are alternately neglected and bounced between agencies, subjected to their respective bureaucratic inscriptions—shows continuities with both colonial and precolonial times. As noted already, the park protects ideas and images as much as ecology. It protects memories of the past on which, to some degree, current political action is contingent (Lye 1997, 2002; cf. Connerton 1989). These memories are embodied in the Batek's behavior and persona; by continuing to persist "as they are," the Batek confirm and reinforce Malay environmental ideas.

The park is fraught with symbolic inversions. On the one hand, the colonial legacy is evident in the ordered, culturalized environment of the park. The enclosure is firmly signified by yellow markers posted at one hundred–meter intervals along the boundary while, internally, there are cleaned, cleared debris-free trails, rope fences, direction pointers, and markers that transform the forest into a park in which visitors may wander in safety. It is a tourist destination. In contrast, there is the naturalizing of its people. Only by embodying a certain image of the wild—whether this is the Orang *liar* of Malay perception, the classical hunter-gatherer of the popular imagination, or the colonials' Sakai—can the Batek fit into the ethos of park management. In so doing, as forest peoples, they are a symbolic Other, the mirror through which Malays see their past and the cognitive threat of the forest. As long as the Batek remain there, the forest is never completely or solely a "park." By remaining there, they symbolically confirm both the appropriateness of management logic (privileging nature over culture) and their own place in and identification with the forest, which in turn is an important part of the imagery supporting their "right" to remain there. Recursively, this cognitive association sustains the wilderness image of the forest—that it is a dangerous place that only a special kind of person, a wild person, can successfully inhabit.

In short, the Batek's hunting-and-gathering way of life symbolically supports the official image of the park as both a protected area and a tourist destination, which in turn supports the continued existence of the forest. However, these associations may not always be self-evident to officials and, as shown in bureaucratic tensions over Orang Asli administration, they are incongruent with the development ethos that operates elsewhere in government. The fissure between the DWNP and the JHEOA reflects some of these symbolic tensions. The JHEOA's assimilation policies attempt to put Semang in their proper place—settled into farming, out of the forest—and erase this primal threat to society. This evolutionism is at core a Malay view, which has become more entrenched in current practice (McKinley 1979). Mobile hunting and gathering seems to exemplify everything Malays find "shameful" or embarrassing about their past.

The DWNP's perspective—and indeed its very existence—on the other hand, is a product of Euro-American perceptions of wildlife. The cracks between the DWNP and the JHEOA, as embodied by the existence and conditions of the Batek, show that this mapping of science onto the local, of colonialism onto Malay, has not been successfully completed. There is no sharp distinction between science and nonscience, between global and local; colonialism has not been banished, just as local ethnoecologies have not been wiped out. Rather, they have been brought together, and the space in the middle, where the ideological braids are woven, is the political and ecological niche of the Batek. As long as the Batek do not need political representation and practical assistance, this is a space for maneuvering that enables them to resist controlled intervention by either the DWNP or the

JHEOA. As such, the orientation of the DWNP (and probably of the JHEOA) also inadvertently serves the Batek's interests and seems to be compatible with their classical mode of life. The Batek are a "boundary" people: they retreat into the forest interior when they wish to place a distance between themselves and threatening Others, and they move out to the forest fringe to take advantage of economic opportunities (trading, tourism, etc.) and when it appears convenient. This boundary niche, of course, is precisely what the Malays saw in the Semang of old and what caused them so much unease. In today's conservation context, this "boundary behavior" is precisely what threatens the presumed order of the park.

Conclusion: People in the Park

Malaysia's Taman Negara is the ritual center of wildlife conservation in the Peninsula. Its success is conventionally assessed by how well it serves its original goal of protecting biodiversity and habitats. Taking an alternative approach, this essay has asked, rather, why the administration of the park, especially with regard to the place of the Batek in it, is the way it is, and this has necessitated a journey through the historical context. For Taman Negara is a product of a profound moment of social change sixty years ago, which led to a radical reorganization of the social-ecological landscape, but without affording clear ideas how those changes would be dealt with in the future. How the national park has survived heretofore is to some degree contingent on the ideas and values that give it meaning.

The colonial identification of forest peoples as noble savages, which was part of a larger conquest narrative, has endured successfully through time. It coexisted with the old fears that forest peoples possessed both human and animal characteristics, and both imageries persist administratively in doubts about the status of the Batek. At the beginning of the essay I asked what were the political consequences of particular representations of the world. The answers from this discussion are counterintuitive. Images of Orang Asli as people who need to be protected from one perspective seem dehumanizing, but from another, it gives them a place in a major conservation area. In one fell stroke, the park enactments assured permanent protection of a sizeable extent of the Batek's territory and thus the reproduction of the Batek's mobile way of life (Lye 1997, 2004). The DWNP's policy position, tolerance as long as things are going well, permitting the Batek to travel freely throughout and in and out of the park without restriction, is congruent with the Batek's desire for cultural autonomy and freedom of movement. Had the park not come into being, it is easy to see what would have happened: their lands would have been claimed by local Malays, opened up, commoditized, and increasingly fragmented.[23] The Batek's capacity to pursue their chosen modes of life and the maintenance of their sociocultural norms would have been seriously compromised.

The current status quo is not, however, without its problems. In common with forestry "narratives" throughout the world (see, e.g., Bryant 1996), there is a tendency to privilege—even inscribe in law—some environmental representations and suppress others. The imagery of wilderness excludes other possible ways of viewing the Batek's hunting-and-gathering way of life. On the one hand, the law protects their right to subsistence hunting, and the Taman Negara administration indirectly protects their dependence on the forest. But this official vision discounts their involvement in commercial activities and does not legitimize other aspects of their environmental relations. One of the latter concerns how their way of life materially maintains the reproductive capacity of the forest. Nor is this lack of recognition the result of an information gap. Long ago, H. N. Ridley (1893) recognized the conservation role played by the Orang Asli in incidentally dispersing seeds of fruit trees throughout the forest (Lye 2004). Hislop (1961) and Wharton (1968), among others, have argued that forest clearance for swiddens enriches browse for wildlife. There is material evidence, in short, that an interdependent relationship exists between biotic life and human environmental relations in the Malaysian forest. These relations actively alter ecological composition and contribute to the apparent diversity of the forest and might thus be incorporated into resource-management programs. But such possibilities are excluded from official representations of the national park. Because the park originated as a wildlife reserve, people can only be seen as an annoyance and a potentially destructive force. The Batek's relationships with the land are constantly under suspicion. Although the Batek are potentially the DWNP's closest conservation ally in Taman Negara (Lye 2005), the constant imposition on them of demands and regulations has created tension and disagreement and has led some Batek to wish the park did not exist. But it is difficult to know how to deal with this.

It is clear that an opportunity has been missed by the DWNP and the Batek to develop a collaborative relationship, sharing different but complementary conservation niches. This missed opportunity is due in no small part to colonial constructions of local identities and Malay environmental perceptions that could not accept forest-based mobility as equivalent in social status to the domesticated world of village and city. Both colonials and Malays were unused to granting political legitimacy to people like the Batek, who desire autonomy and freedom outside the bounds of state. But would creating an institutional framework for collaborative (or participatory) conservation necessarily solve the problem?

As Brosius, Tsing, and Zerner (2005: 9) point out, "collaboration may favor the powerful, rather than the weak, who are unable to set the terms of the collaboration." In comanagement regimes in the Yukon, for example, "attempts to incorporate the lived experiences of First Nations elders and hunters into the language and institutional structures of modern bureaucratic resource management lead inevitably to their compartmentalization and distillation, thus seriously distorting them" (Nadasky 2003: 132). If the official gaze *were* to land on the Batek, however positive in intention, their behavior would be subjected to more attention and control. This might lead to small improvements in everyday park-people relations but would also create more conditions for bureaucratic domination (Hitchcock and Holm 1993). Concern over habitat loss tends to lead to increased institutional levels of governance, as does concern that protected areas may impoverish locals (Wadley, Colfer, and Hood 1997: 264). Reports from elsewhere (e.g., Conklin and Graham 1995), which resonate with the Malaysian context, also support the position that greater public attention, whether for good or bad, sooner or later causes a political "backlash." As I suggested at the outset, bureaucratic reform is unlikely to dislodge underlying perceptions. In the past, the Batek have been viewed both as "embarrassing to the national image" (given their supposedly primitive qualities) and as one of "*our* Orang Asli." The proprietorial attitude is extremely salient. One reason is the ambivalence that I have been pointing to: they are suspected to have wild qualities, to live like animals, to be unable to decide the terms of their future without outside help. In short, the ideological, political, and bureaucratic conditions of Malaysia today do not point to collaboration, comanagement, or participatory conservation as the best outcomes for the future.

The only agency accorded jurisdiction over the Batek is the JHEOA, and its policies are fundamentally incompatible with the Batek's forest orientation. Given the embedded distrust of the Malay state for the nomadic condition, the effect of increased JHEOA attention would be to sedentarize and "modernize" the Batek and, consequently, a loss of place for them. The DWNP, with its interest in promoting better relations with wildlife, is the best-equipped agency to deal with the Batek, but the laws do not permit it to do so. Without fundamental structural reform (see, e.g., Lim Heng Seng 1998), the current management style of the national park may be the key to persistence of the Batek in the wild.

Accordingly, this study offers a critique of conventional neomodernization, which prescribes the state as the agent of social change. What Taman Negara exemplifies is the success of interagency *failure* to meet expected standards of performance. The Batek's persistence and continued well-being is less the product of development intervention than of their ability to move between bureaucratic agencies and their respective systems of thought. In other words, it is because there are limits to the jurisdiction of both the DWNP and the JHEOA that there is room in the middle for the Batek. It is because government agencies do not have a clear idea about what to do with the Batek that the government has managed to protect their interests and, indirectly, the status of the national park.

Notes

This essay is based partially on research funded by the Wenner-Gren Foundation for Anthropological Research, the East-West Center, and the John D. and Catherine T. MacArthur Foundation (1993, 1995–96). Research permissions were granted by the Economic Planning Unit of the Prime Minister's Department, Kuala Lumpur, the State Economic Planning Unit in Kuantan, Pahang, and the Department of Wildlife and National Parks. A postdoctoral fellowship from the Japan Society for the Promotion of Science (1998–2000) made possible the conditions of writing this article and my participation in the Center of Excellence project of Kyoto University's Center for Southeast Asian Studies, which also afforded further opportunities to track developments in Taman Negara. Additional writing support was received from the MacArthur Foundation (1998) and the Resource Management in Asia-Pacific project of the Department of Human Geography, RSPAS, Australian National University (2001). I am indebted to these institutions for their support. For useful comments on earlier versions of this essay, I thank Michael R. Dove, Rosemary Gianno, and Libby Robin. I also received much insight on the history of wildlife protection from H. S. Barlow, Gurmit Singh, Dennis Yong, and, especially, Gordon P. Means, who took the trouble of uncovering insights from the unpublished papers of Paul Means. The rangers and officers of the Department of Wildlife and National Parks, by allowing me access to Taman Negara and letting me watch them work, have also contributed much to my growing appreciation of the problems of park management, as have the Batek, by communicating to me the social dimensions of living in the park. I am grateful to all.

1. *Taman Negara* translates directly as "National Park." The globally established categories of protected areas are enshrined in the World Conservation Union or IUCN's Protected Area Management Categories. These categories range along a continuum of exclusiveness, with the most inclusive categories being areas in which various forms of human activities are permitted (Bates and Rudel 2000; Robin 2000). National parks are only one type of protected area, but for the purposes of this essay, these terms are used interchangeably.

2. The DWNP is an agency of the Federal Ministry of Natural Resources and Environ-

ment. The eastern Malaysian states of Sabah and Sarawak in Borneo manage their own conservation programs and are therefore not addressed in this discussion.

The Batek are an indigenous ethnic minority people of peninsular Malaysia (Orang Asli; Malay for "original people"). They live in Pahang, Kelantan, and Terengganu States. Most of Taman Negara lies in Batek territory. Before large-scale logging and land transformation began in the early 1970s, the Batek territory was contiguously forested; Endicott (2000: 110) estimates that roughly two-thirds of that area is lost beyond regeneration. The total population ranges between 800 and 900. Half of those in Kelantan and most in Terengganu are now semisedentary cash-crop farmers. In Pahang, where the population numbers 450, most are hunter-gatherers who engage in commercial extraction and trade; just over half spend the majority of their time in Taman Negara. A distinctive feature of their way of life is their mobility: they average some two weeks in each location and three or four months in each river valley. The average population per encampment is roughly 38 and, thus, there may be some ten different groups throughout the forest at any one point.

My fieldwork was restricted to the Batek of Pahang. I have conducted fieldwork in Taman Negara on and off since 1993, mainly while living with Batek subgroups. My stay in the park has ranged from several weeks to several months at a time.

3. I am, of course, generalizing here. As Neumann (1998) points out, Euro-American conceptions of the wild are themselves historically contingent. These internal variations and contradictions will not concern us here.

4. There are a lot of gaps in the record: "One of the first things that became evident in studying the Game Department [the former name of the DWNP] in Malaya was the almost complete lack of records and files concerning what had happened in the past" (W. Stevens 1968: 6).

5. Practically all the biographical details that follow are from H. S. Barlow's brief article (2000), little else being available. I have talked with Barlow and gathered additional information from Gordon Means (1999 personal communication, based on personal papers and draft manuscript by Paul B. Means, written between 1960 and 1966). Paul Means, a linguist-missionary who worked with Sengoi-Semai, knew Hubback personally and recorded some of the latter's thoughts about conservation.

6. With, as far as I can tell, one exception: a Jakun leader (Hubback 1932: 1:81).

7. In this essay, I use the terms *aborigines* and *indigenes* interchangeably, although I recognize that there may be analytical and political differences between them.

8. The commission's bibliography (Hubback 1932: 1:421) does include several key volumes on aboriginal tribes.

9. Presumably, Hubback caught the drift of late Victorian enthusiasm for natural history and fear of wildlife extinction (Farber 2000: 87–99). He may also have learnt much during Alaskan hunting expeditions in 1917–18. The British (specifi-

cally, the English and Welsh) did not launch a national park system at home until the 1940s. The Americans are credited with popularizing wilderness protection as a conservation measure (see Robin 2000). However, the wilderness ideals that they espoused drew heavily from English aesthetic traditions (Neumann 1998).

10. In terms of land area today (it has been adjusted since its founding), 57 percent of the park is in Pahang, 24 percent in Kelantan, and 19 percent in Terengganu. As such, the park was founded from three state enactments: the 1938 Kelantan Enactment (No. 14), the 1939 Pahang Enactment (No. 2), and the 1939 Terengganu Enactment (No. 6).

11. His death is shrouded in mystery, as he had fled into the forest when the Japanese invaded Malaya. He was among the handful of colonials (H. D. "Pat" Noone being the most famous) who went underground and were trusted to survive behind enemy lines. Experienced in the forest and known to locals, they were considered capable of organizing a guerrilla force (Chapman 1997). According to one line of speculation, Hubback "was murdered by local inhabitants who harbored a grudge against him for the restrictions that he placed on their hunting activities" (H. Barlow 2000: 23). In another account, uncovered by Paul Means from the Orang Asli of Telom, Pahang, Hubback was murdered by Japanese forces (Gordon Means, personal communication). Neither Hubback's remains nor his collections of papers and artifacts were ever found.

12. There might have been some expulsion of forest marginal communities when park boundaries were readjusted and enlarged after the war.

13. Kuala Tahan, at the confluence of the Tahan and Tembeling Rivers, is a site of some dispute. Currently it is the headquarters of the national park and the main entryway for most tourists. The Malay village of Kuala Tahan, the main beneficiary of park tourism, is located on the opposite banks of the Tembeling (which forms the southeastern border of the park). But the Batek insist that they formerly maintained a permanent base camp at this site and feel displaced by both the park headquarters and the Malay villagers. See Lye (1997: 398–401) for a sketch of the Batek side of the dispute.

14. Contrast this early attitude to shifting cultivation with that toward timber production. The 1937 *Report on Forest Administration for Malaya*, for example, only grudgingly excised forest reserves for Orang Asli: "The aboriginal tribes need protection and assistance, but hardly to the extent of settling them in localities where there is a real danger of a timber famine, to the neglect of other places where the destruction of a few trees is not a matter of great moment" (quoted in Cooke 1999: 44).

15. Related to these sentiments was the colonial denunciation of Malay slavery of aborigines and other modes of sociopolitical control. Dodge (1981) has argued that colonials were not only genuinely shocked by the iniquities of slavery but also used slavery as an excuse to justify the colonial project.

16. As promulgated in the Aboriginal Peoples Act 134 of 1954 (revised 1974).

17. Updated in 1977 with no fundamental change.

18. In the 1990s, the Batek of Taman Negara accommodated to park strictures by moving out to bordering areas (especially on the other side of the Tembeling River boundary) to collect forest products, especially rattan.

19. The park's public profile in the mid-1980s increased following proposals by the federal government to construct a dam on the site and a Jeep track up to Gunong Tahan, both of which were met with concerted opposition from environmental groups. There is a distinct swelling of Malaysian visitors in those years when development proposals are debated in the media.

20. In the years since my primary phase of fieldwork, three new entry points to the park have been established. These developments were discussed as far back as the 1980s, but they did not materialize for fifteen to twenty years. I have not visited these sites. One, Kuala Koh, is within Batek territory in Kelantan and will likely have a significant impact on the people there.

21. Some writers (and most administrators) are still using the derogatory term *Negrito*. I prefer *Semang*.

22. Note, however, that such images are widespread throughout the Peninsula, including among the Batek. Orang Asli (like anyone else) think of neighbors they distrust in such terms and also project like characteristics onto mysterious demonic Others. The point is not that these images are native to the Malays or to specific ethnic groups, but that they reinforce and reflect larger social and political tensions present in society (Hoskins 1996).

23. Enactments to protect aboriginal interests by other means than the park have not been forthcoming (Nicholas 2000) or leave much to be desired (Hooker 1996). All peninsular Malaysian hunter-gatherers except the Batek of Pahang have lost traditional territories and become sedentarized.

THE IMPLICATIONS OF PLANTATION AGRICULTURE
FOR BIODIVERSITY IN PENINSULAR MALAYSIA
A Historical Analysis

To limit future generations to an environment of monotony is to limit their ultimate self-development.—SALLEH MOHD NOR AND FRANCIS NG, "CONSERVATION IN MALAYSIA"

Sacred forests arguably had less relevance in areas of abundant forest and low population density such as peninsular Malaysia (hereafter the Peninsula) than where the demands of larger populations on forest resources called for the protection of specific sites (Gadgil and Guha 1993: 38, 76, 88, 132.) However, the concept of "sacred trees" associated with valuable products like the camphor (*Dryobalanops aromatica*) and *gaharu* (*Aquilaria malaccensis*) was common in Southeast Asia. Taboos, rituals, and elite privileges that imposed a measure of restraint on the harvesting of certain economically valuable species suggest an appreciation of market mechanisms for sustaining low-bulk, high-value production. And the meteoric rise during the mid-nineteenth century of the unprotected gutta-percha (*Palaquium* spp.) as a high-bulk, low-value product in Southeast Asian markets similarly demonstrated indigenous responsiveness to global markets (see Kathirithamby-Wells 2003: 45–48).[1]

Malay readiness to move from foraging to the cultivation of cash crops came spectacularly to the fore with the spread, as in the sixteenth century, of Malabar pepper (*Piper negrum*), followed by the nineteenth-century adoption of neotropical plants. Among the latter, coffee, tobacco, and rubber were in the lead, their production for the market made viable through the simultaneous adoption of neotropical food crops in the form of sweet potato (*Ipomoea baratas*), maize (*Zea mays*), and cassava (*Manihot esculenta*) (Pelzer 1978: 273–74). Less labor intensive, more tolerant of poor soils, and yielding a higher starch output per unit area than rice, the new subsistence crops were readily integrated into the swiddens and *kampung* (village) kitchen gardens of sedentary paddy cultivators (I. Burkhill 1966: 2:168; Crosby 1972: 171–74; Boomgaard 1998; Zaharah 1992: 312–14). The freeing up of land and labor for cash-crop production assisted Southeast Asia's transition into the era of modern international commodity trade, well preceding the emergence of the colonial-backed plantation industry (Kathirithamby-Wells 1995: 35–38).

One of the most successful cash crops of the twentieth century was rubber. A secondary growth "weed" of the Brazilian rain forest introduced in the Peninsula at the end of the nineteenth century, it became part and parcel of a stable "traditional" Malay peasant landscape. But overlooking this ideal pattern of subsistence-cum-cash cultivation that evolved at the level of the peasant economy, colonial policy favored the higher tax yielding but environmentally more disruptive European estate economy.[2]

Following independence in 1957, the colonial legacy of monocultivation was carried over and reached new heights of intensity in rural megadevelopment schemes expressly aimed at poverty eradication. Assisted by massive postwar technological improvements, the expansion of profitable oil palm, less suited to smallholder production than rubber, completed the process of pushing ecologically sustainable peasant agriculture from a core to a marginal position (Kathirithamby-Wells 1995: 39).

The first part of this essay traces the pre–Second World War ascendance of plantations over more environmentally and economically viable systems of resource use. The second part describes the postwar escalation of the trend and assesses its implications for the Peninsula's biological biodiversity.

Nineteenth-Century Monoculture

In the Peninsula as in Indonesia, the demographically adapted practices of shifting cultivators and intercropping conflicted with colonial concepts of commodification for the international market through intensive plantation agriculture (see Pelzer 1978: 44–50; Dove 1985a: 30). Characterized by neat, clean-weeded, even-aged, and regimented rows of export crops, monoculture had its antecedents in regimes of colonial forced cultivation. In the early nineteenth century the practice was adopted by agriculturalists in Singapore as a means of production for the free market. While Europeans established nutmeg gardens, Chinese adopted shifting cultivation for planting gambier, pepper, and tapioca, taking advantage of the vast reserves of forested hinterland to reap reliable returns (Turnbull 1972: 146–55). To contend with rapid tropical soil weathering, exacerbated by monocultivation, they made fresh gardens at the beginning of each cycle of cultivation, gradually extending their activities to neighboring Johor and the west-coast states of the mainland Peninsula.

The Chinese gardens were profitable but within a short time demonstrated the adverse effects of monoculture on the environment. Though the parsimonious recycling of organic waste added nutrients to plants in cultivation, both pepper and gambier exhausted the soil after a single crop cycle of about fifteen years. In the case of tapioca, the soils degenerated even sooner, after about five years (Aiken and Leigh 1992: 53). As a result, the steady wave of clearance for new plantations left behind vast swaths of infertile ground covered in *imperata* (*lalang* or cogon grass) (Jackson 1965: 77–81).

Early twentieth-century development in the Peninsula saw the introduction of the "cheap breakfast": tapioca, sugar, and especially coffee. Encouraged by the government, cultivation became widespread, principally in Selangor and Perak (Voon 1976b: 25). Large tracts of mangrove and freshwater swamps made way for sugar, and steep slopes on Pulau Pinang were cleared for coffee. But beset by problems of organization, falling prices, and disease, neither succeeded as a major plantation crop (C. Barlow 1978: 24; Vincent et al. 1997: 9).

The more successful European experimentation with rubber began with the distribution of seeds by the Singapore Botanical Gardens and improved methods of tapping pioneered by its director, H. N. Ridley, based on indigenous methods of cutting bold incisions (I. Burkill 1966: 1:1170–73). Supported by government loans and 999-year leases, speculators initially hedged their risk by interplanting with coffee, but later, with increased confidence in the crop, they converted to rubber monocultures (Harrison 1985: 107). Rubber, a fast-growing species, typical of secondary, "seral" or subclimax forest of the Amazon Basin, is suited to the Peninsula where the dominant oxisols and ultisols, deficient in chemicals, are unsuited to the continuous cultivation of annuals (Aiken and Leigh 1992: 39–40). Able to thrive on the fired, devegetated, and degraded landscapes prepared for plantation agriculture, rubber was put to intensified production with the input of large capital and cheap labor.

While the corporate plantation investors were largely Europeans, the majority of smallholders were Chinese who switched to rubber in the face of the declining price of gambier (C. Barlow 1978: 30). It also provided them a means of converting to productive use extensive areas of land degraded by tapioca cultivation in Negeri Sembilan (Aiken and Leigh 1992: 53–54).[3] Rubber smallholdings added to the Chinese portfolio investment in tin mining, shopkeeping, and other business ventures. So attractive was rubber as an easy cash crop that the Malay peasantry soon planted seeds they obtained from the estates within the existing mix of nonintensive rural cultivation.

Between 1900 and 1906 the price for rubber in London almost doubled, providing the incentive for a phenomenal expansion in cultivation. Between 1905 and 1907, for example, the area under rubber in the Peninsula more than trebled, from 18,600 hectares to 68,800 hectares (C. Barlow 1978: 26), and it continued to grow, assisted by the construction of roads and railways. The attraction of rubber in colonial Malaya was enhanced by the converging factors of international market demand, a government-supported planta- tion industry, and the bid by the less privileged but robust smallholder sector for a share in the new cash economy (Lim Teck Ghee 1976: 129; 1977: 19). While the carrot—in the form of low land rents, the Loans-to-Planters- Scheme, and the unrestricted entry of overseas financiers—was dangled be- fore European speculators, the stick was reserved for the shifting cultivators, pressured by the government to take up sedentary paddy cultivation without visible market incentives. On the whole, the best lands for "scientific plant- ing" were given out to Europeans and a "no rubber policy" pursued with regard to the peasantry through restrictions on land use (Lim Teck Ghee 1977: 117). These measures merely served to encourage illegal cultivation (Nonini 1992: 75–76).

The Paddy-Rubber Symbiosis

Neither the Chinese monocultivators, nor the Malay counterparts, who raised a mix of crops replicating the vegetational mix of the forest, won the favor of colonial administrators. Smallholder cash cropping, widely associ- ated with shifting cultivation, conflicted with colonial land, agricultural, and forest policies. Little heed was evidently paid to the lone voice of Hugh Clifford, then British Resident of Pahang, who saw in plantation agriculture the seeds of destruction to the fabric of Malay rural life (Barr 1977: 91).

In their encounter with the exotic para rubber (*Hevea brasiliensis*), the Malays were able to draw on indigenous practices of extracting gum and exudates from forest trees, successfully expanded earlier in the nineteenth century for marketing not only gutta-percha but also *jelutung* (*D. costulata*) (Kathirithamby-Wells 2003: 45–48; 98–99). Among the first to take advan- tage of the experimental planting of *getah rambung*, or India rubber (*Ficus elastica*),[4] in Selangor and lower Perak, the Malays readily took to the com- mercially more viable *Hevea brasiliensis* when it was subsequently intro- duced (Lim Teck Ghee 1976: 74, 97n18).

To the bulk of the Malay-Indonesian immigrants (Bataks, Mendeling, Minangkabau, and Javanese), many of whom came from a background of cash-cropping pepper, gambier, and coffee cultivation, colonial perceptions of a rice-cultivating peasantry represented an imagined tradition. Malay resistance to producing rice for immigrant labor in the tin mines and the rubber plantations was expressed in their rapid entry into the rubber market, notwithstanding the lack of government incentives.

Smallholder rubber proved well adapted to the rural economy. Tapping provided a year-round source of income, developed credit for paddy, and reduced borrowing for everyday needs. Compared to the cultivation of surplus rice for the market, the cultivation of rubber was less exacting in terms of labor, capital investment, and market risk (Bauer 1961b: 196–97). The planting of rubber opened a window on the global economy for an otherwise shuttered rural world under colonial rule. At the same time, subsistence rice cultivation provided a stable backup for entering the cash economy and helped planters tide over the interim before rubber reached maturity.[5]

The strong attraction of rubber brought into play the familiar peasant strategy of seeking favorable opportunities by skillfully manipulating government laws and policies. The offer of rent-free tenure to encourage settlement was seized on for planting rubber (Drabble 1979: 73–74; Kratoska 1975: 10). Likewise, the Malay Reservations Enactment of 1913, aimed at protecting *pesaka* (hereditary) lands traditionally dedicated to wet-rice cultivation, merely stimulated the acquisition of new land (*tanah pencharian*) for conversion to rubber planting (Kratoska 1983: 153). In Kuala Kangsar, Perak, between 1900 and 1920 the total area under rubber increased twofold to occupy 31 percent of the land. The 1911 scheme for rent remission on ancestral lands planted with rice failed to elicit wide response and, in Selangor, the peasantry opted to pay higher rents for the freedom to plant rubber (Kratoska 1983: 153). The wholehearted commitment of Malay peasants to the new economy was demonstrated by their conversion to rubber of any land unsuited to paddy or abandoned by European investors. Frequently even village orchards made way for rubber.

While falling world prices after 1913 reached an all-time low during the 1929–32 slump, rubber remained profitable for the peasant smallholder. It offered better cash returns per unit area, a year-round income compared to rice, and it stimulated the rural cash economy (C. Barlow 1978: 40). Cash income from rubber reduced levels of indebtedness, provided attractive collateral for borrowing during hard times, and was an important incentive for Malay smallholders to stake a claim on land outside the Malay reservations (Voon 1976a: 513, 517). Profits from rubber enabled the village heads (*penghulu*) in particular to weather the economic hardships brought about by the

First World War (Shamsul 1986: 29). In the long term, wealth accrued from rubber contributed to the formation of a new elite.

The spectacular expansion of nonestate holdings from a third of all land cultivated in rubber in 1911 to about half of all land cultivated in rubber by 1920 (Drabble 1973: 83; C. Barlow 1978: 26) evinced the success of the Malay struggle to participate in the cash economy. One of the main reasons for the staying power of smallholders against competition from the estates was the easy and natural conditions under which rubber flourished. Planted from seeds picked at random, rubber grew in most flood-free terrain and was ready for tapping after five to six years. The collection process involved simple tools, namely, a tapping knife and cups that the smallholder improvised from discarded cigarette tins. The latex, collected in used kerosene tins, was conveyed on poles or bicycles from the plantations to village backyards, where it was processed into sheets with the use of hand-mangles (Drabble 1979: 73).

Any advantage the corporate estate had over the family smallholder through mechanized processing was lost in larger capital outlay, a considerable proportion of which covered the high cost of immigrant labor. In contrast to the minimum M$110,000 required to bring a two hundred–hectare plot into production, the investment in smallholder planting represented little more than the price of seeds and the productive utilization of off-periods in the paddy cycle (C. Barlow 1978: 29, 39–41). Smallholdings constituted "the most vigorous sector of the rubber industry," able to score a higher annual yield than estates, averaging 500–568 kilograms per hectare, compared to about 454 kilograms achieved by estate holdings (C. Barlow 1978: 81; Drabble 1979: 85).

Ironically, government discrimination against smallholders in the interests of promoting paddy cultivation inadvertently contributed to escalating peasant initiative in the rubber sector simply to stay competitive. By 1921, the smallholder sector, in the Federated Malay States (FMS) of Perak, Selangor, Negeri Sembilan, and Pahang accounted for 33 percent of total output. To restrict production in this sector, with a view to protecting the estate industry at a time of tumbling prices, the government implemented the Stevenson Restriction Scheme (1922–28) (Lim Teck Ghee 1974: 105–8; Bauer 1961c: 243). Though smallholder production in 1921, estimated at between 578 and 1,345 kilograms per hectare, well exceeded output from plantations (C. Barlow 1978: 71, Appendix 3.2), the restriction went ahead based on prejudice and undue criticism of peasant smallholdings as badly kept and prone to generating disease (Sundaram 1988: 70).

The Environmental Implications of Peasant Smallholdings versus Estates

Contrary to received opinion, smallholder rubber was healthier and environmentally better adapted than that of European estates. The unobtrusive rubber trees blended into the patchwork of orchards and anthropogenic secondary forest and scrub, historically evolved to support forest-edge communities, both human and animal (Spencer 1966: 127; Kathirithamby-Wells 1995: 217–19). The small parcels of Malay peasant rubber averaging 1.4 hectares and interspersed with patches of orchard, paddy, and forest offered a diversity of landscape and vegetation. These contrasted with the unbroken, monocultural stretches of European estates of over forty hectares, or with the largely Chinese-owned medium-sized holdings of eighteen hectares. Well into the mid-twentieth century, peasant rubber holdings continued to be planted among tall undergrowth and fruit trees and could hardly be distinguished from the surrounding jungle if viewed from the air.[6] "With the close planting and the tall undergrowth the peasant holding approaches in appearance an abandoned jungle clearing reverting to secondary forest, differing from it only in that there is a single dominant species in the holding instead of the variegated species common in a rainforest. By duplicating jungle conditions in his holdings, the peasant is following sound conservation principles" (Ooi Jim Bee 1959: 146). The agricultural economist P. T. Bauer observed that smallholder rubber, which retained natural undergrowth and heavy leaf fall, took virtually nothing out of the soil. It was even seen as environmentally beneficial through keeping out the pernicious *lalang* from degraded land (Bauer 1961b: 255).

In contrast to the smallholdings that melded into the natural landscape, estate cultivation proved environmentally unsound. The scale and methods of clearance before planting contributed to serious erosion. It involved the firing of many hectares of forest, followed by the uprooting of stumps and remaining roots, ostensibly to eliminate any disease harbored in them. The young plantations presented, according to one observer, "a queer battle-scarred appearance" (McNaughton 1936: 89). The rubber estates, often established on slope land, were first cleared, leveled, and terraced manually before planting. Thus even without the aid of the mechanical diggers and tractors that became available after the Second World War, ground preparation in the early estates wrought considerable damage on the environment. The removal of ground cover reduced the surface interception of rain flow, exposing soils to a high level of weathering, erosion from accelerated surface runoff, and leaching. These processes were exacerbated during the rainy season when the planting took place. The potential stabilization of the soil after planting was further retarded by the misconceived practice of labor-intensive clean-weeding, aimed at reducing nutrient competition (Belgrave 1930: 492). It was estimated that during 1902–39, clean-weeded rubber estates suffered an average annual loss of 7.6 centimeters of topsoil (quoted in Aiken and Leigh 1992: 83). According to a 1940 estimate, clean-weeding contributed to an annual 1.1 billion tons (1,117.6 billion kilograms) of soil washed down the rivers of the Peninsula (Pahang Natural Resource Board n. d).

Rubber research, funded largely by corporate estates until the founding of the Rubber Research Institute, Malaya (RRIM), in 1926, was blatantly biased toward plantations (C. Barlow and Drabble 1990: 202). Despite drawing some 45 to 50 percent of its revenue from smallholders until 1934, a mere 2 percent of its budget for the same year (raised to 10 percent in 1939) was allocated to assisting smallholders. Ironically, the withholding of assistance to smallholders, including the services of Europeans on the Advisory Service (a research unit within the RRIM), guaranteed the integrity of well-tried practices based on traditional knowledge (Bauer 1948: 277).

By the late 1920s the dual benefits of cover crops in the form of legumes for alleviating erosion and for enriching soils began to be widely recognized by European planters (A. King 1939: 139), but it left unchanged the practice of weeding. It was not until 1932 that RRIM—probably observing smallholder practices—marshaled evidence against clean-weeding (Haines 1940: 39). Reduced weeding was adopted during the Great Depression (1929–32) to cut labor cost but, on economic recovery, many estates promptly resumed it (McNaughton 1936: 91–92; Page 1938: 1–3; C. Barlow 1978: 149). The difference between sound smallholder agricultural practices and the methods adopted on corporate estates again came to light when RRIM's research findings shattered the myth linking unweeded peasant smallholdings to "moldy rot disease" (*Malay Mail*, June 19, 1933). Indeed, the investigations suggested that clean-weeding, far from retarding the spread of disease, may actually have contributed to it (Bauer 1948: 58–59).

Similarly, contradicting popular assumption, estates proved to be less productive, in 1929 yielding an average of 459 kilograms per mature hectare, compared to 567 kilograms per hectare by smallholders (C. Barlow 1978: 444–45, appendix 3.2). The difference arose from the greater planting density of an average of eighty-five trees per hectare in smallholdings, which implied more economical land use and lower environmental impact. Furthermore, the solid canopy created by close planting in imitation of the forest, the natural environment of para rubber, enhanced soil conditions. This compared with only thirty to forty trees per hectare planted on estates based on the fallacies that lower density would increase the yield per tree, potentially reduce overall planting and tapping costs, and prevent the spread of disease (Bauer 1948: 56).

Notions about the superiority of the planting techniques adopted on estates, which bolstered imperial ideology, were tenacious. For example, experiments with different planting densities, conducted by the Rubber Research Institute and the Experimental Station of the General Association of the Rubber Planters of East Sumatra, pronounced closely planted and overgrown plots untappable (Bauer 1961c: 241n2). On the contrary, tests carried out during 1936–37 on smallholdings in the Netherlands East Indies and in Sarawak suggested a correlation between greater density and higher yields per unit of land. The evidence, nonetheless, failed to alter existing practice in the peninsular estates due to an unsubstantiated belief that higher productivity on the smallholdings was achieved through overtapping, perceived as injurious to trees (Bauer 1961c: 244–45; Bauer 1961a: 237–40).

Bauer's research efforts in the 1940s on Malayan and Indonesian smallholdings exposed other benefits of close planting (C. Barlow 1978: 40). The denser tree canopy acted as a break on erosion and reduced undergrowth, especially the shade-intolerant *lalang*. Moreover, by inducing high humidity, the close canopy was thought to encourage bark renewal (Bauer 1961a: 240).

In financial terms, the working of rubber smallholdings with surplus household labor contrasted with the heavy dependence of estates on immigrants to carry out their labor-intensive practices. Unlike the case of peasant smallholders, about four-fifths of whom were estimated to have been self-sufficient in rice (Caldwell 1977: 41–42), the demands of immigrant labor on the estates pushed up government rice imports. Environmentally sustainable development, often a cliché in modern parlance, proved nowhere more real than within the sphere of the subsistence-cum-rubber economy of the Malay peasant. It provided a near-perfect formula for incorporating environmentally benign peasant agricultural production into the volatile international rubber market.[7] Peasants worked under a flexible regime of tapping, in accordance with personal needs, circumstances, and market trends.

Although smallholder production is calculated to have dropped from 31.3 percent of total output to 26.5 percent during the first three years of the restriction, it promptly recovered on the revocation of the scheme in 1929, provoking disbelief in official quarters (Lim Teck Ghee 1974: 114–16). Again, with the onset of the Great Depression, smallholders cashed in on the adverse effects of market volatility on the plantation industry, which was forced to cut costs. Under these circumstances west-coast smallholders especially, who were located closer to markets, shifted labor from food-crop cultivation to rubber production, purchasing rice to make up for shortfalls (Bauer 1961b: 194–15). While smallholders were able to adjust output to demand, the same flexibility was not within the reach of estates burdened by high maintenance costs (C. Barlow 1978: 61–62).

The Attrition of Forests by Plantation Rubber

Despite the success of rubber smallholders, restrictions on shifting cultivation and the utilization of Malay reservation land curbed smallholder expansion, while the plantation sector experienced rapid expansion into the foothills. Founded in 1901, the Forest Department became the single agency committed to stemming the tide of forest depletion by the plantation industry.

The expansion of rubber, which thrived in the lowlands up to about three hundred meters, involved the clearance of the dipterocarp forest at its richest, with some forty tree species per hectare. The same lowland forest also contained the most valuable commercial species, including the much sought after *cengal* (*Balanocarpus heimii*) and *meranti* (*Shorea*) (Wyatt-Smith 1995: III-7/7). Thus, in the face of plantation expansion and the absence of sufficient provisions for wildlife and environmental protection, forest reservation was imperative for other than timber needs. It offered the only viable means for biodiversity conservation, even allowing for some silvicultural modification involving the weeding out of the less commercially valuable species. In reality, however, given the economic priorities of colonial development, forest reservation lagged behind agricultural conversion. Not until the slowed pace of plantation development brought by the First World War and the subsequent slump in the price of rubber did reservation accelerate significantly. It doubled during 1914–19 and, again, during 1922–23. Reservation, though at a slower pace thereafter, rose in the Federated Malay States from 13.1 percent in 1923 to 19.8 percent in 1927. By the time the economy recovered in 1932, the Forest Department had achieved slightly more than its 25 percent target (Kathirithamby-Wells 2003: 90, 149–50, 156).

The International Rubber Restriction Scheme (1934–38), which implemented a ban against new planting, put a temporary halt to rubber's invasion of the forests. This was consistent with the government's policy to bolster rice production for increased self-sufficiency in the face of the depression and the sharp fall in tin and rubber prices (Kratoska 1983: 159).[8] The subsequent relaxation in the regulation, which permitted replanting and new planting up to 5 percent of the regulated area, provided a boost to the industry (C. Barlow 1978: 66–67). Overall, between 1929 and 1939, there was an increase of more than three hundred thousand hectares under rubber (C. Barlow 1978: 444–45, table 3.2).

The new wave of tree-crop expansion preceding the Second World War almost matched in magnitude that of forest reservation. In 1932, there were 1.8 million hectares under forest reserve in the FMS (where most of the reservation work had been focused) and just a fraction over 1.3 million hectares under rubber, concentrated largely in the same states and in Johor and Kedah (FD 1933; C. Barlow 1978: 444–45). More than half of the rubber was owned by non-Europeans, chiefly Chinese, in plantations of less than forty hectares. These, together with the European estates, ranging between five and six hundred hectares, but with some as large as twenty-five hundred to thirty-two hundred hectares (McNaughton 1936: 80), formed contiguous blocks of monoculture, contrasting sharply with the dispersed and fragmented patchwork of Malay plantings.

The Plantation's Marginalization of the Smallholder

After the Second World War competition from U.S. synthetic rubber presented a strong case for competitive production at low cost by promoting cost-effective smallholder production. But a number of factors conspired against this. On the metropolitan end, the paramount concern of the British government was the valuable tax on profits earned by U.K.-based Malayan estates (Rudner 1976: 236). Another factor was the drive within the estate industry to increase per-hectare yields in the face of falling prices—a move that dovetailed with the Forest Department's interest in restricting plantation expansion.

As rubber plantings aged and yields declined in the 1930s, the estates proceeded to replant, using cloned seeds, which contributed to both improved yields per unit area and higher total production. There were clear disincentives to smallholder replanting, however. The activity was labor intensive, involving two months of felling with the axe, with an increase in total labor inputs during the first eighteen months, on average sixty to one hundred days per hectare (Ooi Jim Bee 1959: 154–55). Moreover, as many could not afford to lose their cash income during the five-to-six-year period before the new plantations came to maturity, they were reconciled to the practice of allowing self-sown seedlings to take root among existing trees (Bauer 1961c: 254).

For smallholders, as Bauer argued, new planting rather than replanting would have made more economic sense, allowing them to take advantage of the newly introduced high-yielding seeds while continuing to draw some income from declining stands. An important constraint on new planting was the decreasing availability of land along the customary river-valley axis of Malay settlement, leaving the peasantry without the necessary capital and labor for developing the more remote slope lands (Ho 1970: 85). This precluded smallholders from taking advantage of the improved planting materials offered by the government (Rudner 1976: 237), thus nullifying any potential benefits to them from contribution to the rubber tax (Ooi Jim Bee 1959: 153; Caldwell 1977: 41). As the replanted estates based on selected high-yielding seeds gradually gained an edge over the declining smallholder stands (C. Barlow and Drabble 1990: 208), the opportunity was lost forever to combine the inherent environmental merits of smallholder cultivation with optimum yield in a sustainable environment. By 1939 Malaya produced 54 percent of the world output of rubber; but the industry continued to be dominated by plantations of forty hectares or more, occupying in the FMS and the UFMS (Unfederated Malay States) roughly two-thirds and two-fifths of the area, respectively (A. King 1939: 138).[9]

Given the spread of landlessness and shifting cultivation after the Second World War (Kathirithamby-Wells 2003: 268–69; also see below), resettlement and agricultural reform became high priorities for the government. An integral part of the program was to put the economy back on its feet by revamping the plantation industry. In 1955, the replanting program received a boost with a M\$280 million allocation from the International Bank for Reconstruction and Development. This triggered a flurry of activity among estates and the larger smallholdings, but it left behind the majority who owned less than two hectares (Shamsul 1986: 86–87). On the whole, the smallholdings never caught up with the estates' head start, not only in terms of replanting but also of the new technology for rubber processing. By the 1950s, even the estates that had experienced a slow recovery after the war had outstripped smallholder performance (Drabble 1979: 88–89). By 1966–67, about 72 percent of the large estates, but only 54 percent of smallholdings, had been replanted, accounting for the difference in average yield of 327 kilograms and 60 kilograms per hectare, respectively (Drabble 1979: 88).

The halfhearted government effort to introduce replanting in smallholdings can be explained in part by its policy of promoting crop diversification as both a more secure and stable alternative for the peasant sector (Rudner 1976: 244) and a means of regulating rubber production in the interests of the supposedly more efficient estates. The program of commercial crop diversification, proposed in 1938 by Frank Stockdale and shelved during the

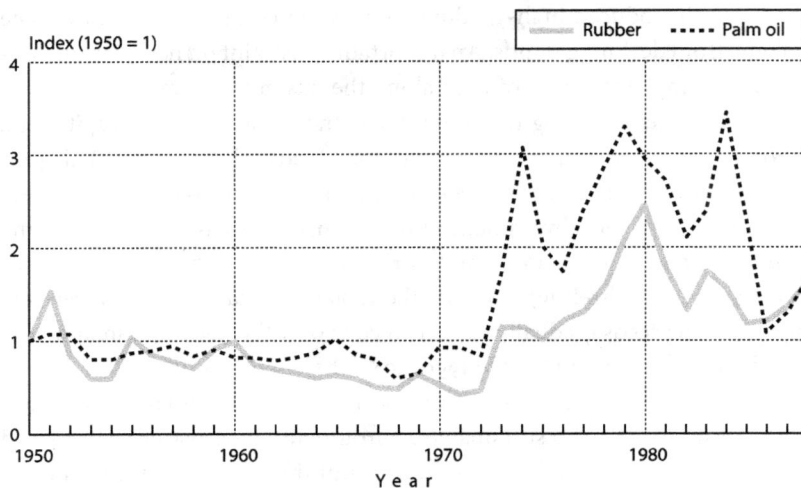

FIGURE 1 Price trends in rubber and palm oil, 1950–90

war, reappeared in the World Bank *Report* of 1955 and then was wholeheartedly embraced by the postwar administration when it faced a downturn in rubber prices in the early sixties (see figure 1).

The crop that emerged as the close competitor to rubber, injecting new life into the flagging plantation industry, was oil palm (*Elaeis guineensis*). First introduced into the Peninsula in 1875 as a decorative plant, the cultivation of the *duro* or "Deli type" of oil palm as a cash crop began in 1926 in Selangor and spread to Johor, Perak, and, to a lesser extent, Negeri Sembilan, Pahang, and Kelantan. The crop's tolerance of wet and swampy conditions allowed planning to extend into low-lying areas unsuited to rubber (Duckett 1976: 176–77). The planted area expanded from a mere forty-nine hundred hectares in 1926 to thirty-nine thousand hectares in 1950 (Ooi Jim Bee 1959: 160–61, Khera 1976: 9, 23–25). Oil palm requires large capital outlay, coordinated harvesting for quality control, and expensive machinery for extraction, all conditions that make it more suitable to large-scale estate cultivation rather than that by smallholders (C. Barlow 1978: 396; Snodgrass 1980: 173). The profitable operation of an oil-palm factory is economically viable only on a plantation comprising at minimum eight hundred hectares (Fryer 1964: 243). Again, plantation rather than smallholder development was favored, as manifested in a 1934 government ruling that permitted only plots of a minimum of eighty hectares to be alienated for oil palm (Ooi Jim Bee 1959: 161).

Under the postwar program of diversification the area under oil palm expanded rapidly and, by 1973, covered 274,898 hectares, with a little more

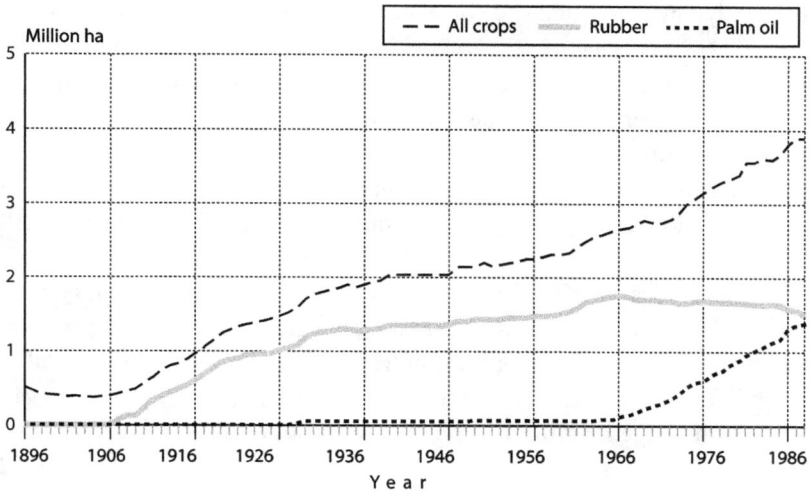

FIGURE 2 Trends in land use in peninsular Malaysia

than one half of this under non-Malaysian ownership (Courtenay 1979: 139, 148). Conversely, the area under rubber shrank during the same period from 809,716 hectares to 607,287 hectares (Snodgrass 1980: 173; see also figure 2).[10] Simultaneously, cocoa and coffee, which, unlike oil palm, were well suited to smallholder cultivation, found their way into rural home gardens but, under competition with established producers in West Africa and South America, neither gained importance (Khera 1976: 22).

The New Economy, Land Development, and Mega Plantations

By independence in 1957, mechanization and land fragmentation under the green revolution had undermined smallholder paddy farmers. The declining London price for rubber (see figure 1), with a sharp drop in 1960 from around 300 cents to around 150 cents per kilogram, further aggravated the plight of the peasantry (C. Barlow 1978: 93, Table 3.3). The declining yields of smallholdings versus the improved productivity of estates provided ample justification for the newly independent government to chart a new direction for the peasant community.

Overlooking the proven economic initiative and competitiveness of the Malay peasant, early nationalists saw commodity production for the capitalist market, with its violent swings of boom and bust, as detracting from the ideal of a stable, protected rural environment (Rudner 1976: 245). The benign Malay leadership at the helm of the independent government similarly subscribed to realizing "the ideal of self-supporting Malay communities" originally envisaged under colonial rule (Kratoska 1983: 166; Voon 1976a: 511–12). Neither the government nor the Malay leadership factored into its formula for rural betterment the reality of a robust peasantry, which time and again had tenaciously ridden the competitive market economy, however meager and uncertain the returns. Contrary to the nationalist ideal, in the postwar economy, tree-crop cultivation under a regime of state intervention and protection irretrievably gained dominance over paddy (Vincent and Ali 1997: 152–53).

The critical issue of landlessness faced by the newly independent government of Malaya stemmed from demographic growth. A rise in 2.5 million during 1957–70, outstripping previous rates of increase, brought the total to 8.8 million people by the end of this period. Of this number, 4.6 million were Malays, the large majority of whom were rural peasants (Sundaram 1988: 326). In giving priority to solving landlessness, the Ministry of National Resources acted on the premise that Malaysia was essentially a primary producer, with land as its main asset (Tunku Shamsul and Perera 1977: 10). The National Rural Development Council, formed in 1960, functioned as "the nerve center for the rural offensive" (*Straits Times*, 1960). Following independence, the shift of government support from the corporate plantation to the smallholder sector was inspired by the Bumiputera (Sons of the Soil) policy, underscored by new research orientations sympathetic to rural interests (C. Barlow and Jayasurija 1986: 651).

As a preliminary measure toward upgrading rubber smallholdings, grants were made available for replanting holdings of six hectares and under with high-yield clones through the Rural Industrial Smallholdings Development Authority (RISDA) (Ooi Jim Bee 1976: 246; Rudner 1976: 259). The sale and subdivision of a large number of European estates after independence also favored smallholders. By 1961, the total area under smallholdings exceeded that of the estates (Rudner 1976: 256, 258), and by the following decade, their total output had also improved. The average yield per unit area, however, was distinctly lower than that of the estates, which were privileged by early planting (C. Barlow 1978: 115–16, 444–45, appendix 3.2).

Despite the progress that independent smallholders made toward adopting new technologies, they were outpaced by the ambitious government-sponsored "group smallholdings" under the Federal Land Development Authority (FELDA). As the largest of the national development agencies, FELDA, founded in 1956, aimed to solve Malay rural poverty through the integrated development of smallholdings. Its activities, expanded under the New Economic Policy formulated in 1971 following the race riots of May 13, 1969, spearheaded the largest and most ambitious development scheme in the

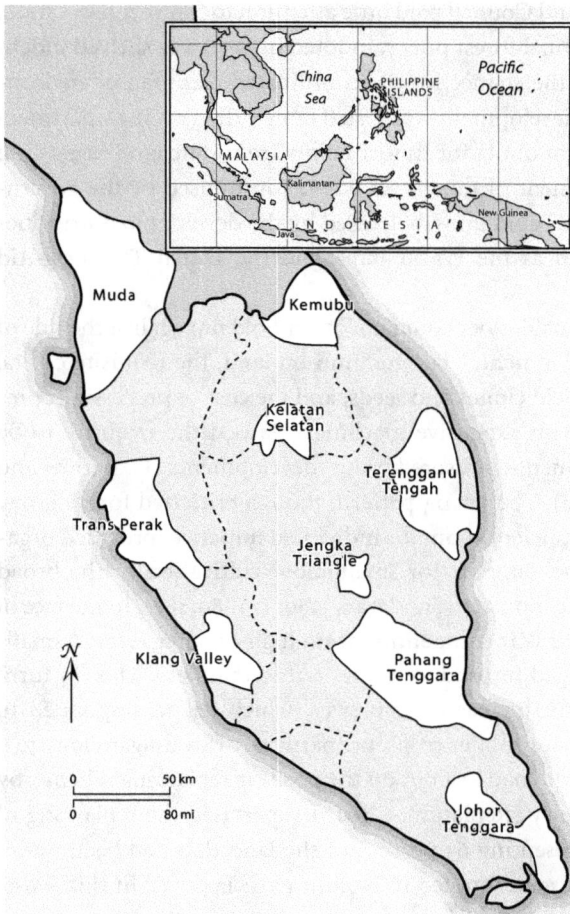

MAP 1 Land Development Schemes in Peninsular Malaysia

region (see map 1). For the period 1971–90, it foresaw the planting of 1.7 million hectares of rubber, oil palm, and other crops (GM 1976: 210).

FELDA's program for so-called group settlement had two objectives. First, it sought to open up new areas of primary forest for *integrated settlement*, with individual plots forming, where possible, an estate complex with central processing and marketing facilities. Second, it aimed to convert through *fringe development* programs holdings of a size deemed uneconomic into larger, viable units (Tunku Shamsul and Perera 1977: 12–13; Snodgrass 1980: 177–78).[11]

The scale of land clearance associated with rural development under the post–Second World War government struck at the heart of forest planning, crucial for environmental protection. In the absence of a formal land-use policy, the National Land Council paid little attention to forest conservation. The question of a national forest policy, mooted in 1959, was shelved indefinitely (GM 1968), and the chance for the coordinated planning of environmentally sustainable development was lost. Though the Forest Department set an annual 30 percent quota for timber felling from state land forests (FD 1959: 65), the enforcement of this target was compromised by the government's commitment to rapid growth, backed by the development priorities of "aid" agencies such as the World Bank and the Dutch Technical-Aid Mission (FD 1972: 90).

The independent smallholder could no longer hold out against the tide of capital-intensive development. The premium on land, the requisite capital investment for high-yield clones and seeds, and the cost of processing competitive high grades with expensive machinery caused the majority to be absorbed into government-sponsored group developments (C. Barlow and Jayasurija 1986: 651–52). The FELDA pattern, though criticized for its heavy subsidization and insufficient room for individual initiative, provided organizational and financial support for smallholder cultivation in the broad framework of the estate industry (Snodgrass 1980: 179–80, 182). In the face of the post–Second World War competition from high-yielding estates, smallholders had been obliged to produce more, rather than less. This, in turn, had led to the need for increased farm sizes, which FELDA responded to with a shift away from nonrubber crops, primarily rice (Sivalingam 1993: 71). For the smallholders who had lost out on the postwar replanting scheme, by 1973 the Rubber Industry (Replanting) Board supervised the replanting of 493,000 hectares, representing 63 percent of the land that had been under smallholder rubber in 1946 (quoted in Nonini 1992: 124–25). In this sense, state aid fulfilled an obvious need.

The total area under FELDA schemes dedicated to oil palm doubled during 1966–67, and by 1973, it was more than twice the size of that under rubber (Khera 1976: 145). By 1975, rubber and oil palm occupied 4 percent and 13 percent, respectively, of the land area in the Peninsula (Duckett 1976: 176). Smallholder oil-palm output in FELDA land schemes increased from 16 percent of scheme totals in 1966 to 46 percent by 1988 and, by 1987, oil palm under FELDA and FELCRA (the Federal Land Consolidation and Rehabilitation Authority) accounted for 38 percent of the country's total production, of which 86 percent came from smallholdings (Vincent and Hadi 1991: 16).

The larger integrated settlement schemes for new planting were nonetheless beset by problems, many of which were environmental in nature. Between 1956 and 1990, FELDA alone had developed over 800,000 hectares of some of the best forest land in the country (Aiken and Leigh 1992: 66).

Furthermore, not all the land cleared was necessarily planted. During the Second Malaysia Plan (1970–75), for example, land cleared by FELDA and other agencies tripled, covering 799,600 hectares (Hurst 1990: 121, Table 3.3). But of this total, only 353,411 hectares, or less than half, were actually developed by the various agencies. The remainder was abandoned, usually after logging (GM 1976: 128), leaving extensive scrub (Vincent and Hadi 1991: 8).

Other shortfalls in land-development targets may be attributed primarily to two factors. First, the loss of independence and initiative among settlers who now, rather than "participants," have been turned into "recipients" (Vincent and Hadi 1991: 20). Agencies provided too much aid and assistance, leaving a lack of opportunity for individual initiative under the cooperative systems of management. Second, the radical transformation of the Malay peasant landscape (Hurst 1990: 65–66). The development agencies took up contiguous blocks of agricultural land for the schemes based on inaccurate and unreliable land-classification surveys and failed to integrate environmental dynamics based on physical characteristics such as slope, accessibility, and infrastructure (Joseph 1964: 276–77).

The once self-reliant Malay peasant, who earlier had successfully adapted commodity production to customary traditions of land use, was now made to rely on externally derived, capital-intensive, and often ill-suited methods and tools. Land-development schemes helped raise the incomes of many (C. Barlow 1978: 264–65), but the heavy social cost invited the following critique: "The settlers are not even *participants* but are mere *recipients* of the government development projects . . . the settlers find that they can no longer work as and when they feel they like to [*sic*]. . . . There is therefore that 'cultural lag' between modern technology represented by the scheme itself and the settlers themselves with their traditional values, beliefs, and ways of life. For any land development project to be successful it must be able to bridge this gap" (Mansor bin Ahmad, quoted in C. Barlow 1978: 240).

The Environmental Impact of Megaplantations

Post–Second World War rural development uprooted peasant cultivators from a settled and familiar environment and transplanted them to an alien pioneer landscape to which many were unable to adapt. The deleterious environmental impact of large-scale tree-crop cultivation, made evident on the early European estates, was magnified in scale in the land-development schemes. It is calculated that agricultural area rose from 21 percent of the country's land area in 1966 to 35 percent in 1982, an increase of 68 percent in just sixteen years (H. Brookfield and Byron 1990: 44). The total area under rubber and oil palm rose spectacularly from 1.9 million hectares in 1970 to

TABLE 1 *Cultivated Area by Crop, Peninsular Malaysia, 1970–90 (in thousands of hectares)*

Year	Rubber	Palm oil	Paddy	Coconut	Cocoa	Other	Total
1970	1,724	222	524	214	3	221	2,908
1975	1,695	569	596	233	18	248	3,359
1980	1,697	907	529	245	42	167	3,587
1985	1,663	1,232	465	220	101	215	3,896
1990	1,534	1,662	465	220	148	215	4,244

Source: Prime Minister's Department for 1970 and 1979; Kementerian Pertanian for 1980, 1985, and 1990. See Vincent, Ali, and Chang 1997: 152.

2.8 million hectares by 1985 and 3.1 million hectares in 1990 (Courtenay 1979: 135; see table 1 and figure 2). At 3.4 million hectares in 1994, tree crops occupied 25.9 percent of the land area, compared to 5.9 million hectares of forest, or 44.8 percent of the land area (Ministry of Science, Environment, and Technology, Malaysia [MOSTE] 1997: 64). Overall, forests constituted less than twice the area under tree-crop cultivation, and the latter was dominated by large blocks of monocultivation that had suppressed in extent the biologically richer patchwork of mixed cultivation.

The problem of forest depletion and the corresponding loss of biological diversity cannot be overemphasized. Conservationists in Malaysia, while agreeing that land development had a vital role to play in the national program of socioeconomic development, voiced concern over the exposure of soils to heavy rainfall when vegetation was cleared in development schemes ("Editorial" 1975). Runoff from rain, which is increased by about 60 percent on cleared land, exacerbated rain splash and gully action and contributed to erosion, siltation, flooding, and water pollution. Surveys described the whole of the Peninsula as bearing "at least a moderate risk of soil erosion" and identified at least three of the major land development schemes (the Jengka Triangle, Kelantan Selatan, and Terengganu Tengah development schemes; see map 1) as located in areas in which erosion was potentially "very high" (Morgan 1974: 92, 104). Floods of the magnitude experienced in January 1971, and comparable to the "Great Flood" of 1926, though rare, demonstrate potential vulnerability to meteorological perturbation. The government focused on raising production, profits, and the GNP, but it seemingly ignored the environmental and social implications of its development schemes (Bruenig 1985: 23). In 1974 the Environmental Quality Act was passed but did not provide for an environmental impact assessment (EIA) in land development. When finally incorporated into the Revised Act of 1985, there were serious shortfalls in implementation.[12]

Short-term political goals, which compromise environmental safeguards in land development, can only be understood in the context of national and state politics. As Dorothy Guyot has demonstrated with reference to Terengganu, rural-development schemes play a crucial role in politics. In the two years leading up to the 1964 elections, the United Malay National Organization (UNMNO), then pitted against the powerful Pan-Islamic Party (PMIP), rushed to implement land-development schemes to garner votes, resulting in the selection of unsuitable sites for many of them. Every district needed its scheme, and an estimated 20 percent were poorly sited on sandy ridges and steep slopes. Similarly, election fever fueled the grandiose East-West highway project, linking Kuala Terengganu with Butterworth, to serve the interests of certain individuals in developing oil-palm estates, following immediate profits from logging (Guyot 1971: 372–73).

The Third Malaysia Plan (TMP; 1976–80), which had a whole chapter titled "Development and the Environment," marked a tuning point in official policy (GM 1976: 218). The TMP envisaged "cooperation between the Federal and State Governments in ensuring that, as far as possible, man's activities were [sic] in balance with his environment" (GM 1976: 219–20). It drew attention to potential environmental problems arising from land and natural-resource development. The TMP's policy of environmental protection was fully expounded in the 1977 National Forest Policy (NFP). Its central aim was to encourage the sound utilization of forest land by development agencies through proper coordination and planning.

Implementation of the NFP, however, proved fraught with difficulties. In contrast to the land classification survey that had assigned only residual land to protection forests, the NFP took a major step forward in declaring that "Protective Forests should never be alienated save in the most exceptional circumstances and after the most careful objective study" ("National Forest Policy" 1980: 4). This clause, evidently unacceptable to state governments that jealously guarded their rights over land, was duly dropped in the revised NFP of 1992 (see GM 1995). State control over land meant that the enforcement of national policy was contingent on the initiative of individual states, which in many instances remained wanting.

A flagrant disregard of environmental safeguards was evident in the conversion of forests for plantations. The largest was the Jengka Triangle scheme (1968–80) covering a triangular area of some 121,781 hectares between the towns of Jerantut, Temerluh, and Maran. About 77 percent of the development was dedicated to creating the largest oil-palm estate in Southeast Asia, while the rest was allocated to rubber (GM 1966: 78). The scheme involved the excision of 41,654 hectares, or 75 percent, of the Jengka Forest Reserve, consisting of small patches of swamp forest interspersed with heavily timbered hill slopes of near-climax dipterocarp forest (Wikkramatileke 1972: 486, 488). The 1970 *Land Capability Council Report* used for the Jengka Triangle scheme was biased toward economic interests, well served in the absence of the environmental impact assessment made mandatory only in 1985 (Teh and Tunku Shamsul 1992: 84). In the planting of rubber and oil palm, the Land Conservation Act of 1960, which prohibited the cultivation of slopes more than eighteen and a half degrees in areas proclaimed as hill land, was contravened (Senftleben 1978: 81; Teh and Tunku Shamsul 1992: 98–100). The problem was exacerbated by the opening of access roads that left bare and erodable surfaces. Erosion and widespread gullying was especially severe during the intermediate stage between forest clearing and the planting of cover crops (Teh and Tunku Shamsul 1992: 106). The potential for massive erosion under these conditions is estimated at 24.9 tons per hectare per year on slopes of thirteen degrees (quoted in MOSTE 1997: 78).

Adding to the problem of erosion in the Jengka Triangle was river pollution. Development in the Jengka Traingle scheme proceeded in disregard of existing evidence in oil-palm and rubber plantations of extensive turbidity and *E. coli* contamination of streams exceeding world health standards (Teh and Tunku Shamsul 1992: 98). In the early phase, untreated industrial effluents from oil palm– and rubber-processing factories were discharged directly into the streams and rivers (Teh and Tunku Shamsul 1992: 102). The Jengka development scheme displayed all the symptoms of environmental degradation, including erosion, flooding, and water pollution associated with many of Malaysia's rural development projects (Soong et al. 1988: 1–2).

By 1990 Terengganu had set aside some sixty-six thousand hectares in Ketengah, east of the East Coast Range, for conversion to oil palm. The area, with soils ranging from marginal to poor in quality and prone to erosion if exposed, was selected notwithstanding the advice of the Natural Resource Section of the State Economic Planning Unit (SEPU/UPEN) that it be retained under natural cover (WWFM 1983: 49). This was a clear example of development initiatives proceeding in disregard of the ecological realities that actually govern economic productivity.

Kelantan launched a similarly ambitious project. To relieve pressure on the lowlands where about 80 percent of the state's population lived, it proposed agricultural conversion in the southern hill forests in the Lebir and Ninggiri Valleys. During the 1980s, prime forest land in the lower altitudes were logged, making way for rubber and oil palm and simultaneously displacing the area's remaining fragments of the aboriginal Batek (Endicott 1997: 32–33). The development agencies KESEDAR (Kelantan Selatan Development Authority), FELDA, and FELCRA opened up huge estates of oil palm and rubber (WWFM 1991: 2:12, 79, 82; see also Kathirithamby-Wells 2003: 354–55).

Also ill conceived was a UNDP (United Nations Development Program) supported project for agricultural development in southwest Kelantan (WWFM 1991: 2:13). The project encompassed a corridor along the Lojing highway and subsidiary development in the Cameron Highlands and contravened the recommendations of local agencies like SEPU and MARDI (Malayan Agricultural and Rural Development Institute) (WWFM 1991: 2:195–97; see also Kathirithamby-Wells 2003: 308–11; 404–5).

Between 1966 and 1975, agriculture expansion roughly matched deforestation, but that correspondence subsequently broke down. In 1990, the area under agriculture was estimated at 4.5 million hectares (Vincent and Hadi 1993: 455–56). With the exception of some diversification to cocoa, the emphasis in commodity production remained on oil palm, with serious environmental implications. Soil erosion through the process of mechanical clearing, followed by burning, remained a problem (Ng Siew Kee, quoted in Vincent and Hadi 1991: 18).

Furthermore, chemicals and pesticides used to make up shortfalls in soil fertility, partially as a consequence of erosion, continued to have adverse effects on water quality. About 90 percent of the agricultural inputs used, of which 75 percent were herbicides, were for rubber, oil palm, and paddy. Due to the difficulty of excluding agrochemicals from modern farming methods, herbicides and insecticides affected surface runoff and water courses, reducing the diversity of aquatic life (Sham Sani 1993: 14). In Terengganu, for example, air-breathing anabantids and catfish were reported to have replaced the carp species *Cyprinidae* (WWFM 1983: 49). The rubber- and oil palm–processing industries contributed as much as 35 percent of the nation's industrial pollution (US-AEP 1989: 3–4). Oil-palm mills, usually located near water courses to draw water for processing, discharged about two thousand kilograms of effluent for every thousand kilograms of crude oil produced (Vincent and Ali 1997: 328). By 1982, only 51 of 186 rubber factories had introduced any form of effluent treatment (H. Brookfield 1994: 275–76).

Biodiversity Loss

Arguably, plantation cover can substitute for forest cover in that it retards soil erosion and provides organic nutrients for the soil. It does not, however, approach the complex biodiversity of the tropical forest canopy. Estimated to have 2,650 species of trees alone (Ng 1988: 104), the Malaysian rain forest has more species per unit area than any other forest in the world. Within the species-rich Malesian region, the peninsula has been singled out as having a key role to play in conserving the global heritage of biodiversity. Trees of ten centimeters (in diameter at breast height) and larger, in just fifty hectares of the Pasoh Forest Reserve in Negeri Sembilan, a typical example of lowland rain forest, represent 660 species in 244 genera, out of the 2,830 species and 532 genera recorded in 1989 for peninsular Malaysia (Soepadmo 1995: 21). The presence of rare species further contributes to the heterogeneity of the lowland forest (Ng 1988: 105). Dependent on this type of forest habitat are some 50 percent of the Peninsula's mammal species, as well as an equivalent percentage of resident birds (Wells 1999: xxiii). This, coupled with the fact that most species are present at very low densities and are randomly distributed, makes for an extinction-prone environment. Any reduction of forest implies the elimination of an undetermined number of species (Soepadmo 1995: 30).

Despite its exotic origins, the cultivation of *Hevea braziliensis* on the scale and manner of the large European estates is little different from other forms of monoculture. Rubber estates contain only one species, with no indigenous faunal association and, by definition, they are incapable of supporting a biomass even remotely similar to the native forest. Only when gradually integrated into regenerating secondary forest, as in the indigenous combination of smallholdings and shifting cultivation, can rubber contribute to the composite structure of the tropical forest. Before the introduction of oil palm, rubber estates commonly included pockets of primary or secondary forest on low-lying areas that were left unplanted. These offered a limited refuge to a variety of wildlife. Further, the stable crown of the rubber tree provides nesting sites for birds and squirrels. Contrastingly, oil palm, with its moving fronds, is less successful in harboring arboreal species. Furthermore, in comparison with rubber's annual supply of hard-coated seeds, oil palm kernels provide a year-round supply of desirable food for mammals, which, in the long term, proves detrimental. For in the uniform environment of the oil-palm estate the predominance of one particular food source contributes to the dominance of a narrow range of vertebrates, including rats, and their predators, namely, snakes, hawks, owls, and wild cats. The extensive damage to oil palm caused by the rats (*Rattus* spp.) leads to the use of poisons that can prove destructive to other animals. Thus, as more of the patches of secondary forest covering swamps and low-lying areas in rubber estates were converted to oil palm, the threat to wildlife progressively intensified (Duckett 1976: 177–81).

According to a survey conducted during 1966–68 by the Colombo Plan ecologist W. E. Stevens, the land capability classification had made no provision for protecting some 80 percent of wildlife in the Peninsula that are confined to the lowlands below six hundred meters above sea level (W. Stevens 1973: 6). The clearance of wildlife habitat was compounded by the lack of legal provisions for protecting existing reserves (with the exception of the Taman Negara National Park covered by a special act). Consequently, many mammals, including the Sumatran rhino (*Dicerorhinus sumatrensis*), suffered extinction in the course of land development for plantation agriculture (W. Stevens 1973: 42, 46).

In 1962 the Game Department made a proposal to upgrade Kerau Reserve to a national park, extending its western boundary to include salt licks essential for the survival of *seladang* (Malayan gaur or *Bos gaurus hubbackki*). Instead, the area was excised for rubber planting under FELDA. Similarly, the spread of cultivation in Sungai Cior, Perak, originally a *seladang* reserve with salt licks, resulted in the loss of many animals (Kitchener 1961a: 199).

Land-development schemes frequently either destroy or fragment elephant territory (Mohd 1968: 45). Extensive land clearance in Johor and Pahang, in particular, endangered elephant populations (Mohd 1977: 6–7). Displacement by logging and land development led to their convergence on oil-palm estates in search of food (Modh 1988: 262; WWFM 1991: 2:202). Consequent crop damage in turn resulted in the shooting and culling of animals, substantially reducing their population. During this period the Game Department resorted, in the name of crop protection, to the periodic culling of herds, shooting an average of fifteen elephants per year. Between 1960 and 1968 alone, the elephant population for the Peninsula fell from 692 to 486 (W. Stevens 1968: 12). Moreover, the encroachment by elephants and other big herbivores on plantations, in combination with the improved infrastructure associated with land development, drew hunters and trappers closer to their prey.

Lessons Learned?

Forest management, though not entirely supportive of the conservation of biodiversity due to selective logging, protected some of the biologically rich lowlands as reserves in the face of the land-hungry plantation sector. A reservation policy, well entrenched before the onset of the acceleration in development after the Second World War, provided the only basis for negotiating forest retention in the face of pressure for degazettment, logging, and development.

Ironically, the threat of plantations to biodiversity was first noted by Ridley, who as the director of the Singapore Botanical Gardens, contributed most to developing rubber as a plantation crop in the Peninsula. He observed in 1922 that since the introduction of rubber cultivation in the Peninsula, a certain number of species had either become very rare or had been entirely exterminated (Ridley 1922: xiii). This trend escalated in the postcolonial period with the reduction in the extent of total forest cover, especially in the biologically rich lowlands. Though in 1994 some 58 percent of the land area was still under natural forest cover (MOSTE 1997: 64), only 38.9 percent of that classified in the Forest Inventory II of 1981 as "virgin forest" (read "primary forest") was particularly significant in biodiversity terms (see Vincent and Ali 1997: 117, table 4.2). Furthermore, bearing in mind that the inventory, like the preceding Inventory I, used the 1,000-meter boundary to distinguish between the lowland and the hill forests, what primary forests still remained represented largely hill (300–780 meters) and upper dipterocarp forest (780–1,250 meters). There was little primary forest in the altitudinal zone below 300 meters, although that zone is most important in terms of the diversity of lowland species such as birds (Wells 1988: 171). Also significant is that the overall 18 percent decline in forest loss between the two inventories, as well as the related decline in biodiversity, resulted largely from the conversion of forest to plantation rubber and oil palm (Brown, Gillespie, and Lugo 1991: 114–15).

In 1990, the Permanent Forest Estate of Peninsular Malaysia, at 4.8 million hectares, only marginally exceeded the total area of 4.2 million hectares under cultivation, in which plantation rubber and oil palm accounted for 3.2 million hectares (Vincent and Ali 1997: 152). With no guarantee against the degazettment of areas in the Permanent Forest Estate, only the wildlife parks, reserves, and patches of virgin jungle reserves, constituting a mere 7,400 hectares (MOSTE 1997: 55), provided partial amelioration for the massive loss of biodiversity caused by plantation agriculture.

Against the background of past shortfalls in environmental protection, the end of the twentieth century witnessed a sea change in official policy. One of the fundamental driving forces was the growing appreciation of the Peninsula as a plant genetic reservoir, pivotal for pharmaceutical research and biotechnology.

In the field of agriculture, the increased resistance of plant pests to insecticides resulted in a reappraisal of the role of predators and parasites, among them birds and insects for crop protection (Myers 1979: 65, 72–73, 125). Apart from a growing awareness of the environmental and health risks of industrial chemicals, natural methods of pest control offered the advantage of reduced costs, as demonstrated in the successful biological control of rodents in oil-palm estates and paddy lands. In addition to attracting rats, oil palm contributed to the Peninsula's colonization in the 1960s by the barn owl (*Tyro Alba*). The provision of nesting boxes for barn owls in oil-palm estates encouraged their breeding, and their consequent colonization of the Peninsula reduced crop damage and chemical use (Wells 1999: 434, 437).

Improved environmental awareness influenced other agricultural policies such as the discouragement of chemical yield stimulants by rubber smallholders (Snodgrass 1980:174–75; Aiken, Leigh, and Leinbach 1982: 265). Furthermore, a growing appreciation that environmentally friendly practices such as intercropping offered soil protection and the market advantage of diversification resonated with the wisdom of earlier peasant practices (see Nazaruddin 1990: 65). During the 1970s, FELDA experimented with the planting of cash crops such as cocoa, maize, sorghum, and fruit trees and the interplanting of grasses with oil palm for cattle rearing as a means of promoting subsistence farming alongside plantation crops (Wikkramatileke 1972: 495). This program formed part of a broader plan to boost the initiative and participation of settler communities through the decentralization of estate management (Boon 1971: 75, 81). By now the Rubber Research Institute had likewise rid itself of some of its earlier prejudice against smallholder practices. It recommended not only interplanting, especially in fringe schemes, but also denser planting to increase revenue and employment for settler farmers (Lim Teck Ghee 1976: 286, 288–90, 292–93).

Agricultural improvements incorporating the wisdom of indigenous practices remained largely tentative but gained new relevance in the context of the UN Convention on Biological Diversity in 1992. Biologists emphasized the merits of introducing orchards and semiwild vegetation around the peasant homestead. Other than relieving the monotonous plantation landscapes, such groves and patches of vegetation would provide fruit and firewood and harbor owls and hawks for biological pest control (WWFM 1983: 50–51). The integration of environmentally friendly practices such as these was recognized as offering the best hope for reducing the environmental impact of the plantation industry.

By the end of the millennium other positive signs were apparent. The rising trend toward industrialization and the development of off-shore oil and gas, in combination with the program for in situ development and crop intensification (GM 1991: 94–95), suggested that any significant increase in the area under plantation agriculture was unlikely. Where new developments were undertaken, the Seventh Malaysia Plan (1996–2000) envisaged that "environmental and conservation considerations will increasingly be integrated with development planning" (GM 1996: 589). The gradual and progressive shift of policy, planning, and legislation toward the integration of environmentally sound practices into plantation agriculture (GM 1996: 609–10) represented a major advance, albeit one still heavily circumscribed by political and economic exigencies.

Conclusion

Plantation agriculture, more than any other single phenomenon, has contributed to the transformation of the landscape in the moist tropics. Perceptions of modern development, whether by colonial or independent regimes, have been singularly concerned with restraining traditional modes of nonintensive cultivation well adapted to preexisting environmental and demographic conditions in favor of intensive agriculture. Such policies have had profound environmental implications, with severe biodiversity loss in peninsular Malaysia, a region endowed with one of the richest ecosystems on the planet.

Smallholder rubber played a significant role in the Peninsula's entry into the twentieth-century global economy. Unlike coffee cultivation, which expanded at the expense of forest in Brazil, Central America, and the uplands of Java (Topik and Clarence-Smith 2003: 311–13), in the Peninsula the adaptability of rubber to the settled rural landscape promoted by government restrictions on shifting cultivation addressed the tension between development and environmental protection. Rubber was successfully adapted to Malay peasant agriculture involving modest outlays of capital and labor and within flexible regimes of tapping and marketing. Its success, mirrored by the more sustained participation of the Indonesian rubber smallholder into the present,[13] demonstrates that environmental and economic imperatives are not mutually exclusive. In the Peninsula, by contrast, the plantation-biased government policy, originating in the colonial period, undermined the survival of environmentally sound smallholder practices. British efforts to discourage smallholder rubber cultivation, like the French disparagement of similarly successful indigenous coffee cultivation in the Cameroon (Ackert 2003: 306), manifested the dualism implanted in colonial economies between a perceived "unscientific" indigenous cash-crop production and "scientific" capital-intensive plantation agriculture. The colonial vision of the Malay peasantry as strictly paddy cultivators, one that overlooked their traditional participation in the market economy, resulted in the ultimate triumph of the intensive, land-hungry plantation industry. The outcome was rural landlessness and poverty.

Under the postcolonial national government committed to the economic uplift of the Malay rural poor, contiguous group settlements involving massive forest clearance fast superseded the patchwork of family smallholdings dedicated to subsistence-cum–cash cropping. The megaplantation schemes, premised as much on political considerations of vote-buying as the dictates of development aid agencies, militated against the survival of the independent smallholder, particularly when rubber gave way to oil palm. Oil palm's proneness to pest problems and the more complex processing technology associated with the industry rendered the crop less suited to smallholder production.

Malaysian land development arguably achieved the fastest pace of poverty eradication in the developing world, but it escalated biodiversity loss through excluding biologically sound options and safeguards, as visibly evident in vast swaths of abandoned land, for example. The history of smallholder versus plantation agriculture in the Peninsula offers a salutary comment on the need for a nuanced partnership between indigenous genius and Western science and capital for environmentally sensitive development in the tropics.

Notes

1. Gutta-percha was traditionally molded by the Malays into handles for chopping knives, swords, and krises. With European settlement it was ingeniously designed into handles for riding whips that found a popular market but became a product of international importance for lining telegraph cables (I. Burkill 1966: 1:1652; Kathirithamby-Wells 2003: 45–46).
2. An estate was officially defined as "land contiguous or non-contiguous with rubber . . . under single ownership" and having a minimum size of one hundred acres or forty hectares. The same criterion was generally adopted for other crops (Fryer 1964: 228–29). Throughout this essay, the terms *estate* and *plantation* have been used interchangeably and in contrast with the term *smallholder*.
3. In a similar scenario, para rubber took up land exhausted by tobacco cultivation in east Sumatra (Pelzer 1978: 52).
4. *Ficus elastica*, found from the eastern Himalaya to Java, is a source of rubber, traditionally used for lining receptacles made from bamboo and other materials. When subsequently discovered by Europeans and used for erasing pencil marks, it got the name "India rubber" (I. Burkill 1966: 1:1023–24).
5. The unavailability of sufficient flat land and water rendered shifting cultivation the predominant method of subsistence production in precolonial times. Colo-

nial laws, aimed at Malay sedentization, focused on wet-rice cultivation but failed in the main to raise production beyond the subsistence level (Sundaram 1988: 55–56). Also see note 9.

6. H. Brookfield, Potter, and Byron (1995: 121) note, however, that relative to rubber planting by the Dayak in Borneo, the rubber holdings of the Malays were characteristically regular, as they were not intermingled with other trees to produce a multistoried forest. The latter situation evidently pertains to a much later period than that observed by Ooi Jim Bee (1959).

7. According to Lord Keynes, "There is, on average, some date in every year at which the price of rubber is approximately double its price at some other date in that year" (quoted in Bauer 1961b: 185).

8. Between 1920 and 1940 Malaya is estimated to have imported about 65 percent of its rice annually (Kratoska 1983: 159).

9. The UFMS comprised Kedah, Perlis, Kelantan, Terengganu, and Johor, all of which passed from Siamese to British protection in 1909 and received British advisors during the 1920s and 1930s.

10. During 1966–70 the annual growth rate of rubber acreage was only 6.9 percent, compared to 24.1 percent for oil palm (Khera 1976: 22, table 1).

11. Whereas in FELDA's rubber schemes settlers were allocated individual and separate holdings, in oil-palm schemes the settlers participated as cooperative shareholders (Khera 1976: 158).

12. To avoid upsetting ongoing schemes and those already on the drawing board, the act was only partially implemented during the course of the Fourth Malaysia Plan (1981–85) (Aiken and Leigh 1982: 266–67).

13. In Indonesia, the second-largest producer of rubber in the world, about 75 percent is cultivated by smallholders (Dove 1993c: 136). Apart from its association with a low incidence of poverty, as in Sumatra (C. Barlow 1991: 90, Table 5.4, 93–94; Booth 1991: 55), rubber has done less damage environmentally than subsistence annuals based on shifting cultivation (Barlow 1991: 95).

RUBBER KILLS THE LAND AND SAVES THE COMMUNITY
An Undisciplined Commodity

The first two chapters of this volume—Lye's analysis of conflicted state policies toward the Batek in Taman Negara National Park, and Kathirithamby-Wells's analysis of state favoring of plantation over smallholder models of rubber cultivation—examine dichotomizing policies toward nature and culture. The reality of relations between society and environment is more complicated than these policies, which illustrate the contradictions of modern environmentalism. (cf. Slater 2003). Another example of this complexity and contradiction comes from the Dayak smallholder rubber producers of Borneo.[1]

The Contradiction

Kantu' tribesmen in West Kalimantan (map 1) say that when they plant rubber (*Hevea brasiliensis*) in their swiddens or swidden fallows, the land thenceforth becomes *tanah mati* "dead land," in implicit contrast to the "living land" of the rice swidden agricultural cycle.[2] This indigenous characterization of the consequences of rubber cultivation as "killing" is puzzling, because it seems to conflict with what otherwise appears to be its benign impact. From an indigenous perspective, rubber is an economic boon that mediates between subsistence priorities and market needs; from the perspective of social scientists and ecologists, it is one of the best economic adaptations to the Southeast Asian rain forests yet devised (Dove 1993b; Gouyon, de Foresta, and Levang 1993). From the perspective of many environmentalists, the rubber silvicultural system is so admirable that it is touted as not merely a complement to but even a replacement for swidden agriculture, and the wood of rubber trees became one of the first "eco-labeled" tropical timbers (Gordon 1993). How can this contradiction, of a crop that both nurtures and kills the land, be explained?

MAP 1 Kantu' Territory in West Kalimantan, Indonesia

To begin with, the problem of cross-cultural translation of a metaphor like "dead land" must be confronted. Obeyesekere (1992: 175), in his critique of the perception that the indigenous peoples of the Pacific viewed European explorers as "gods," has called this the "problem of literalizing tropes." He argues that any such language must be interpreted in a "context of utterance" (1992: 175). There are examples of interpretations of images like the killing rubber from elsewhere in the world. For example, Keller (2008) has observed a similar system of belief in Masoala, Madagascar, in which the death of the banana tree is seen to be a prerequisite to life and local values, whereas the eternalness of the moon is associated with lifelessness and the extralocal values of conservationists. Slater (1994) has reported on the contradictory folk images of dolphin-human spirits in the Brazilian Amazon, which she says reflect Amazonians' own mixed reactions to the ongoing destruction of the rainforest and the resulting transformations in the social as well as physical landscape. In contrast, it will be suggested here that the killing of the land by rubber in Kalimantan reflects not unresolved tension regarding the changes in society and the environment brought about by globalization; rather it reflects the cultural construction of a tension that manages these changes.

The Kantu' characterization of rubber lands as dead reflects a pervasive ideology of exchange involving both society and the environment. Rubber is not part of this system, and accordingly it "kills" exchange and thus the land. Exchange cannot, however, be equated with sustainability, nor can we conclude that rubber cultivation is destructive—for in many respects it is just the opposite of this. Land planted with rubber is seen as being forever dead, in fact, because it is seen as being potentially alive forever with rubber. The opposition of dead and alive must be interpreted in terms of an opposition between two distinct transactional orders (Bloch and Parry 1989): one focused on household reproduction and involving the subsistence cultivation of annual food crops, the other focused on the production of commodities for the market and involving the cultivation of perennial tree crops. I will show that the cultural distinction between the two systems in terms of alive and dead land has helped the Kantu' and other smallholders participate in global commodity production with minimal economic and political risk. In this respect, their rubber smallholdings are distinguished from other types of smallholder commodity production, as well as estate systems of production, that carry much greater risks.[3]

Recent efforts to apply Foucault's work on the concept of governmentality to the field of conservation and development may help us understand this contradictory image of smallholder rubber. Scholars like Agrawal (2005) and Sivaramakrishnan (1999) have drawn on this concept to examine the way in which marginal forest communities are involved in natural resource management by the central state (see also Li 2007). In particular, they have suggested that government programs purporting to empower such communities have potentially had the effect of locking them even more firmly into central state rule by means of daily mechanisms of self-discipline. A similar self-disciplining also can be achieved by involvement in state circuits of production of commodities like rubber.

The system of smallholder rubber production to be examined here historically escaped from this sort of central penetration and control. Critical to this escape was the simultaneous involvement in distinct commodity and food production sectors, the two transactional orders referred to above. The conception of the land as being killed by the rubber was essential to the construction and maintenance of these two distinct orders. Governmentality operates through the embedding of structures of central authority in the daily routines of production. The cultural reification of rubber production as killing makes these wider structures of power explicit and, thereby, easier to isolate and mitigate. The insight that this case study offers into relations between global commodity production and local subsistence production is relevant to ongoing efforts to involve forest communities in production for global markets, typically with little thought being given to the challenge of articulation between the market and subsistence spheres.

The Kantu' and Kantu' Rubber Cultivation

The bulk of rubber in Borneo is produced in tiny gardens of a hectare or so, by smallholders like the Kantu' (Dove 1993b; Gouyon, de Foresta, and Levang 1993). The product of their cultivation—the pressed but usually uncured rubber slabs—is traded or sold to Dayak, Malay, or Chinese traders in the interior, who in turn sell them to traders on the coast. Most smallholders have, and have had for centuries, so-called composite economies: they cultivate food crops—usually by extensive swidden agricultural technology—to meet their subsistence needs, while gathering or cultivating export commodities like rubber to meet their market needs.[4] In the case of the Kantu', an Ibanic-speaking Dayak people, the primary subsistence crops are dry rice (as well as some swamp rice) and a wide variety of nonrice cultigens, including a second cereal, maize, a number of different cucurbits, and a number of different tubers, including cassava and sweet potato and, to a lesser extent, taro. Rice is the primary starch staple and the focus of each meal, supplemented with one or more of the nonrice cultigens as a "relish." When the rice crop fails to meet subsistence requirements (as it may several years in ten), the market crops are sold to buy rice, and if that fails, then cassava and other tubers become the starch staple for the duration of the famine period. Para rubber (*Hevea brasiliensis*) is generally the most important market or cash crop for the Kantu', with pepper (*Piper nigrum* L.) being a distant second.

Rubber, a native of the Amazon basin, was first planted in the interior, tribal territories of Borneo toward the end of the first decade of the twentieth century. It was adopted with remarkable speed by local communities throughout this region (not only in Borneo, but also in Sumatra and in peninsular Malaysia). Whereas it was initially propagated as an estate crop in the former Dutch and British colonies, within two decades of its introduction it had become predominantly a smallholder crop. Smallholders held over 84 percent of Indonesia's rubber acreage in 2004, and they produced more than 80 percent of its rubber (Government of Indonesia 2004: 1, Table 1). This success is all the more notable because it occurred among culturally and politically marginal peoples who have been labeled as resistant to innovation and development by both colonial and postcolonial governments, and because these governments did nothing to support this adoption and a great deal to hinder it (Dove 1996).

There are two principal reasons for rubber's success in the face of these obstacles. The first involves the serendipitous fit of rubber into a tradition of gathering natural forest rubbers (Dove 1994). During the second half of the nineteenth century, booming international markets brought new players and more intensive systems of exploitation to bear on these rubbers, which had been internationally traded for more than a millennium; at the same time, the gradual evolution of the tribal rubber gatherers toward more sedentary settlement patterns and more intensive agriculture was transforming the niche into which the forest rubbers had formerly fitted. The exotic *Hevea* suited the transformed niche better than the native forest rubbers, and as an introduced and thus "domesticated" species by definition, it was protected against some of the efforts of opportunistic outsiders to dominate the extraction of the "wild" forest rubbers. The second reason for rubber's success involves the way in which it complements the indigenous system of swidden agriculture (Dove 1993b). The respective agronomies of rubber and swidden rice permit them to be jointly cultivated with little competition for either land or labor.[5] The advantage of this joint cultivation is that it significantly reduces the measure of agricultural risk borne by the small farmer: rubber tapping helps the smallholder survive periodic swidden failures due to environmental variables, and swidden cultivation helps the smallholder survive periodic collapses in rubber prices due to market dynamics.[6]

Despite this historical and economic complementarity, swidden land planted in rubber is said by the Kantu' to be dead. Rubber appears to be the first, and so far the last, cultigen to have this impact. No swidden crop, nor any aspect of the swidden cycle, is ever said to kill the land, not even when opening primary forest. To understand why the land is dead under rubber, it is first necessary to understand how the land otherwise is alive.[7]

Social and Ecological Exchange

The life of the land, among the Kantu' and many of the other groups in Borneo that adopted rubber at the beginning of the previous century, pertains to its role in pervasive relations of exchange in which not only land but also people participate. The most fundamental principle underpinning these relations is that of reciprocity.

The operation of the principle of reciprocity can be seen, first, in the Kantu' belief that wealth or bounty bears a cost. This is aptly illustrated by tribal beliefs concerning the "mast fruiting" of the trees in Borneo's rain forest. Every three to five years, many of the rain forest dipterocarps bear a heavy crop of fruit (Curran et al. 1999; Dove 1993a).[8] Borneo's tribespeople welcome mast fruitings as an opportunity to increase the personal consumption of, and trade in, the forest fruits. But they also believe that the increased consumption of the mast year entails a cost:[9] thus the Kantu' say that the spirits of the forest demand "sigi' kolak mata mensia [one basket of human eyes]" during a mast-fruiting year (i.e., the spirits demand human dead in sufficient numbers to fill a basket with their eyes).[10] A corollary to this principle that bounty bears a cost is the reverse one that cost yields a bounty: the most important statement that most animist tribespeople of Borneo can make to their gods involves the sacrifice of one or more domestic pigs. Such a sacrifice entails a significant economic cost to them, but it is one they justify by saying that any family making this sacrifice is certain to reap better harvests as a result, or at least better harvests than those families that do *not* make the sacrifice.[11]

The converse of the spirits giving the Dayak a mast fruiting in exchange for a human sacrifice is their withholding the mast fruiting because, in effect, such a sacrifice has not been made. The Kantu' say that their relations with the spirits are governed by a number of proscriptions and prescriptions, and prominent among the former are sexual prohibitions (against adultery, illegitimate birth, and incestuous unions). The Kantu' say that when they honor these sexual proscriptions, the spirits will assist them in enjoying the fruits of the land and the forest, and when they do not they will not.[12] Thus the Kantu' say that because of repeated violations of these sexual proscriptions mast fruitings are less frequent today than before, and when they do occur they are still less abundant than in earlier times (see also Curran et al. 1999). The only way to avoid this withdrawal of the spirits' favor, in the event of a sexual delict, is to make a compensatory offering or sacrifice to the spirits. Formerly, the lives of the offending couple themselves (i.e., those who committed the sexual delict) were sacrificed to the spirits.[13] In contemporary times, domestic pigs are sacrificed instead. For a serious sexual delict, seven pigs should be sacrificed: one each at the tribal territory's oldest *rian* (durian, *Durio zibethinus* Murr.), *engkabang* (illipe-nut, *Shorea* spp.), *tapang* (a bee tree, often *Koompassia excelsa* [Becc.] Taub.), and *buah* (any other forest fruit) in the territory, and one each in the *rumah* [longhouse], on the *tanah* [ground], and at the *sungai* [river].

The second and related dimension of the principle of reciprocity in Kantu' society is that too much bounty can be a bad thing. That is, there is a point at which the accumulation of material wealth—whether due to the direct efforts of humans or the bounty of nature—violates indigenous norms for re-distribution and reciprocity and, indeed, is made possible only by such violation. Thus the majority of swidden rice surpluses are typically dis-tributed each year, from households enjoying surpluses to those households experiencing shortfalls. As a result, any long-term accumulation of rice stores by a given household is suspect, since it implies that requests for rice to tide others over harvest shortfalls have been denied.[14] The assumed immoral origins of such accumulation are reflected in explanations of the perceived material wealth of the Catholic Church in West Kalimantan: thus the Kantu' say that the loft of the Catholic mission nearest to their territory is inhabited by a *naga* (dragon) that excretes gold coins when it is fed human dead. The implication, therefore, is that only such a gross violation of the moral order as feeding one's dead to a dragon could produce the kind of accumulated wealth that they associate with the Catholic Church.[15]

THE CREATION AND DESTRUCTION OF BOUNTY

The overaccumulation of bounty can be avoided by its distribution or de-struction. As the story of the Catholic dragon suggests, there is a linkage between destruction (i.e., death) and creation (i.e., of wealth). This linkage, and the principle of cycling and recycling that underlies it, is central to the cosmology of many Dayak groups (and, indeed, to the traditional cosmolo-gies of peoples throughout Southeast Asia).[16] An example is a myth held by the Ngaju, a people of south-central Kalimantan, concerning the Tree of Life, a central symbol of their cosmology: "When the Tree of Life bears its fruit . . . it does not simply stand in solitary height, for then its time has come. The birds fly to it. They peck at the fruit of the Tree, they destroy it, and from the destruction and self-destruction arises the whole cosmos" (Schärer 1963: 128).

The image of epochal fruiting in this myth resembles—and, indeed, per-haps derives from—the phenomenon of mast fruiting discussed earlier. The stochastic character of production underlying both mast fruiting and Ngaju cosmology has an ecological basis: mast fruiting is an adaptation to the generally impoverished environment of the tropical forest, and the more extreme the impoverishment in particular regions, the more extreme the mast (i.e., the less frequent but the greater in magnitude the mast [Curran et al. 1999; van Schaik, Terborgh, and Wright 1993: 358]). There is increasing evidence that patterns of perturbation of this sort—stochastic or catastrophic change as opposed to gradual change—characterize forest history and, indeed, the natural history of the earth's environment in general (see chapter 1) and should be recognized as the basis for human management (Berkes and Folke 1998; Gunderson and Holling 2001; Haila and Dyke 2006). There is increasing evidence, that is, that periodic disturbance in forests and other ecosystems is both predictable and integral to sustainability.

Creation and destruction are linked in a cycle, and anything that interrupts or resists this cycle is seen as a violation of the cosmological order. This is reflected in the traditional Kantu' *pantang* [proscription] against the use of Borneo's foremost timber, from the ironwood tree (*Eusideroxylon zwageri*), in house construction.[17] The Kantu' rationale for this proscription is that the durability of this wood and anything constructed from it surpasses the life span of the people using it. In light of the current analysis, the problem with such long-lived construction is that it represents a bounty taken *from* nature with no guarantee of when it will return *to* nature. The cycle of destruction and creation, the cycling between nature and culture, thus is broken by such construction.[18]

Another example of such beliefs is reported for the Merina of central Madagascar: Bloch (1989: 179–80) describes a proscription among the Merina against planting trees whose life span surpasses that of humans: he interprets this proscription in light of the tension in Merina society between the transitory life of the individual and the eternal life of the undifferentiated descent group. The former, which Bloch glosses as "individuation," opposes the latter because it terminates with death. The problem posed by individuation is resolved by the prescription that all individual property must be distributed before death. The problem with planting long-lived trees is that they interfere with this cyclic relationship between the individual and the descent group. As Bloch says (1989: 179), such planting "would therefore compete with descent in an absolute way. It would render something of the individual permanent." Bloch's explanation is evocative of the principles elucidated here concerning the need to distribute rather than accumulate wealth and the need to destroy so as to create. Long-lived trees in Borneo present a challenge to the moral order and the moral cycle not because something has been destroyed forever but, counterintuitively, because something has been *created* forever.[19] Permanent life presents as much of a problem to this moral ecology as permanent death.[20]

The principles of this cosmological order are well applied in the farming system that still persists throughout the interior of Borneo and the rest of Indonesia's outer islands: swidden agriculture. The cyclic linkage of creation and destruction is central to this system: the near-instantaneous destruction of the forest by fire creates the swidden, whereas the gradual destruction of the swidden, through both natural and anthropogenic processes of afforestation, recreates the forest.[21] An emphasis on distribution versus accumulation also is central to the swidden system. The successful operation of the swidden system among a people like the Kantu' is made possible only through the regular inter-household exchange of labor and grain. The existence of environmental constraints on the timing of planting, weeding, and harvesting obliges swidden farmers to exchange labor with one another to cope with these labor bottlenecks; and the inherent unpredictability of the environmental conditions that determine the swidden harvest obliges each year's successful swidden farmers to distribute grain to the less successful as an investment against the time when the former will be less lucky (Dove 1988: 162). The accumulation versus distribution of grain stores is socially sanctioned because it is seen as evidence of the violation of this principle. Finally, the principle that bounty bears a cost also applies to the swidden system. The Kantu' regard the outcome of the swidden cycle, the harvest, as the product of an agreement between themselves and the spirits: if the Kantu' perform the requisite rituals and offer sufficient sacrifices to the spirits, they expect this to be reflected in a bountiful harvest. The Kantu' quite explicitly say that whatever they expend on ritual sacrifice will increase their favor with the spirits, and this will in turn increase their swidden returns sufficient to, minimally, surpass their ritual expenditure (Dove 1988: 151).

Exchange in Rubber Cultivation

The principles of exchange just discussed, as applying to the system of swidden cultivation, do not, for the most part, apply to the system of rubber cultivation.

BOUNTY WITHOUT COST OR DISTRIBUTION

The Kantu' do not believe that the bounty produced by rubber cultivation bears a cost in the way that the bounty produced by swidden cultivation does. When rubber was first introduced to the Dayak early in the twentieth century, they feared that the rubber trees would eat the spirit of their swidden rice, and they moderated their commitment to rubber in response. This represented a cultural statement about the conflict between two different transactional orders (Dove 1996; cf. Bloch and Parry 1989: 23–24): one, the swiddens, oriented toward the long-term subsistence needs of the group; and the other, the rubber, oriented toward the short-term market needs of the individual. However, the system of beliefs that dominates the system of swidden agriculture, regarding the ritualized exchange of bounty between spirits and humans, is absent from the system of rubber cultivation. That rubber lies outside this belief system is reflected in the fact that rubber is not one of the trees mentioned earlier to which pigs are sacrificed when a major sexual delict occurs.

Associated with the lack of exchange between humans and spirits in rubber production is a lack of exchange among humans. Little or no inter-household reciprocity or exchange is involved in providing the inputs to rubber cultivation (the foremost of which is labor) or consuming the out-puts (sheets of rubber), in sharp contrast to swidden cultivation.[22] Unlike swidden cultivation, there are no ecological imperatives for such exchange in the case of rubber: the seasonal variables and labor bottlenecks that con-strain swidden production are absent from rubber production. There also is greater commoditization in rubber cultivation of inputs such as labor, and greater separation of the laborer from the product of his or her labor, than is the case in swidden rice cultivation. Thus while sharecropping is unknown in swidden cultivation, it is commonly practiced in rubber cultivation, as in the *bagi dua* (split-in-two) system, wherein laborers who have no rubber trees of their own tap the trees of others in exchange for a half of the yield.[23]

BOUNTY WITHOUT CREATION AND DESTRUCTION

Just as there is no socio-economic exchange in rubber cultivation, so is there no ecological exchange: there is no automatic cycling back and forth be-tween rubber and forest. Unlike most of the economic species planted by the tribal peoples of Borneo in their swidden sites, rubber does not succumb to the secondary afforestation that naturally occurs there; rather, it *becomes* the succession.[24] In its native habitat in South America, rubber is a denizen of mature forest, not younger successions (Dean 1987: 60). Rubber trees are long-lived: individual rubber trees can have a productive life of thirty to forty years. The life of the rubber garden is not limited to thirty to forty years, however, because naturally grown seedlings typically replace the planted trees with a second generation of trees. Further iterations of this natural process make the productive life of rubber groves potentially open-ended. In practice, however, the productivity of older rubber groves tends to decline. Gouyon, de Foresta, and Levang (1993: 194) note that the impact of shade on seedling growth will result in a drop in tree density in the third and higher generations, as the spacing of the rubber trees moves from the artificial spacing of five hundred per hectare in rubber groves toward the two to three per hectare that naturally obtains in its native habitat in the Amazon, which eventually makes the rubber grove unprofitable to tap. And indeed, it is not uncommon for Dayak to clear older rubber groves for new swiddens.[25] Such a reentry into the swidden cycle does not alter the status of the land as *tanah mati*, however. The Kantu' say that land planted in rubber is still *mati*, dead, even if the rubber itself dies.[26]

The nonparticipation of rubber land in these ecological and socio-economic exchanges is reflected in tenurial practices—which is the clearest practice behind the rhetoric of the "dead land." For example, after a Kantu' *bilek* (household) undergoes partition, its swidden forest land will continue to be held in common (*kuntsi*) by the resulting households, at least until the children of those carrying out the partition mature and begin to make their own swiddens (Dove 1985b); any rubber groves, however, will be divided. Similarly, if a household moves out of a longhouse, its forest lands must be returned to the longhouse to be held and used in common, but not its rubber groves, to which the departed household can continue to exercise exclusive proprietary rights. The adoption of rubber thus resulted in the development of a special tenurial category for it, distinct from ordinary swidden regrowth.[27]

Further insight into the meaning of the label "dead" for land under rubber is provided by the use not of the Kantu'/Iban term for death, *parai*, but the Malay term, *mati* (A. Richards 1981: 210, 254–55; Wilkinson 1959: 2:749–50).[28] The use of a Malay versus a Dayak term reflects the fact that unlike the locally oriented transactional order of swidden production, the transactional order of rubber is more oriented toward markets and the wider, traditionally Malay-speaking society. And whereas the Malay term *mati* does have the literal meaning of "death," it also has a figurative meaning of "fixed" (Wilkinson 1959: 2:749–50), and it is this latter meaning that dominates its uses in Kantu'/Iban (A. Richards 1981: 210). The phrase *tanah mati* thus refers to the way in which the combined ecological and social consequences of planting rubber "fix" the land in the context of a wider swidden landscape that is by implication *not* fixed. Nor, counterintuitively, does this fixing of the land mean its death in a literal sense. Indeed, land planted in rubber is dead forever because the rubber is, at least potentially, alive forever; just as the accumulation of wealth from rubber trees is unend-ing because of the lack of social exchange in its cultivation. Rubber is judged not in terms of the values of life or death, therefore, but in terms of the cycle between life and death, creation and destruction, field and forest. Unending life without death jeopardizes this value, just as does unending accumulation without distribution. The former interrupts the ecological cycle of creation and destruction, the morality of exchange between nature and culture; whereas the latter interrupts the economic cycle of interdependence and reciprocity, the morality of exchange between people.

COMPLEMENTARITY IN POLITICAL RATHER THAN
AGRONOMIC TERMS

The relationships between systems of cash cropping and systems of swidden cultivation have not been well interpreted, even by such respected scholars as Pelzer (1945) and Wolf (1982). Both argue (Pelzer 1945: 24–25; Wolf 1982: 330) that cash crops like tobacco, pepper, and coffee complemented native swidden systems during the colonial era—Wolf (1982: 330) even uses the term *symbiosis* to describe the relationship between tobacco and swidden cultivation—whereas rubber and other perennial tree crops competed with and disrupted these swidden systems.[29] Thus Pelzer (1945: 25) writes: "Although this modification of the old economy [by planting perennials in swiddens] has led to permanent gardens of rubber, coconut, and benzoin trees and thus to a partial abandonment of shifting cultivation, where pepper and coffee supplement the traditional crops of the shifting cultivator no fundamental change takes place, because *these* [added] gardens are not permanent." Based on Pelzer's research, Wolf (1982: 330) writes of Deli, Sumatra:

> In this area, the Dutch had long grown tobacco on plantations, which developed in symbiosis with the slash-and-burn agriculture of local Malay and Batak villagers. The plantation took over the labor of burning off the covering vegetation. It then raised the first crop, tobacco. When productivity decreased in the second year, the plantation opened a new field, allowing the villagers to take over the tobacco plots in order to raise food. When rubber was introduced in 1906, this symbiotic relationship came to an end. Rubber trees were a perennial crop and could not be alternated with annuals. Instead, rubber cultivation, carried on by imported Javanese and Chinese laborers, now engulfed the subsistence plots of the native population.

Wolf's view is partially correct, at least in narrow agronomic terms. Crops like tobacco, pepper, and coffee have a highly deleterious impact on the land. They rapidly extract its nutrients, requiring its abandonment after just a few years.[30] Pelzer and Wolf are right, therefore, in thinking that this pattern of land use in some sense resembles and thus complements swidden cultivation. The same fundamental ecological principle underlies both systems of cultivation: cultivation exhausts available nutrients and leads to a proliferation of weeds and pests, which obliges a cessation in cultivation until such time as a weed- and pest-free nutrient base is restored.[31]

The ecological principles underlying rubber cultivation are quite different. In rubber cultivation, the decline in productivity of older gardens is much more gradual and it is not driven by a decline in the stock of available nutrients or a rise in weeds and pests; rather, it is driven by the natural progression of older gardens toward a lower density of rubber. Whereas in the case of tobacco, pepper, and most swidden crops, the decline in productivity and eventual abandonment is driven by a fast shift to a younger, depauperate, pest-ridden succession, in the case of rubber the process is driven by a slow shift to an older, richer succession. There is a clear difference between the two processes: whereas the former briefly interrupts and holds in abeyance natural processes of vegetative succession, the latter enhances them; whereas land subject to the latter process becomes "fixed" or "dead," land subject to the former process remains unfixed or alive. In these senses, rubber does indeed seem to be "competing" with swidden cultivation, as Pelzer (1945: 25) and Wolf (1982: 330) suggest; but this competition is not necessarily malign.

Despite the seeming competition between rubber and swidden cultivation, their combined practice has been extraordinarily successful, as measured by the number of people involved and their contribution to the international rubber trade. This success is explained by the fact that the combination of rubber and swidden cultivation has been both politically and economically empowering (Dove 1993b, 1994, 1996). This empowerment is a function not just of the relationship of the rubber system to the swidden system but also of the significance of this relationship in the wider context of colonial and postcolonial political-economic relations. In this wider context, any symbiosis between a system of cash cropping and a system of subsistence agriculture bears a risk: it may be seized upon by superordinate political and economic authorities as an opportunity to piggyback a centralized regime of extractive agriculture on a local regime of subsistence agriculture, so that the latter subsidizes the costs of the former. This is precisely what happened with tobacco and swidden cultivation in Deli, Sumatra via a sort of agronomic governmentality: the daily pattern of smallholder agriculture was altered by the introduction of annual crops (tobacco, pepper, coffee), with the result that the patterning of smallholder production in time and space was implicitly reordered toward the priorities of the state.

The colonial tobacco planters and swidden cultivators of Deli did not contribute equally to their so-called symbiosis. Wolf (1982: 330) implies that the swidden cultivators benefited because "the plantation took over the labor of burning off the covering vegetation"; but burning is actually the briefest stage by far of the swidden cycle, taking as little as an hour or two to complete. The slashing and felling of the forest that precede the burn, in contrast, are onerous tasks; and by integrating their tobacco into the Deli swidden cycle, the planters could avoid expending this labor themselves. This arrangement was ironic, in that colonial planters were in effect drawing a labor subsidy from a traditional system of agriculture that was otherwise fiercely criticized and suppressed by the colonial establishment. This irony reached a peak in the famous *taungya* system of British Burma, in which colonial teak cultivation piggybacked on indigenous swidden cultivation. Bryant (1994) has exposed the dual and paradoxical function of *taungya*, noting that it was intended to extract an economic surplus from the swidden peoples—in the form of assistance with the replanting of teak—at the same time as it was intended to eliminate swidden cultivation by gradually putting all swidden lands under state teak plantations.[32]

Although the Burmese swidden cultivators ultimately did not succumb to the *taungya* system (Bryant 1994: 249–50), the potentially debilitating effects of such subsidies of state enterprise are well documented in Geertz's (1963) study of the piggybacking of colonial and postcolonial sugar cane cultivation on peasant wet-rice cultivation in Java.[33] Where there is a threat of the extraction of such subsidies, a system of cash cropping that offers—by means of its complementarity—an easy avenue into local systems of resource use is not empowering; and one that does not offer this complementarity, like rubber, *is* empowering. In this sense, the lack of complementarity between rubber and swidden cultivation, and the supposed death of the land under rubber, indicate social and ecological relations that are sustainable from the local perspective.[34] And the presence of such complementarity indicates the reverse.

THE RUBBER-SWIDDEN FIT AND THE BENEFIT OF TENURE

The *lack* of complementarity in the foregoing sense between rubber and swidden cultivation is related to the development of policies by colonial and postcolonial political and economic authorities to support the development of a state-led plantation rubber sector as opposed to the smallholder rubber sector. The existence of this smallholder system beyond the control of central political-economic authorities and in direct competition with their planta- tion system had a predictable impact on the way in which smallholder rubber has been viewed by these authorities. Colonial authorities believed the rubber smallholdings to be diseased and overexploited (Dove 1996: 40–

41), and colonial and postcolonial authorities alike dismissed them as *hutan karet* (rubber jungles) (Gouyon, de Foresta, and Levang 1993: 182).[35]

This state emphasis on the wild and unworked character of the small-holder rubber is no accident: it reflects tension over tenurial issues. In the colonial as well as the postcolonial state systems governing Kalimantan, land involved in swidden cultivation cycles has been accorded little or no tenurial recognition. In contrast, some recognition *is* accorded smallholder rubber plantings, however disparagingly, as reflected in the sarcasm of the "rubber jungle" epithet.[36] Indeed, the fact that rubber plantings win this recognition from the state has become a principal motivation among smallholders for planting the crop.[37] It is perhaps inevitable that a production system that wins such recognition from the overarching state system, and that fits well into this wider system, will *not* fit well into a local subsistence system like the swidden system of the Dayak. That is, it is probably inevitable that what thrives in the wider political-economic system will be "dead" in the local one. Opposition to the local system was probably a prerequisite, by definition, for success in the extralocal system.[38]

The local distinction between swidden and rubber cultivation is reproduced in extralocal commercial forest exploitation. The swidden-cultivating peoples of Kalimantan have been bedeviled over the past several decades by the explosive growth of logging concessions, *hak pengusahaan hutan*. Notwithstanding the fierce opposition between the logging concessions and the swidden cultivators, both rely on the same ecological dynamic, namely, a natural cycle of artificial forest opening and natural forest closure. They differ in the greater magnitude of destruction that the logging concessions precipitate compared with swidden cultivation, because of the scale of clearing involved, the associated disturbance of the soil, the road building, and disinterest in promoting the afforestation needed for subsequent cycles, as well as having conjured versus legitimate rights to the forest. Notwithstanding all the ills that attend logging concessions, however, Dayak activists say that they are less worried about them than about the *hutan tanaman industri*, or commercial tree plantations, including oil palm. The Dayak say that whereas the logging concessions will eventually go away and may also offer opportunities in the interim for village employment and participation (compare McCarthy 2006), the tree plantations may stay forever (Stepanus Djuweng, Institute of Dayakology Research and Development, Pontianak, West Kalimantan, personal communication, July 14, 1994). Whereas the logging concessions are part of a forest cycle, the tree plantations are fixed, like the rubber gardens. This analogy is reflected in the fact that many of the tenurial conflicts that arise in the development of timber plantations involve locally owned smallholder rubber gardens, whereas most conflicts in the development of logging concessions involve locally owned swidden lands. Whereas the fixedness of the rubber gardens protects local landowners in the face of state encroachment, therefore, the fixedness of the timber plantations dispossesses them.

Exchange between People and Trees

An important element in much of this discourse about exchange is a metaphorical relationship between people and trees.

PEOPLE-TREE METAPHORS

Metaphoric linkages between people and trees, or more generally between people and plants, afford many societies a mechanism to conceive of and structure social relations (Rival 1993; for examples from the Indo-Malay region, see James Fox 1971 on the Rotinese of Eastern Indonesia, Sather 1990 on the Iban of Sarawak, and Sugishima 1994 on the Lionese of Central Flores).[39] A traditional Javanese example of this is the folk saying that a king is to his people as the tiger is to the forest (Moertono 1981: 22).[40] De Boeck (1994: 461) relates a similar example from among the Aluund of southwest Zaire, who say that the people are to the king as animals (frugivores) are to a fruit tree.[41]

One of the most important metaphoric linkages of people and trees involves a comparison of age.[42] The dimension of age is implicit in most discussions of trees and other perennials in Indonesia: they are colloquially called *tanaman umur panjang* (long-lived plants) in Indonesian. The significance of this life span lies in its comparison with the human life span, as illustrated in the proscription by the Merina of Madagascar of planting long-lived trees (Bloch 1989)[43] and by the Kantu' of using long-lived timber. In some cases, that a given tree exceeds the human life span is accepted and constructively utilized: it is claimed that this is why durian trees (*Durio zibethinus* Murr.) are planted in West Kalimantan as tenurial markers (Charles Peters, personal communication, New York Botanical Gardens, July 13, 1994). The significance of the life span of a planted or "worked" tree exceeding the human life span is illustrated by the case of rubber. The Kantu' say of land planted in rubber: "tanah mati s'umur idup [the land is dead for as long as one lives]."[44] In this context, the meaning of *s'umur idup* is actually, and counterintuitively, "for *longer* than [one] lives"; which means forever.[45] The significance of this open-ended life of the rubber grove, in comparison with the finite life of humans, is reflected in the fact that rubber occupies an entirely distinct transactional order within the household, separate from that of the annual food crops.

The comparison of age between people and trees is often part of a wider belief in the shared interest and welfare of people and trees. Stories like the following, from the former Banjar kingdom on the southeast coast of Borneo, are common: the sixteenth- and seventeenth-century Banjar court chronicles tell of a battle in which the king's standard fell to earth and was reerected with a branch of the *jingah* tree (Ras 1968: 175–76, 432, 438).[46] This pole took root and grew; and then, whenever it lost a branch, the imminent death of a member of the royal family was foretold.[47] A similar identification of person and tree was found traditionally among some of the Dayak groups of southwest Borneo, as described by Frazer (1951: 790):[48] "Amongst the Dyaks of Landak and Tajan, districts of Dutch Borneo, it is customary to plant a fruit-tree for a baby, and henceforth in the popular belief the fate of the child is bound up with that of the tree. If the tree shoots up rapidly, it will go well with the child; but if the tree is dwarfed or shriveled, nothing but misfortune can be expected for its human counterpart." One of the closest identifications between persons and trees was achieved in the traditional treatment of the dead among some Bornean tribes: there are several nineteenth-century accounts of burying the dead in apertures in living trees, which then grew around the bodies (A. Maxwell 1992: 8–9).[49]

The identification of people and trees may also be achieved at the level of the entire cosmos. The conception of the cosmos as a "tree of life" is common throughout Indonesia (indeed, throughout the world).[50] I mentioned the Ngaju Tree of Life earlier, along with the belief that its cyclic creation and destruction betokens the creation and destruction of the cosmos. The key elements of this vision—the linkage of creation and destruction, and the related concept of an exchange between life and death and between culture and nature—are reflected in the Kantu' beliefs that when the forest spirits provide excessive bounty as in a mast fruiting, they will exact a price in human dead and, conversely, that when humans engage in excessive consumption by committing sexual delicts, they must atone by making sacrifices to the spirits.[51]

THEORIES ABOUT PEOPLE-TREE METAPHORS

The identification of people with trees has implications for attempts to denaturalize the tree as a conceptual structure, as in Bouquet's (1995) analysis of four "trees"—Dubois's family tree of *Pithecanthropus erectus*, Haeckel's phylogenetic diagram of human evolution, the Tree of Jesse showing Christ's earthly ancestry, and River's genealogical tree for anthropological inquiry.

Bouquet (1995: 44) suggests that an unthinking commitment to treelike conceptual structures is reflected in the fact that evolutionary theory contested the biblical creation story but borrowed the same genealogical motif to represent its own version: "This replication of the genealogical motif seemed to me a very clear example of how scientific discourse is rooted in what I will call local cultural discourse." She cites Bourdieu's (1977: 38) critique of anthropology's preference for official or logical, as opposed to practical, genealogical trees;[52] and she also cites Deleuze and Guattari's (1987) discussion of tree-based versus rhizome-based models of reality (Bouquet 1995: 42, 51).

Deleuze and Guattari's (1987: 18) discussion is sufficiently interesting to merit citation at some length. They write:

> It is odd how the tree has dominated Western reality and all of Western thought, from botany to biology and anatomy, but also gnosiology, theology, ontology, all of philosophy. . . . The West has a special relation to the forest, and deforestation; the fields carved from the forest are populated with seed plants, produced by cultivation based on species lineages of arborescent type; animal raising, carried out on fallow fields, selects lineages forming an entire animal arborescence. The East presents a different figure: a relation to the steppe and the garden (or in some cases, the desert and the oasis), rather than forest and field; cultivation of tubers by fragmentation of the individual; a casting aside or bracketing of animal raising, which is confined to closed spaces or pushed out onto the steppes of the nomads. The West: agriculture based on a chosen lineage containing a large number of variable individuals. The East: horticulture based on a small number of individuals derived from a wide range of "clones." Does not the East, Oceania in particular, offer something like a rhizomatic model opposed in every respect to the Western model of the tree?

Deleuze and Guattari are correct in suggesting that there are important cultural and political differences—beyond the obvious agronomic ones—between systems of production based on vegetative propagation and systems based on reproduction from seed.[53] They are wrong, however, and themselves fall victim to Orientalism, when characterizing the rhizome/seed distinction as one of the East versus the West. There are rhizomes in Western history just as there is seed in Eastern history—and the same holds true for trees, as is shown by the evidence from Kalimantan presented earlier.

Whereas efforts like those of Bouquet, Bourdieu, and Deleuze and Guattari to raise our consciousness about the use of embedded tree models are instructive, any suggestion that such models have simple roots in Western culture ignores the environmental present and past of both the East and the West. The discussion here of the identification of trees and people in Kalimantan reflects the fact that most of human history, in most parts of the world, has involved an intimate relationship with trees. The antiquity and ubiquity of this relationship is reflected in the fact that the oldest trade good in the world may well be tree sap. Camphor (*Dryobalanops aromatica* Gaetn. F.), dammar (gum from a variety of dipterocarps, especially of the genus *Shorea*), "dragon's blood" (*Daemonorhops* Blume, spp.), gum benjamin (*Styrax* spp.), and various pine resins (especially from *Pinus merkusii*) are the oldest trade products of Southeast Asia; and they fit into a trade niche that was originally created for the even more ancient traffic in the tree saps of the Middle East, the fabled frankincense (*Boswellia* spp.) and myrrh (*Commiphora* spp.).[54] The value of such ancient tree products notwithstanding, there is often great ambivalence in conceptions of trees, which is reflected in the way different societies view forests even today. In many contemporary hunter-gatherer societies, the forest is conceived of as an ever-giving parent (Bird-David 1990); whereas in many agricultural and state-based societies, the forest is conceived of as the principal obstacle to progress and the refuge of spirits, criminals, and political enemies. These sentiments are reflected in the common Javanese term for development, *babad alas*, which literally means "to clear the forest."

THEORIES ABOUT THEORIES: TRUNK TO TIP AND
SAP VERSUS TREE

It is illuminating to compare tree metaphors in the East and the West, as James Fox (1988) does in an analysis of descent "trees" in Roti. He notes that the meaning of the image of the tree is far from given. He writes, "Trees present protean forms. They can grow up but they can also grow outward. Some, like the waringin, can grow up and then set down new roots. Hence, tracing relations using the image of a tree can take many forms" (James Fox 1988: 16). He illustrates this point by noting that whereas descent is conceptually represented by a tree in Roti, as in the West, the directional orientation of the two images is not the same. Thus, he writes of Roti, "Here we have the conception of an ancestral tree, which is the opposite of that of the West: instead of being read from the top down, as a form of descent, it is read from the bottom up as a kind of ascent—from trunk to tip" (James Fox 1988: 9).[55]

A difference in metaphors is also evident in the case of rubber and its moral ecology. From the Dayak perspective, as I have noted, this morality is articulated in terms of the killing impact on the land. From the perspective of the Western-derived plantation complex, the morality of production is characterized very differently, with a focus on the sap or latex itself, which receives no attention whatsoever in Dayak cosmology. For example, in the colonial plantations of Indochina, Woodside (1976: 210, quoted in Murray 1992: 60) says that the sap was called "white gold" by the planters, whose workers countered by calling it "white blood." Shifting to the Western hemisphere, Schultes (1987: 93–94), following the lead of a Brazilian novelist from the 1940s, calls the sap the "blood of the gods." The prominence of blood in these metaphors seems to have something to do with sacrifice, the sacrifice of humans in particular. Trees have sap, only humans have blood. Thus, whereas the Dayak metaphor focuses on the cost to the land, these Western metaphors seem to refer to the cost to human workers, to coolies. This is a historically accurate reflection of the difference in modes of production: the smallholder tribal mode of production exploits the land, the plantation mode exploits human labor.

Conclusions

To recapitulate, the Kantu' of West Kalimantan say that swidden land planted in rubber trees becomes dead, which is initially puzzling because the combined cultivation of swiddens and rubber groves has proved to be one of Southeast Asia's most sustainable natural resource management systems. An analysis of the ethnography in one rubber-producing region, in West Kalimantan, suggests that the meaning of "dead land" is based on indigenous processes of social and ecological exchange, from which rubber is excluded. Three principles underlie this exchange: (1) bounty carries a cost; (2) bounty, or wealth, should be distributed, not accumulated; and (3) the creation and destruction of bounty are linked. None of these principles apply to the system of rubber production, which carries no such costs, in which distribution is not necessary, and in which there is no similar cyclic linkage between resource creation and destruction. These differences result in a lack of complementarity between rubber and swidden cultivation at the local level. But at the extralocal level these same differences and this same lack of complementarity become not a liability but an asset: the lack of complementarity hampers access to the local subsistence base by extraction-oriented state elites; at the same time, the lack of complementarity facilitates access by local communities to powerful national and international tenurial and resource regimes. That this lack of complementarity frustrated the centralized state control of production that was achieved with so many other tropical commodities helps explain the history of state antipathy to smallholder production like that of the Kantu'.

Rubber cultivation seems at first glance to exemplify the insidious extension of the global political-economy into marginal regions. An exotic imported to Southeast Asia from the New World, rubber in the early years of the twentieth century spread with remarkable speed through the interior of Borneo, Sumatra, and the Malay Peninsula, becoming an integral part of the daily production routines of millions of smallholders. It seems, thus, like an archetypal case of indirect means of disciplining marginal peoples; but what actually took place was far more complicated. As discussed here, the Kantu' and other smallholders drew on indigenous cultural values to mark off the rubber production sphere and limit its wider impact on their society. This is reflected in the statement about rubber killing the land, which essentially represents a folk commentary on the disciplining effects of rubber. By making these effects explicit, and by commenting on their impact on traditional production values, their impact was circumscribed.

These insights into the history of Dayak involvement with rubber are especially relevant today because of the interest in mitigating deforestation and promoting afforestation that is being driven by worries over global climate change. These worries are feeding into the explosion of interest in establishing oil palm plantations in the tropics, initially to produce edible and industrial oils, but now also out of interest in bio-fuel production and carbon sequestration (Potter 2008: 70; see also Acciaioli 2008; Cooke 2002; Sandker, Suwarno, and Campbell 2007; Sheil et al. 2009). The historic case of rubber shows that global claims about the benefits of oil palm cultivation misapprehend the way that oil palm, in practice, complements or conflicts with local systems of production and environmental relations. The case of rubber also shows the importance of considering smallholder alternatives to plantation-type production, something that has been completely ignored to date in oil-palm development in Southeast Asia.[56]

Beyond oil palm, this analysis has several more general lessons for policy makers seeking the sustainable management of natural resources. First, there are no strictly local solutions. The swidden-rubber system of the Kantu' described here, for example, is Janus-faced: it is oriented toward both local and extralocal needs, constraints, and opportunities. In an interconnected world, sustainable systems will not just partake of these interconnections but will emphasize them; they will concentrate on balancing the costs and benefits of these connections in a way that is sustainable for the local community. This balance necessarily involves both the political and the economic arena: Dayak successes with rubber came at the expense, and in spite of the opposition, of a state-supported plantation sector.

Second, there are no static solutions to sustainability: in a world of change, sustainability is, in part, all about incorporating change into what already exists, so as to protect existing advantages and diminish existing disadvantages. Thus the swidden-rubber system of the Kantu' is not traditional, if by this term we mean to signify something from an unchanging past. It is as new to Kalimantan as the *Hevea* rubber on which it is based. The change implied in the phrase *tanah mati* reflects the changes in Dayak society that accompanied the introduction of *Hevea* rubber.[57] This was not simply a passive response to the introduction of rubber, but an aggressive pattern of adoption designed to take advantage of a new element on the resource landscape. And further changes in the rubber smallholder system are occurring today (Penot 2007).

Third and finally, there are no preordained outcomes. Smallholder rubber systems like that of the Kantu' predominated in Indonesia and Malaysia throughout most of the twentieth century, but a completely different estate model was a strong competitor in these two countries and actually prevailed in Indochina for many years. In addition, in the Malay peninsula, matrilineality and Islam led to the development of a smallholder system more dominated by gender politics than was the case in Borneo (Kato 1991: 144–45; 1994: 147–48, 157).

Notes

I first carried out research in the interior of Indonesian Borneo (Kalimantan) from 1974 to 1976 with support from the National Science Foundation (Grant #GS-42605) and with sponsorship from the Indonesian Academy of Science (LIPI). I gathered additional data during six years of subsequent work in Java between 1979 and 1985, making periodic field trips to Kalimantan, with support from the Rockefeller and Ford Foundations and the East-West Center, and with sponsorship from Gadjah Mada University. A subsequent series of field trips to Kalimantan, beginning in 1992, was supported by the Ford Foundation, the United Nations Development Programme, and the John D. and Catherine T. MacArthur Foundation, with sponsorship by the National Planning Board (BAPPENAS) and Padjadjaran University. An initial analysis of this data was supported by a fellowship from Yale University's Program in Agrarian Studies. An earlier version of this analysis was published in Dove 1998.

1. See also Endah Sulistyawati's essay in this volume for a later restudy of the same group of Dayak, the Kantu'. See Dove (2011) for a book-length analysis of the history of involvement of Borneo's native peoples in production of commodities —including rubber—for global trade.
2. Whereas the Kantu' use the term *tanah mati* (dead land) for rubber gardens, they do not use the term *tanah idup* (living land) for swidden land. In this opposition, swidden land is the privileged register: rubber land needs to be differentiated from swidden land, but not the reverse.
3. Recent work that is relevant to the drawing of such distinctions is that on shade-

grown versus non-shade-grown coffee (Bray, Sánchez, and Murphy 2002; Gobbi 2000; Vandermeer and Perfecto 2005).

4. Other examples of this combination of market- and subsistence-oriented agricultural activities from Indonesia are swidden agriculture and rattans in East Kalimantan (Fried 2000) and swidden agriculture, coffee, and dammar (resins) in Sumatra (Michon et al. 2000).

5. But see Peluso (2009) for a case in West Kalimantan where competition for land eventually arose, leading to the near-disappearance of rice.

6. Not all of Indonesia's smallholdings conform to this pattern. Pelzer (1945: 24) describes in Sumatra a pattern of historic adoption of rubber by smallholders that was based not on combining market-oriented rubber and subsistence-oriented rice but rather on a complete and one-way shift from the latter to the former: "A shifting cultivator, may, over a period of years, convert his old ladangs into rubber gardens, for example, until he is assured of a sufficient income from his permanent tree crop to buy his food from the outside and no longer requires a ladang for the production of food." This pattern of agricultural transformation was historically not common in Kalimantan among Dayak like the Kantu', but it is common among Malays (Islamicized Dayak), for whom it can be an identity marker.

7. See Penot (2007) for recent developments in smallholder rubber cultivation, including the adoption of some estate practices.

8. The evolutionary purpose of mast fruiting is to overload seed predators during the mast years and thereby enhance the prospects for seed survival, while denying predators food during the intervening years and thereby suppressing their population growth. Mast fruiting in Borneo is triggered by dry air masses, caused on the western side of the island by "a chance fluctuation in the subtropical monsoonal circulation system" and on the eastern side by the El Niño–Southern Oscillation phenomenon (Ashton, Givinish, and Appanah 1988: 61).

9. On ambivalence regarding bounty, see Sugishima (1994: 159) on the Lionese of central Flores: "Witches are also feared as the providers of 'extraordinarily large harvests' (kesu), since their souls (ana wera) are contained in these. Those who eat of such crops will become incurably ill and die."

10. The kolak [basket] that is mentioned here is a regional unit of standard measure of volume (Wilkinson 1959, 1:607, 621), which averages 5.9 liters among the Kantu'. Based on an average volume for the human eye of 6.37 cubic centimeters (J. Scott Kortvelesy, personal communication, 1996), one kolak could hold approximately 157 eyes or the eyes of 78.5 persons. Compare the belief among the Dogon of Mali that spirits of the bush "will exchange eyes with humans, rendering them blind" (van Beek and Banga 1992: 67). An alternate version of this belief is that the spirits only demand a "basketful of human eyes" when the mast occurs two years in a row. The principle is the same in either case, however: unusual bounty is thought to bear a cost that is commensurate with its benefit.

11. On the economic fortune of one household relative to another, see Weber's (1958: 271) comment, cited in Jane Schneider's (1990: 27–28): "[T]he fortunate is seldom

satisfied with the fact of being fortunate [but] needs to know that he has a *right* to his good fortune. He wants to be convinced that he 'deserves' it, and above all that he deserves it in comparison with others. . . . Good fortune . . . wants to be 'legitimate.' " Jane Schneider (1990: 278) adds that, bound by this ethic, the members of small communities anticipate spiritual danger from "overstepping" their legitimate bounds.

12. On the morality of exchange, Jorgenson (1989) writes that the Pwo of Thailand periodically thank the forest for the bounty that it provides them by turning over to the forest, and to forest animals, an entire swidden, complete with standing crops. Jane Schneider (1990) suggests that this sort of "equity consciousness" with nature spirits was common in the Western world as well before the advent of capitalist relations of production.

13. It is said that formerly the offending couple would have been buried in the ground with a live bamboo stake driven through them, the subsequent growth of this stake intended to serve as a warning sign to would-be transgressors. On the underlying principle of reciprocity, see Clay (1991: 266), who writes that the Tukano of Brazil "believe that if fish are taken from restricted areas, the ancestors of the fish will take infant children—one child for one fish."

14. Rice stored for more than a few years goes bad and becomes unfit to eat, which makes its accumulation all the more immoral. Thus to say that a household has such large and long-standing rice stores that the rice has gone bad is the ultimate Kantu' expression of both farming success and social failure.

15. The belief that material accumulation, as opposed to distribution, is the work of the devil is familiar from works such as that of Taussig (1980) on Peruvian mining.

16. Compare the complex of Indo-Malay myths regarding the rice goddess Dewi Sri, in which the afterworld is associated with life, the death of the goddess is associated with the birth of rice, mortuary shelters are associated with rice barns, and so on (Stutterheim 1956; Carpenter 1989).

17. This proscription was specifically applied to the use of ironwood for *tiang* (long-house pillars), *atap* (roof shingles), *tangga'* (longhouse ladders/stairs), and *perau* (canoes). The Kantu' subgroup with whom I worked, the Melaban Kantu', still honored this proscription in its entirety in the 1890s. By the 1930s, they were honoring it for the longhouse pillars but not for the roof shingles, for which they had started using ironwood. By the 1950s, they were using ironwood for all parts of the longhouse.

18. Compare van Beek and Banga's (1992: 69) analysis of the dichotomy between the fixed village and the "moving" bush among the Dogon of Mali: "Anything in the bush moves and changes, in any season—sand dunes, gullies, trees and rocks. Only the village stays put as the only fixed point in the Dogon ethnogeography, inhabited by a series of succeeding populations (Toloy, Tellem, Dogon). They are the areas of stability. However, they also represent stagnation, the places where the forces of the bush whither away: life and death, wisdom and knowledge coming from the bush are applied in the village, but used up and worn down in the process."

19. The threat to the moral order in the case of the Merina is based, in part, on the fact that the enduring trees are individually owned. When trees are owned not by individuals (or individual households) but by larger groups, their value within this morality of exchange may be reversed. Compare Sather (1990: 20) on an Ibanic group in Sarawak: "Thus, tree rights transcend the basic units of everyday social life—the household and longhouse—and in doing so, reflect a sense of deeper historical connection, linking individuals and families through ties of ascent to past generations of household members, regional pioneers and long-house and household founders, thus reinforcing their membership in a wider regional society encompassing the whole of the upper and middle Paku."

20. The phrase *moral ecology* is based, in part, on Scott's (1976) concept of moral economy. According to Scott's work, the moral economy is one that guarantees basic subsistence, often through social investment as opposed to extraction of agricultural surplus. I suggest that the moral *ecology* is one that guarantees the basic sustainability of both society and environment through an investment in exchange relations of great time-depth and spatial breadth.

21. This linkage between creation and destruction is also reflected in the pattern of forest land tenure, in which the destruction or clearing of the primary forest is the basis for creating rights to the succeeding regrowth (Dove 1985b).

22. The income from rubber tapping *may* be used to support systems of reciprocal exchange: the most common example is the use of rubber income to purchase commercial *arak* (distilled spirits) for use in ceremonial feasts, which are part of a wider regional system of exchange (Dove 1988). However, if rubber income is *not* used for this purpose, there are no social sanctions; whereas there *are* sanctions against holding back the surplus produce of the swidden rice system from this system of ceremonial exchange.

23. The extralocal orientation of the transactional order of rubber cultivation is reflected in the use of an adopted Malay expression, *bagi dua*, "split in two," for rubber sharecropping, instead of a native Kantu' expression.

24. Rubber thrives well amid the natural afforestation that takes place on fallowed swidden plots (Rambo 1982: 282). Indeed, this afforestation, which was accepted by smallholders but initially fought by the estates in an ill-conceived clean-weeding policy, does not threaten but actually enhances the productivity of the rubber, by increasing shade, air temperature, and humidity in the smallholding, all of which promote quicker bark renewal after tapping (Bauer 1948: 58; Pelzer 1978: 283).

25. This might explain the rubber-clearing that Pelzer (1945: 24–25) observed: "In Bengkalis in the East Coast of Sumatra a few years ago, rubber gardens occupied so large an area that there was actually a scarcity of land suitable for the making of new ladangs for food crops, and rubber trees had to be cut down." It might have been that Pelzer saw the clearing for swiddens of old and unproductive rubber groves and misinterpreted this as evidence of land scarcity. Another possibility is that the Bengkalis people initially planted larger amounts of rubber than needed to strengthen land claims (cf. Gouyon, de Foresta, and Levang 1993: 192–93) and

subsequently, once their tenure was secure, cleared some of the very same rubber for swiddens. But the fact that smallholders such as the Kantu' tend to plant rubber on land that is not ideal for swidden agriculture generally minimizes the necessity of clearing rubber for swiddens (Dove 1993b: 141–42).

26. The Kantu' say that this same rule applies to land planted in fruit trees: it remains *mati* (dead) even if the fruit trees themselves die.

27. The adoption of rubber also contributed to some historic changes in the tenurial status of swidden regrowth, involving the progressive shift of the balance of control from the longhouse to the household (Dove 1985b).

28. Nor, in talking about the "dead land" of rubber, is the Kantu'/Iban term *bunuh/bunoh* (to kill [the land]) ever used (A. Richards 1981: 56).

29. See Sherman's (1980: 127) critique of Pelzer.

30. The Kantu', for example, say that pepper removes all the *lang* (scent) from the land and that land used for pepper cultivation is thereafter *kusi* (barren). The exactions of pepper cultivation are reflected in the short productive life of pepper gardens: the average Kantu' pepper garden can be cultivated for a maximum of seven to eight years, after which it must be abandoned.

31. The similarity in this respect between pepper cultivation and swidden cultivation is reflected in the fact that the Kantu' do *not* say that land used for pepper is dead, but only that it is barren.

32. Pelzer (1945: 31) writes of the articulation between colonial Burmese teak forestry and indigenous swidden cultivation: "It is ingenious in that it recognizes the habits of the shifting cultivator and fits them into a modern forest economy instead of persecuting the cultivator and forcing him to give up his traditional culture." He does not explicitly address the issue of who benefits from this ingenuity (aside from the negative benefit for the swidden cultivators of not being persecuted), but his summary of the *taungya* system implies that it is the government forester: "The taungya forestry system is an excellent way of getting pure stands of commercial timber. The typical forest of the tropics is expensive to exploit because of the large variety of trees which it contains, many of which have little or no commercial value" (ibid.).

33. The cultivation of sugar cane fit so well, agronomically, into the cultivation of wet rice fields that—in the context of the colonial and postcolonial prioritizing of sugar production—peasant wet-rice production wound up subsidizing state sugar production, which contributed to the former's historic failure to develop, its "involution" (Geertz 1963).

34. Conversely, the so-called complementarity of a crop like pepper with swidden cultivation indicates locally *un*sustainable social and ecological relations (see also Jürg Schneider 1995: 31–32).

35. When estate owners eventually abandoned their destructive clean-weeding policy and began to emulate the smallholder practice of allowing spontaneous afforestation to engulf their rubber groves, their view of the jungle underwent a predictable reversal: the negatively loaded phrase *rubber jungle* became the positively

loaded phrase *rubber forestry*. Pelzer (1945: 78) writes as follows: "There are two principal means of protecting the soil between tree crops: by allowing the growth of indigenous plants or weeds (in other words, the development of a naturally mixed cover), or by establishing a pure cover of leguminous plants, many of which came originally from the New World. Rubber planters in Malaya seem to have had more success with the first method, which has been called 'the forestry method of cultivation' or 'rubber forestry.' "

36. This transition was one of the factors that drove the replacement of indigenous Bornean latexes (such as *gutta percha* and *jelutong*) in the markets with the introduced Para rubber. Exploitation of the native latexes, which relied on the same principles of destruction and creation, reciprocity and exchange as swidden agriculture, did not win any tenurial recognition from extralocal authorities, and this left the native tappers constantly open to challenge by state elites (Dove 1994).

37. See Gouyon, de Foresta, and Levang (1993: 192–93) on the situation in southeast Sumatra: "Most farmers in the area lack official land titles; yet rubber contributes to family wealth by bearing witness to land occupancy. An area covered with rubber is usually regarded by local land right as belonging to the planter, and as such can be inherited or sold. Planted land can also be claimed as an individual asset in case of conflict over land property with the government or estate companies. In some areas, smallholders are planting rubber as fast as they can to occupy an uncultivated area before it is seized by such external bodies." See also Dove (1985c) and Heersink (1994: 54n24).

38. See Peluso (2009) for an analysis of the role of rubber in local, interethnic land struggles.

39. The role played by trees in social relations, especially gender relations, has been studied by Rocheleau and Edmunds (1997), Schroeder (1999), and Schroeder and Suryanata (1996).

40. A still current example of the metaphoric linkage of people and plants in Indonesia is called *ilmu padi*, which literally means "rice knowledge" but which Echols and Shadily (1992: 220) translate as "the ability to be well educated but humble." A common example is the aphorism, "Makin berisi, makin berunduk [The fuller it is, the more it bends over]" (Schmidgall-Tellings and Stevens 1981: 121). This statement refers literally to the agronomic fact that the fuller a rice panicle is, the more it bends over. Metaphorically, it refers to the norm within Javanese culture that the more one has to say, the quieter one is.

41. Another example comes from the Kissi of Guinea, who use the phrase *single forest* as a metaphor of political solidarity (Fairhead and Leach 1995: 63).

42. The comparison of tree age and people age is part of a wider set of metaphorical relations in which plant imagery is used to refer to human growth and reproduction (see Bourque 1995 on the Andes). The reverse also occurs: thus Visser (1989: 82–83, 95–97) describes how the Sahu of Halmahera in Eastern Indonesia use metaphors of human reproduction to characterize the cultivation and maturation of rice plants.

43. The principle in the Merina belief that long-lived trees should only be associated with long-lived social groups is also illustrated among the Huaorani of the Ecuadorian Amazon by the association of enduring peach palm groves with long-term local endogamous groups, on the one hand, and on the other the association of short-term manioc gardens with ephemeral exogamous marriages and political alliances (Rival 1993: 648–49).

44. The fact that this status is unchanged by the death of the rubber trees themselves affirms that it is not the life of the trees that is referred to here, but the life of the planters.

45. See Wilkinson (1959: 1:407, 418, 2:1265) for the Malay phrase *s'umur idup* (literally, "for the age of life").

46. This has been variously identified as either the fig tree (*Ficus benjamina* Linn.) or as what the Malays call the *rengas* tree, which refers to a group of trees (mostly *Anacardiaceae*) with an irritating sap (Ras 1968: 175–76, 176n5, 543; Wilkinson 1959: 2:964–65).

47. A desire to avoid this sort of identification is found in practices like that of the Kuna Yala of Panama, who will not use *Tachigalia versicolor*, which dramatically flowers once and then dies, for fear that they would follow suit (Archibold and Davey 1993: 55).

48. Thongmak and Hulse (1993: 162) similarly report that among the Karen of highland Burma and Thailand, after the birth of each child the afterbirth is placed in the crook of a branch of a large tree, which exemplifies life and longevity; thereafter, the health and well-being of the child is linked to this tree.

49. Some Iban groups link the welfare of the human soul with the welfare of a mythical plant. D. Freeman (1967: 323n) writes: "It is believed by the Iban that every soul has a counterpart or *ayu*, which takes a plant-like form. These *ayu* grow on a mythical mountain where they are tended by celestial shamans. An *ayu* is said to be in mystical symbiosis with the individual to whom it belongs, and its appearance is supposed to reflect his state of health." Sather (1990: 37) similarly writes of the Iban: "Every living human being is represented, not only by a conjunction of body (*tuboh*) and spirit or soul (*semengat*), but by an invisible plant counterpart, the *ayu*, which symbolizes human life in its vulnerable, mortal aspect."

50. An example is the Javanese *gunungan* (literally, "mountain-shaped" [Horne 1974: 225]), a shadow puppet that depicts all manner of life grouped on a giant tree.

51. The relationship between humans and spirits, and between culture and nature, is addressed explicitly in the symbolism of Dayak headhunting. Davison and Sutlive (1991: 191, 203) write as follows: "Trophy heads are like the fruit of the forest: they are gathered, and their seed is planted to provide the Iban with sustenance. . . . In symbolic terms, to take heads is to gather the fruits of the forest: Iban warriors must go headhunting, to bring back the fruit of the mythical *ranyai* palm and thereby supply their community with the means for their continued existence, namely rice seed and children." In headhunting, therefore, human death is linked

to the life of both crops and people; and just as the spirits are seen as taking the eyes of humans in exchange for giving them fruit, so is headhunting seen as taking human heads *as* fruit. The image of fruiting, and fruit gathering and consumption, constitutes a key linkage between people and trees, between human culture and spirit nature.

52. Bourdieu (1977: 38) writes: "The genealogical tree constructed by the anthropologist, a spatial diagram that can be taken in at a glance, *uno intuitu*, and scanned indifferently from any point in any direction, causes the complete network of kinship relations over several generations to exist as only theoretical objects exist, that is, *tota simul*, as a totality present in simultaneity."

53. More compelling is Velásquez Runk's (2009) analysis of the rhizomatic cosmology of the Wounaan of Panama versus the arboreal vision of western conservation.

54. Compare Nguni (2009) on the cognitive, cosmological, and psychochemical values of trees depicted in southern African rock art.

55. See also Mosko (2009) on the metaphoric implications of the base, body, and tip of the yam plant in the Trobriand Islands.

56. As Penot (2007: 579) notes, rubber is also being examined for its carbon-fixing potential.

57. This is not to suggest that Dayak society was "traditional" and autochthonous before the coming of Para rubber. On the contrary, Dayak society experienced transformations analogous to those experienced with Para rubber during the prior boom in native forest latexes in the mid-nineteenth century (Dove 1994). Before that Dayak society experienced changes of other sorts. There was always, thus, a *history*, as opposed to an "unchanging past."

II

COMMUNITY RIGHTS DISCOURSES THROUGH TIME

ADAT ARGUMENT AND DISCURSIVE POWER
Land Tenure Struggles in Krui, Indonesia

Since the end of the twentieth century, the topic of "communities and conservation" has been a key issue in international debates on conservation. Absent for decades from these debates, communities only came under public scrutiny in the 1980s (West 2005), which was accelerated in 1989 when the Coordinating Body of Indigenous Organizations of the Amazon Basin (COICA) made an appeal to conservation practitioners. COICA called for an active involvement of the indigenous peoples in conservation projects "in the defense of our Amazonian homeland" (Chapin 2004: 19). This appeal sparked enthusiastic discussions and collaborations between indigenous people and international conservation organizations. Within a few years, indigenous peoples and traditional communities gained stature in conservation discourses and were deemed to symbolize the European ideal of the "ecologically noble savage" (Conklin and Graham 1995; Redford 1991). The political fecundity of linking communities and conservation resulted in ambitious programs, such as the World Bank's Integrated Conservation and Development Programs (ICDP). The limited indigenous input and bias toward them in conservation agendas, however, soon exposed the ICDPs to severe criticism and led to their eventual decline (Chapin 2004). Before long, scholars and activists began questioning other community-based conservation projects, problematizing the usage of the term *community*, as well as notions of access and rights, and accountability and effectiveness (Neumann 1997; West 2005).

By the end of the "communities and conservation" honeymoon, indigenous peoples and cultural communities had gained a position of continued valorization, particularly due to the efforts of environmental movements and organizations (Brosius 1999b; Dove 2006).

These movements deploy the image of indigenous peoples and traditional communities as stewards of nature and emphasize their connection with the natural world to challenge their marginalization, as demonstrated in Indonesia (Afiff and Lowe 2007; Tsing 2003; Li 2000), Mexico (Jung 2003) and the United States (Pulido 1998), among other places. Such "strategic essentialism" (Brosius 1999b), at times oversimplified for the benefit of national technocracies and bureaucracies, was a global phenomenon precisely because it had political capital. In 2000 indigenous peoples took an active role in the World Park Congress in Durban (Brosius 2003). Contesting many basic assumptions about conservation and the roles of Western science, they injected their own discourse about themselves into the "regime of truth" that has governed conservation debates (Escobar 1995). The Durban congress demonstrated that indigenous peoples and cultural communities have become active agents in the discursive arena of conservation and development.

What follows is a narrative of similar challenges and triumphs faced by a customary community in Indonesia. It demonstrates how customary communities and the concept of *adat* (customary law) have resurfaced as a key element in the discourse of conservation and natural resource management in Krui, Indonesia, in the late 1990s, in line with global trends in transnational symbolic politics.

For more than five generations, villagers in Krui have cultivated *damar* trees (*Shorea javanica*) in customary lands bordering Bukit Barisan Selatan National Park, on the southwest coast of Sumatra. When an incision is made in its trunk, a *damar* tree produces high-quality resin, a commodity traded and used for marine caulk, incense, varnish, paint, and batik making since the seventeenth century (Boomgaard 1998a). *Damar* tree stands managed by Krui villagers mimic the function, structure, dynamics, and diversity of natural forest, hence the use of the term "forest gardens." *Damar* forest gardens provide the villagers not only with significant cash income but also with hydrological services for irrigating paddy fields (Michon et al. 2000).

In 1994 the Ministry of Forestry extended the Bukit Barisan Selatan National Park beyond its original boundaries demarcated in the 1930s by the Dutch colonial government, trespassing onto customary land where *damar* forest gardens were situated. This unilateral act led to unhappiness within the local communities. A coalition of researchers and nongovernmental workers who had been active in Krui determined to assist the villagers in reclaiming rights to their forest gardens. Their strategy embodied elements of strategic essentialism. They represented the villagers as an *adat* community who, through its forest gardens, has been successful in maintaining the ecological integrity of the customary land surrounding the national park, making them eligible, indeed, for the national environmental award, the Kalpataru. Winning this high-profile award created more political space for negotiation for the community. Since then, the so-called *adat* argument has assumed a central role in the community's strategy to reclaim ownership of their forest gardens. I define the *adat* argument as the deployment of *adat*-related concepts in the discursive arena of conservation and natural resource management, particularly in terms of asserting *adat* land tenure rights and *adat* customary rights, to legitimize land ownership claims.

Invoking *adat* to serve political objectives has had a long history in Indonesia. Accounts from the colonial period demonstrate how the colonial state oscillated between abandoning and reinstating *adat* institutions as the local governing bodies. In this essay, I explore the contemporary invocation of *adat* by various actors to restore ownership of *damar* forest gardens to the villagers. I contrast this phenomenon with historical accounts that chronicle the deployment of *adat* by both the colonizers and the colonized. I suggest that the *adat* argument contains discursive power that, at times, can bring about unexpected results.

This essay is based on research carried out in Indonesia during the summer of 1999. I conducted numerous interviews with local villagers, nongovernmental activists, funding agency personnel, research agency staff, as well as government officials involved in policy making for the *damar* forest gardens. Formal documents, informal notes, and transcripts of village meetings complement the primary data gathered.[1] In the first section of the essay I give a brief review of the evolving definition of *adat* and the discourse of the *adat* community in natural resource management. Next, I explore the landscape of Krui, the deployment of *adat* institutions during the colonial period, and their weakening role in postcolonial Indonesia. In the third part of the essay, I look into the increasing importance of the *adat* argument in today's debate over community-based conservation and natural-resource management, paying close attention to how *adat* is represented differently by different actors and how the strategy of deploying *adat* argument has created tensions within the Krui community. I conclude with the argument that the discursive power inherent in the *adat* argument is accessible even to the weakest party in a struggle, as represented by the *damar* farmers in Krui.

The *Adat* Community as a Loaded Concept

Adat is a heavily loaded concept. In its most general sense, *adat* refers to the customs and practices of Indonesia's various ethnic groups. It encompasses customary laws (*hukum adat*), ritual conventions, marriage rules, kinship systems, methods of conflict resolution, rules for resource ownership and utilization, and other formally articulated norms and ideas (Moniaga 1993; Fried 2003; Tsing 1993; Zerner 1994a). *Adat* as a source of law was one of the most debated issues in Dutch colonial politics. Cornelis van Vollenhoven, a famous legal scholar and an influential professor at the Leiden University Faculty of Law, in 1906 published a detailed study of Indonesia's customary laws in his magnum opus *Het adatrecht van Nederlandsch Indië* (*The Adat Law of the Netherlands Indies* [van Vollenhoven 1918–33]). This was followed by numerous books and articles, which were widely read and debated by Dutch legal scholars of the period (Burns 2004). Vollenhoven's crusade for inclusion of indigenous legal concepts in the colonial legal system instigated the famous debate between Leiden and Utrecht University scholars, focusing on *adat* land rights (Burns 1989, 2004; Djalins 2007). After van Vollenhoven's premature death in 1933 the debate waned, and the colonial administration was able to acknowledge *adat* land rights only nominally.

Colonial ambivalence on *adat* land rights is still echoed in Indonesia's legal system today. For example, colonial wording on the conditionality of *adat* land rights was repeated almost verbatim in Indonesia's 1960 Basic Agrarian Law, as stated in Article 5: "*Adat* applies to agrarian matters *unless it conflicts with national and state interests*, Indonesian socialism, and legislative regulations, in which case the national law provisions prevails" (MacAndrews 1986: 22; emphasis mine). It is further limited by the requirement that "*adat* property rights shall be adjusted to the national law and interests, and shall not be in conflict with such national laws and regulation" (MacAndrews 1986: 22).

With its rising importance in the international discourse of conservation and natural resource management, 'community'—like *adat*—has also become a loaded concept. In conservation studies, 'community' often refers to "a bundle of concepts related to space, size, composition, interactions, interests and objectives" (Agrawal and Gibson 1999: 633). Current literature on conservation and resource management approaches community from one of three ways: as a spatial unit, as a social structure, or as a set of shared norms, all of which are problematic (Agrawal 1997). In the discourse of natural resource management, community has undergone a major evolution: from the culprit in environmental degradation to a key interest group.[2]

In the past two decades in Indonesia, *adat* community has become a powerful concept because it articulates extremely well to the prevailing transnational symbolic politics surrounding indigeneity and natural resource management. In the early 1990s a boom in media reporting on global warming, deforestation, declining biodiversity, and the impact these processes have on indigenous groups attracted the attention of an international audience.

For example, the struggle of the Kayapo Indians in Brazil and the crusade for the Penan of Sarawak both generated tremendous support from numerous nongovernmental organizations, politicians, and policy makers, as well as celebrities (Conklin and Graham 1995; Brosius 1997b). Although Indonesia's indigenous communities do not fit all of the characteristics of those in Latin America, Australia, or other regions, the opening created by this transnational symbolic politics focused the nation's attention on the plight of marginalized, mostly rural *adat* communities in Indonesia. This is illustrated in the ability of the Aliansi Masyarakat Adat Nusantara (AMAN, an umbrella organization for *adat* communities in Indonesia) to successfully challenge the state's derogatory term *masyarakat terasing* (isolated communities) and replace it with the more politically correct *masyarakat adat* (*adat* communities) in addressing their members. Thus, despite some downsides to this transnational trend,[3] it has raised global interest in alternative modes of conservation and natural resource management as practiced by indigenous and customary communities alike.

The Discourse on the *Adat* Community in Indonesia

Research on the indigenous management of natural resources has generated considerable interest in Indonesia. Research on *repong damar* (dammar plantations) in Lampung (Mary and Michon 1987), *tembawang* (managed forests) in West Kalimantan (Padoch 1995; Salafsky, Dugelby, and Terborgh 1993),[4] and swidden agriculture (Dove 1983b, 1985d, 1993a, 1993b) are frequently cited by environmental activists advocating for *adat* communities (see, e.g., Moniaga 1993; Heroepoetri and Aliadi 1994). Inevitably perhaps, given the numerous and diverse ethnic groups and *adat* communities in Indonesia, there is no single, officially acknowledged legal definition of the *adat* community (Safitri 1999). The earlier-mentioned NGO, AMAN, defines *adat* community as "a group of people who live together based on a shared genealogical ancestry, within a specific geographical area, with a unique value and socio-cultural system, exercising autonomy with respect to its land and natural wealth, and regulating and managing its livelihood through use of *adat* law and *adat* institutions."[5] Another NGO, the Agrarian Reform Consortium, based in West Java, defines *adat* law as a set of unwritten regulations sustained and honored by the *adat* community (Latin 1999). This definition notwithstanding, the Consortium fails to present an agreed upon definition of *adat* rights, particularly *adat* property rights, perhaps due to the fact that "*adat* rights" is an ambiguous term with different meanings for different actors.

Communal *adat* rights are often mistakenly contrasted with individual rights. Moniaga (1993) presents *adat* rights as "community-based" and refers to them as *hak ulayat*. Moniaga cites Zerner's definition of *hak ulayat* as "mutual or joint rights of a group of inhabitants occupying a specific piece of land" (Moniaga 1993). In contrast to the prevailing wisdom that *adat* rights are communal, Dove argues that peasants' rights to land and to natural resources in practice involve both individual rights (rights of ownership, or *hak milik*) and communal rights (rights of avail, or *hak ulayat*). He writes, "In fact, the two types of rights have occurred congruently, as part of a single land tenure system, at many times and in most parts of Indonesia. . . . Among the contemporary tribal Kantu's of West Kalimantan, for example, individual households hold rights of ownership or 'primary' rights to fallowed swidden land, while the long-house holds rights of avail or 'residual' rights to the same land" (Dove 1985c: 266). A misunderstanding of *adat* community rights is apparent in Indonesia's 1960 Basic Agrarian Law, which addresses these rights in completely separate articles, and in discussions among development officials, "which make clear that a given area or people is assumed to be subject to rights of avail or rights of ownership but certainly not both" (Dove 1985c: 266).

Only recently has the understanding of *adat* and *adat* rights taken a different trajectory among Indonesian scholars. Zakaria (1999), who is acutely aware of this misunderstanding and its repercussions for local community struggles, argues that for a long time rights of avail (*hak ulayat*) have been misunderstood as rights of communal use (*hak pemanfaatan bersama*). He suggests that rights of avail actually reflect the fact that a community living in a particular territory has sovereignty over it (Zakaria 1999: 130–31).[6] Thus, he continues, the regulations and acknowledgment of *adat* property rights should not be interpreted solely as rights of communal use, as is often the case in the state's acts and regulations. They should be extended to include private rights (*hak milik*), which are indeed regulated and managed by *adat* law (Zakaria 1999: 130–31).

A redefinition and reconstruction of what *adat* is and what it means is not only the work of present-day scholars. In the province of Maluku, the local community itself has redeployed *adat* regulations for natural resources exploitation—the *sasi*—to contest the state's control over resources (Zerner 1994a). In Zerner's words, "*adat* or custom, rather than being museumized as a cultural curiosity or an object of touristic concern, is being used as a discursive projectile carrying community claims to resources and territories into a hotly contested political-economic arena" (Zerner 1994a: 1115). Is this approach problematic, as Zerner (1994b) argues? Indonesian *adat* advocates might disagree. Ruwiastuti, a noted advocate for agrarian reform and for the acknowledgment of *adat* rights, stands by the *adat* argument approach. She argues that as long as *adat* is "understood as an entity that is alive, dynamic and growing in parallel with the transformation experienced by its community, then revitalization of the *adat* community is a legitimate, deliberate effort to counter attempts to thwart the advancement of the community" (Ruwiastuti 1999: 10). In other words, if the revitalization of *adat* and the *adat* community helps the community overcome obstructions to its advancement, then it is a legitimate endeavor. This philosophy is widely embraced by advocates of *adat* communities, including those involved in helping the Krui communities to reclaim ownership of their forest gardens.

The Landscape of Struggle

For decades Indonesia's New Order regime policies on *adat* communities systematically weakened the fabric of these societies, marginalizing their position vis-à-vis the larger society and diminishing their share of benefits from their natural resources. This was demonstrated, for example, in the replacement of traditional governance structures throughout Indonesia with the uniform *desa* system in 1979—wherein a homogeneous, government-run system of village administration was imposed on the entire country—and the freezing of *adat* community rights to collect forest products in corporate timber concession areas (Kato 1989). Stripped of their authority over many of the community's affairs, most notably its territorial claims, *adat* institutions in Krui were already starting to lose their relevance with the villagers when NGOs and other sympathetic actors came to help revive them.

THE MAKING OF THE KRUI COMMUNITY

Krui is located on the southwestern coast of the province of Lampung in West Lampung District. Nestled in between the heavily forested Bukit Barisan Selatan National Park and the Indian Ocean, it comprises three subdistricts (*kecamatan*): North, Central, and South Pesisir (map 1). The people who inhabit the Pesisir area are identified as the Abang or Abung people (LeBar 1972), who belong to the Peminggir *adat* group.[7] The *adat* system in Lampung comprises several layers of kinship; the smallest is called the *marga*, a distinct genealogical unit with a specific territorial division (Kingston 1991) led by a leader called *sai batin*. A *marga* may be spread over a number of dispersed villages and is the fundamental pillar of *adat* institutions in Krui. According to oral history, the ancestors of the Krui migrated from the southern part of Pesagi Mountain called Sekala Begrak (also known as Sekala Berak or Sekala Be'rak) about 450 years ago (LeBar 1972; Hadikusuma 1988). They established sedentary villages along the coastal plains,

MAP 1 Krui, Lampung, Sumatra

practiced swidden agriculture, grew pepper and coffee, and gathered forest products, especially *damar* resin and rattan (Michon et al. 2000). Later, to maintain a continuous supply of resin, the villagers started to domesticate and cultivate *damar* trees.

The *damar* forest gardens in the Pesisir area can hardly be distinguished from the natural forests in the adjacent Bukit Barisan Selatan mountains in terms of diversity, structure, dynamics, and functions. Yet an examination of the structure and composition of the damar forest demonstrates that it has been shaped by generations of the villagers' stewardship. The forest canopy and understory is a complex mixture of trees, herbs, and shrubs, with *damar* (*Shorea javanica*) as the dominant species (Michon et al. 2000; Torquebiau 1984). The villagers call these forests *repong damar*,[8] a term scholars translate as "*damar* forest gardens" or damar agroforests. Researchers and scientists consider *damar* forest gardens to be evidence of a sophisticated local knowledge that has successfully replicated the complicated reproduction of a dipterocarp forest (Suwito 1998; Mary and Michon 1987; Torquebiau 1984). The *damar* forest gardens help protect the soil and maintain the hydrological functions that irrigate the adjacent paddy fields (Mary and Michon 1987; Michon et al. 2000). The *damar* forest gardens are also very successful, certainly when compared to monocultural plantations, in maintaining a relatively high biodiversity (Michon et al. 2000).[9]

Since the initial settlement of Pesisir, *damar* production has been a major economic activity in the area,[10] especially after the trees began to be cultivated. Sources indicate that the earliest cultivation of *damar* took place between 1870 and 1885 (Michon et al. 2000).[11] By the late 1930s, *damar* production was booming and 80 percent of the *damar* produced in Krui was coming from cultivated trees (Michon et al. 2000). By 1995, *damar* production in Krui reached ten thousand tons, providing a yearly income averaging $683 per hectare (Michon et al. 2000: 175). As a *damar* agroforest includes fruit trees among its species, earnings also come from fruits, which offer substantial income to the villagers (de Foresta and Petit 1997).

The cultivation of *damar* is important for the villagers not only for its economic value but also for its social role. According to Krui custom, a *damar* tree planted in a swidden field represents a family's claim to the cleared land (Lubis 1997). Without planted *damar* trees, cleared land left fallow will return to early stages of forest succession without any marks of previous use or ownership; and according to the *adat* in Krui, such land can be claimed by others. Rights to the *damar* forest are managed according to the customary (*adat*) laws and institutions. Established *damar* forest gardens are passed down to the eldest son, who is responsible for managing them and supporting his siblings until they become independent (Lubis 1997; Suwito 1996). Because ownership of the land remains in the family, a *damar* garden eventually becomes the common or joint property of an extended family. The tenurial rules governing expansion and the opening of virgin forest for *damar* gardens were laid down by the *marga* and its leaders, but they changed dramatically when Dutch colonial influence reached Lampung.

THE *ADAT* COMMUNITY IN THE COLONIAL CONTEXT

Dutch colonial influence in Krui is not well documented, but some data on Dutch-Lampung interaction are available from the Van Royen ethnological report of Lampung (Kingston 1991).[12] Historically, external powers such as the Sultanate of Bantam or the Dutch colonialists greatly influenced the power and authority of *marga* leaders, whose influence rose and fell according to the political needs of these outsiders (Hadikusuma 1988). In the early nineteenth century, after the return of South Sumatra to the Dutch following the British interregnum, the *marga*'s role as an administrative apparatus was alternately abandoned and reclaimed depending on the colonial need to maintain control over the natives. In 1857, under pressure from its officials in the field, the Dutch colonial government terminated the role of the *marga* leaders as official representatives of the government, a role that had been instituted by Thomas Stamford Raffles, the governor of Java during the

British interregnum (Kingston 1991: 28). In its place, the Dutch appointed their own village heads. This transfer of power profoundly affected the land-tenure regime in the region, as Kingston (1991: 28–29) explains:

> Under the new system, the government only recognized the land claims of individual villages up to 6 km from the village and 3 km from the *umbil*, temporary settlements established by the villages that were opening up virgin territory for swidden cultivation. The land in-between the village claims, formerly the common *marga* territory, now came under state control. This was a critical decision with far-reaching implications, for the Dutch had abolished, at least on paper, the authority over land use formerly exercised by *marga* leaders and thus the basis of their power and wealth. Not surprisingly, this elevation of the village head's status, who had no standing in *adat*, vis-à-vis the traditional *adat* leaders unleashed intrigues detrimental to effective administration.

The colonial government eventually restored the *marga* leader's authority, but the hiatus in their control of forests and waste lands had enabled immigrants to inhabit these lands, thus complicating the local land-tenure and property-rights system. This resulted in so many disputes that in 1874 the government reversed its decision. Once again the *marga* was stripped of its jurisdiction over land outside the village's immediate boundaries (Kingston 1991: 29). It took another fifty years before the Dutch once again reinstalled the *marga* system to serve its political needs.

During the upsurge of nationalism and activism inspired by Indonesia's communist movement in the late 1920s, the Dutch attempted to revive the *marga* and to create a stable *adat* community system. Their objective was to obstruct the nationalist movement by reemphasizing people's subservience to tradition and by reinforcing regional differences (Kingston 1991: 25). Many in Lampung fiercely resisted such efforts through a strong anti-*marga* campaign. The leftist organization Comite Tani Lampung (CTL) led an effort to portray the colonial revival of *adat* institutions as an "an attempt to turn back time and place the people under an autocratic, exploitative system that primarily served the interests of the government and the elite at the expense of the people" (Kingston 1991: 29). Kingston's assessment is not supported by local accounts, however, which give a different picture of the conflict (Bukri et al. 1981). Local accounts represent the reestablishment of the *marga* system as colonial submission to the local system. In other words, the local view was that the *marga* was so rooted in their lives that the Dutch had no choice but to acknowledge it and reinstall it as part of their governance system. This decision was officially announced in the Native Municipality Ordinance of the

Outer Islands (Inlaandsche Gemeente Ordonantie Buitengewesten, IGOB) in 1928. Bukri et al. (1981) suggest that the anti-*marga* campaign waned after the *marga* leaders were reinstalled.

The strategy of indirect rule enabled the community in Krui to expand their *damar* gardens without much conflict with the Dutch. In 1935, the Dutch gazzetted five blocks of land as a closed game reserve in the Bukit Barisan Selatan mountains, bordering on numerous villages in the Pesisir area.[13] Known by the villagers as BW (from *Boschwezen*, a Dutch term for forested area), this became the boundary between closed forest under government stewardship and "waste lands" that, although not officially acknowledged as such, was left under *adat* jurisdiction. The villagers cite this "understanding" as the basis for their tenurial claims and insist that they have always respected this boundary when opening new *damar* forest gardens, except for some minor encroachments about two decades ago due to the increasing pressure for land (Mary and Michon 1987; Lubis 1997).[14]

This brief history reveals multiple interpretations of *adat* by the locals and by outsiders. As I will demonstrate in the following paragraphs, various interpretations of *adat* and *marga* in the colonial period resonate with contemporary interpretations and reconstructions by farmers, elites, and external parties such as research institutes, nongovernmental organizations, and the state.

In the postcolonial era, *adat* has undergone numerous reinterpretations by the Indonesian government. As the fledgling government tried to establish its presence throughout the country, the national authorities gradually transformed colonial governance inherited from the Dutch into an "Indonesian" system. In 1952, the *marga* system reestablished by the Dutch in 1928 was replaced by the *kenagarian* system. In 1979, the uniform *desa* system finally replaced the *kenagarian* system with the enactment of Law No. 5/1979, completely stripping the *marga* and *adat* leaderships of any remaining authority iand power.

CONTEMPORARY KRUI, NGOS, POLICY MAKERS, AND THE STATE

During Indonesia's New Order era, *adat* communities and their rights were largely marginalized by a regime that prioritized economic development almost at any cost. During the tenure of the New Order regime, the area surrounding the Bukit Barisan Selatan National Park was targeted for various purposes. In 1982, the government allocated fifty-two thousand hectares of forest along the outside border of the national park, land overlapping with the locals' customary land, to commercial forest concessions, without the locals' knowledge or consent. As was common practice, the allotment of concessions was based on secondary data on maps without ground checking (Kusworo 1997). When the commercial value of logging waned, the concessions were abandoned. The Ministry of Forestry, created in 1983 after a split from the Ministry of Agriculture, then converted the concessions into limited production forest (*hak pengusahaan terbatas*, HPT)[15] based on the Consensual Forest Land Use Agreement (*tata guna hutan kesepakatan*, TGHK) of Lampung in 1992.[16] The objective was to create a buffer zone around the national park, ignoring that this area was in fact the site of the villagers' *damar* forest gardens.

Such unilateral zoning incited tensions between the state and the villagers. When the Ministry of Forestry demarcated the new boundary with concrete markers in 1994, the villagers regarded this as a threat to their customary land rights. In that same year, the Penengahan and Pahmungan villagers—alongside other villagers in the surrounding subdistricts—began the long struggle to defend their ownership rights to the *damar* forest gardens, both independently and with the support of nongovernmental organizations and sympathetic advocates. The villagers made consistent, clear, and almost unanimous demands to the government: the boundaries between the villages and closed-forested lands should be returned to the old *Boschwezen* boundaries of 1935, and the *adat* land tenure rights should be acknowledged. This struggle attracted support from NGO activists and researchers who had already been working closely with the community.

Although studies of the *damar* agroforests in Krui started as early as the 1930s, contemporary studies of their ecology and silviculture date only from the mid-1980s.[17] These later studies portrayed the villagers and their forest gardens as exemplars of indigenous knowledge and sustainable natural resource management. Putting the villagers and their forest gardens at the forefront of scientific discourse in agroforestry created an alternative to the state's discourse, which frequently presents indigenous resource management regimes as primitive and backward (Dove 1985a; Tsing 1993; Fried 2003). The study of the Krui *damar* agroforests and related farming systems was also supported by the International Center for Research in Agroforestry (ICRAF), an institute under the umbrella organization of the Consortium of Governments in Agricultural Research (CGIAR). The minister of forestry eventually requested ICRAF to assist in the formulation of a policy on the Krui *damar* agroforests.

In the early 1990s, the Ford Foundation in Indonesia started a social forestry program on the islands beyond Java with the goal to "promote collaboration among the government, academic institutions, and nongovernmental organizations in rethinking the respective roles of government and villagers regarding sustainable forest management in the Outer Islands of Indonesia" (Ford Foundation 1998: 3). Close cooperation with numerous research institutions introduced the foundation to the research on the Krui *damar* forest gardens. It discovered that research findings on the gardens were not informing government policy. In line with the goal of its program, the Ford Foundation then gave grants to a number of research institutions and NGOs already working in Krui, such as LATIN, based in Bogor, West Java, and WATALA, based in Bandarlampung, the capital of Lampung Province. By 1995 these and other institutions, including ICRAF, had become more actively involved in the Krui case. To ensure cooperation among the different entities working in Krui, the parties set up an informal consortium, Team Krui.[18] Team Krui would go on to foster dialogue between local community representatives and the government, to forge a win-win solution to the Krui case, and to guarantee the sustainable management of its forest resources.

Reviving the *Adat* Community

Increasing contemporary pressure on the *damar* forest gardens came not only from the central government through its zoning policy but also from the local government, which planned to establish and expand palm plantations along the coastal area in Krui. In response, Team Krui came up with a plan to help the villagers reclaim ownership of their forest gardens. A number of strategies were deployed to strengthen the villagers' claims. During discussions with the local peoples, the strategy of using the *adat* argument surfaced and eventually dominated if not overshadowed the strategy of emphasizing the local community's success in nature conservation. This section of my essay presents the ways in which different actors reconstructed and used *adat* community rights to influence the behavior of others and/or to maintain their position in the land rights struggle. To understand this dynamic, we first need to understand how the *adat* argument was adopted, reshaped, and challenged by the local farmers in their own discourse.

ADAT ACCORDING TO THE VILLAGERS

The strategy the farmers deployed to reclaim ownership of their *damar* forest gardens evolved through a number of different stages. In the earlier stages, the *adat* argument was not deliberately invoked to justify their demands; instead, the farmers emphasized their long-term management regimes for maintaining the forest gardens. The first official conference on *damar* agroforests took place in 1995.[19] In this important meeting, the farmer representatives objected to the "forest" status of *damar* groves in Krui as perceived by the forestry agency and external observers.[20] They argued that what the outsiders saw as natural *damar* forests were actually cultivated *damar* gardens, because the trees were deliberately planted by their predecessors or by themselves, consequently farmers should not be subjected to taxes when extracting resin and other products from the forest gardens (de Foresta 1996). Legal recognition of the *damar* trees as farmers' gardens would cancel out the tax obligation. The farmers' attempts to challenge the common misconception about *damar* gardens elevated them to active players in the emerging debate about their gardens. Their testimony showed how the official discourse framing the *damar* trees as "forests" instead of as "gardens" obscured their property rights claims and burdened them with unjustified taxes.

During the course of their struggle, the local farmers used a variety of arguments to reject efforts to undermine their claims on the *damar* forest gardens. A fascinating glimpse of this rejection is revealed in a December 1996 transcript of a meeting between a subdistrict chief (*camat*) and the villagers. The meeting was intended to introduce the villagers to a plan to open an oil palm plantation in the area and to register existing property rights as the first step to realize the plan (Suwito 1996). Despite the chief's attempt to convince the villagers that the meeting was benign in its intent, the villagers approached the meeting with suspicion. This was expressed in the seemingly innocent questions they posed to the subdistrict chief, which proved to be an effective strategy, especially considering that the subdistrict chief had deliberately brought in journalists to prove his benevolent intentions. One farmer questioned the chief's knowledge of forest conservation, citing the central government's encouragement to maintain forest cover like that of the *damar* vegetation. Another farmer rhetorically inquired about the impact of a monocultural palm oil plantation on biodiversity. He expressed his confusion about the logic of the subdistrict chief's argument that a monocultural plantation would bring prosperity to the villagers. He contrasted that argument with what the local agriculture extension (*penyuluhan*) had taught him: namely, that the benefits of having diverse species in the forest-garden system were superior to having a monoculture system. He went on to assert that an agroforestry system such as *damar* forest gardens could provide a more diverse source of income than a monoculture, in addition to providing hydrological services for the paddy fields at the same time. Other arguments presented by the farmers revolved around the role of *damar* forest gardens in preventing erosion and landslides given the steep topography of the area, and in providing a buffer zone for the national park. Throughout the question-and-answer session, the villagers added various conservation and ecological arguments, on top of economic arguments, to reject the palm plantation. Hardly any references were made to the indigeneity of their system, to tradition, or to *adat* institutions to justify protecting their forest gardens.[21]

Conservation arguments were consistently used to challenge the local government's efforts to create an oil palm plantation, but over time the farmers also started to deploy the element of *adat* in their discourse. In a letter sent to the West Lampung district chief (*bupati*) in 1996, five self-help community groups (*kelompok swadaya masyarakat*) protested a plantation company's expropriation and cutting of the trees and plants in their *damar* gardens. These groups proclaimed their affiliation to Marga Ngaras, one of the sixteen *margas* in Krui. They deliberately expressed their appreciation of the district chief's "deep understanding of the culture and custom (*budaya* and *adat*) of the Marga Ngaras community and the rights of the community to the *repong damar* inherited from our predecessors," which the chief had expressed during an earlier meeting with them. In this statement, the villagers portrayed respect for *adat* and culture as a virtue that a highly educated gentleman such as the district chief would naturally possess, which provided the farmers with a discursive setting to justify their claims to land based on the *adat* argument. In claiming that their proof of ownership lay in its acknowledgment by *adat* (*pengakuan secara adat*), the villagers represented *adat* as simply the ownership regime granting legitimacy to their claim.

The farmers did not always use the a*dat* argument to counter outside pressures. On a few occasions they presented *adat*-related institutions simply as the background to their struggle, as was shown in a letter sent by the Marga Malaya members from the North Pesisir subdistrict to the governor of Lampung in August 1996. This letter, which was also copied to the Ministry of the Interior, the Ministry of Forestry, and the Ministry of Agriculture, stated that after an *adat* meeting,[22] the villagers had unanimously agreed to reject an oil palm-plantation project in their village. To increase the pressure on the governor, another letter was sent to the Minister of Forestry in November 1996 demanding a decree that the area in and surrounding Malaya be closed for oil palm-plantation development. The letter included an attachment in which the villagers presented the economic, ecological, and social arguments for rejecting the oil palm plantation, including a citation of Act No. 12/1992, which gives farmers the freedom to choose what crops they want to plant, in this case, *damar*. In a letter sent to the Ministry of Forestry dated November 15, 1996, the local Malaya farmers made the following statement in support of their arguments, in part: "(1) The *damar* cultivation is a continuation of sustainable forest management inherited from our predecessors, who started from planting coffee, pepper, cloves mixed with *damar* trees; (2) Based on Act No. 12/1992, farmer households are allowed to pick the crops they want to plant; and (3) *Damar* gardens are important to create a green vista, to improve economic standing, and to maintain environmental sustainability, especially to sustain fresh water sources and to prevent erosion and landslides."[23] This letter shows that the farmers were acutely aware of the power of invoking conservation, biodiversity, and indigeneity. The role of *adat* was not emphasized and was only reflected in the *marga* leader's signature as one of the representatives of the local farmers.

Using the *adat* argument to claim rights to land was not a universally accepted strategy among the sixteen *margas* in Krui. The members of the Marga Penengahan-Laay wrote a letter on June 6, 1997, to the Ministry of Forestry that differed from the strategy of the other villagers. The letter claimed that the *marga* territorial forest in Penengahan no longer existed; hence the villagers demanded that individual ownership (*hak milik*) of the *damar* forest gardens be acknowledged instead. This demand clearly revealed the villagers' understanding of the differences between communal *adat* rights and individual ownership.

For several years after the farmers began their activism in 1992, arguments that used *adat* to reclaim rights to *damar* forest gardens remained an ambivalent element in their strategies. This, however, changed over time. The arrival of Team Krui and the group's strategy to nominate the Krui community for an environmental award did strengthen the position of *adat* in the farmers' discourse. The nomination to win a high-profile environmental award convinced the farmers of the discursive power of the *adat* argument. They gradually began to include it in subsequent campaigns to reclaim their forest gardens.

A STRATEGY GONE AWRY? KALPATARU AND THE
RE-CREATION OF THE *ADAT* COMMUNITY

The Kalpataru is a prestigious environmental award given annually by the president of Indonesia in recognition of contributions to the betterment of the environment. Organized by the State Ministry of Population and Environment, the award is meant to encourage the public to be more involved in the causes championed by award winners (Zerner 1994a). If the Krui community won the Kalpataru, this would give them the upper hand when meeting the Ministry of Forestry at the negotiation table. Based on this rationale, Team Krui decided to nominate the Krui community in a three-step strategy: (1) Getting scientific recognition of the *damar* forest gardens' ecological and economic viability; (2) getting institutional recognition that *damar* forest gardens are agroforests resulting from deliberate planting and management by local farmers; and (3) getting political recognition through the Kalpataru award for the local community.

When in 1997 the nomination was accepted, Team Krui debated who should be the recipient of the award. The farmers were considered too general a category to satisfy Kalpataru recipient requirements. Deliberations resulted in a suggestion that the award should be given specifically to the *adat* community of Krui, whose members resided in the three subdistricts in West Lampung. The *adat* community would be represented by its *adat* leaders, in this case *marga* heads. This decision was supported by several factors: the intense discussions of *adat* community rights among NGO and research institute members; the Ford Foundation's interest in social forestry for local communities; and the expectation that the high-profile status of indigenous and traditional communities in conservation discourses would continue for years to come. It was unforeseen, however, that this decision would represent the *adat* community—and not the *damar* farmers themselves—as the stewards of the *damar* gardens.

One of the most prominent farmers from the Krui community was also a *marga* chief, so he was appointed to represent the sixteen *margas* to receive the award. His dual role as both a *damar* farmer and a *marga* chief proved crucial in alleviating some disgruntled voices among the farmers who felt deprived of their rightful credit for the *damar* forest gardens' success. Aware of other potential conflicts that might arise among the *marga* leaders, Team Krui asked the appointed person to get signatures of approval from the rest of the *marga* leaders.

The national acknowledgment of the Krui achievement brought pride to the Krui community and an incentive to revive *adat* institutions. Shortly after the award ceremony, the sixteen *marga* leaders set up a foundation called Yayasan Penyimbang Marga (I will refer to this as the Marga Foundation).[24] Following the foundation's establishment, the sixteen *marga* leaders sent a letter to the Minister of Forestry, reasserting their demand for the return of the limited production forest boundaries to the original 1935 *Bosch-wezen*. Members of Team Krui observed this development with mixed feelings. Their major concern was that a foundation for *marga* leaders might not be participatory enough to encourage community involvement on the ground, especially when more than half of these *adat* leaders no longer lived in the area.[25]

Dissatisfaction among the *damar* farmers started to spread concerning the undeserved credit that the *marga* leaders had enjoyed, leading them to set up their own organization, Persatuan Masyarakat Petani Repong Damar (I will refer to this as the Farmers' Association) to promote *damar* cultivation. To assert their own role, the farmers' association sent a separate letter to the Ministry of Forestry to demand the removal of the 1994–97 forest boundary, and its return to the original *Boschwezen* marks, so as to return the area they claimed as *adat* land. Despite efforts to distance itself from *marga* and *adat* institutions, the farmers' association could not completely avoid deploying the *adat* argument.

The Kalpataru award clearly bolstered the prestige of *adat*, given that diverse elements in the local community began to invoke *adat* more explicitly. Both the Marga Foundation and the Farmers' Association were describing the *damar* garden lands as "*marga*-owned *adat* land" (*tanah adat milik marga*). Responding to this development, the government also began to deploy *adat* in its discourse. *Adat* had emerged, therefore, as a space of contestation for the government, Team Krui, *marga* leaders, and *damar* farmers.

THE MINISTERIAL DECREE AND THE *ADAT* COMMUNITY

Following the prestigious Kalpataru award, the villagers' struggles attracted more attention from policy makers. After a field trip to Krui and impressed with Team Krui's work, the minister of forestry responded to the villagers' demands for ownership of the *damar* forest gardens, by inviting ICRAF to help formulate a policy for the *damar* forest gardens. In June 1997, on behalf of the Krui communities and Team Krui, ICRAF proposed six points to the Ministry of Forestry, none of which mentioned the *adat* community explicitly: (1) The government should move the forest boundary markers to the original boundary of the Bukit Barisan Selatan National Park; (2) the government should not tax commodities from the *damar* forest gardens as they are not natural forest products; (3) the government should allow the farmers to continue maintaining their forest gardens; (4) there should be no restriction on the harvesting and marketing of timber from the *damar* forest gardens; (5) the farmers should be allowed to pass forest garden ownership to their descendants; (6) the government should formally recognize the *damar* agroforestry as a legitimate management system (Sirait 1998b: 2). Thus began the long process of formulating the unprecedented "Ministerial Decree for an Area with a Special Objective" (Surat Keputusan Menteri Mengenai Kawasan Dengan Tujuan Istimewa, most popularly known as SK KDTI).

In October 1997, ICRAF submitted a written memo with four further policy recommendations for the ministerial decree: (1) It should be general, but specific in addressing certain areas; (2) it should recognize the *damar* agroforest as a legitimate management system; (3) it should be specific in its evaluation of the *damar* agroforest communities; (4) it should be considered an intermediary strategy preparatory to another more specific policy being issued to regulate the area within the boundary of the *adat* forest (Sirait 1998a: 6).

ICRAF suggested that land ownership recognition should be granted to the *adat* community, echoing Team Krui's strategy. The Ministry of Forestry turned this suggestion down, citing the lack of a legal basis to justify a transfer of forest ownership from the state to the *adat* community. An extended consultation with the state minister for land and agrarian issues confirmed this legal argument. Within a very narrow legal corridor, the most the minister could offer was to change the designation of the forest in order to "protect" the forest gardens from future conversion plans by the local governments. The minister suggested changing the designation from that of a state forest area into an "area with a special objective" (*kawasan dengan tujuan istimewa*, KDTI). Additionally, for *damar* forests gardens located outside the state forest area, the villagers could petition for certificates of ownership (*hak milik*) from the National Agrarian Agency (Badan Pertanahan Nasional, BPN) (Sirait 1998a).

Team Krui realized that it could not push for a more progressive policy, particularly because the cabinet was ending its term in a few months, and there was no guarantee that the incoming minister of forestry would be willing to offer any concessions to the farmers. So Team Krui accepted the minister's offer. The remaining negotiation was limited to determining whether rights would be granted to individuals or groups. Sirait writes that "following a lengthy explanation of the social organizations within *damar* forest gardens presented by LATIN . . . it was agreed that individual contracts (with the farmers) was [*sic*] unrealistic and that agreement with each social group, or *marga*, was preferable" (Sirait 1998b: 7). The Ministry of Forestry eventually decided that the rights to manage the *damar* forest gardens should be awarded to the heads of the *margas* as representatives of the community.

The Ministry of Forestry's decision to grant the status of special objective area to the Krui community reflected a sincere effort to involve the local community in forest and natural resource management, yet a deeper analysis of the decree reveals institutional and bureaucratic rigidities. A shift in the state's perception of forest-dependent communities appeared to have inspired the ministerial decree. No longer were forest-dependent communities generally seen as enemies of forest sustainability. Instead, there was an acknowledgment that the community had developed a management system that sustained the ecological integrity of the forest (Ministry of Forestry 1998, consideration b and c). In the contract between the Ministry of Forestry and the local communities, however, the ministry listed its rights vis-à-vis the local communities, the most important of which was to "define the (management) guidelines, [and] implementation of the guidelines . . . for *repong damar* management" (Ministry of Forestry 1998, chap. 5, art. 3, point 3: 9). Thereby the state effectively denied the sophisticated ecological achievement of the local communities. By insisting on its right to produce management guidelines, the government virtually superseded the local communities in determining the appropriate silvicultural techniques for the *damar* agro-forests. The Ministry of Forestry assumed the role of developer and possessor of knowledge concerning forest management, in the process pushing aside local knowledge and techniques.

The Ministry of Forestry also listed its obligations to the local community as including "guidance" (*bimbingan*), agricultural "extension" (*penyuluhan*), "training" (*pelatihan*), and "the granting of facilities" (Ministry of Forestry 1998). These terms further underscore the asymmetrical relationship between the government and the local communities. Again, the government assumes it possesses more knowledge of *repong damar* than the farmers, hence the guidance and extension, demoting the farmers to a marginal position. This language represented the farmers as needing help and assistance, despite their having managed the forest gardens sustainably for generations. This language also expressed mistrust of the resilience and flexibility of the local community to cope with future challenges to the management of the *damar* forest gardens.

Repeated references to the *adat* community might appear to reflect the state's increasing confidence in forest-dependent communities. Again, unpacking the language reveals a more complex story. The state defined the *adat* community as a "traditional community that is still bounded in a *Gemeinschaft* with intact *adat* institutions, defined legal boundaries, adherence to customary law by the community members, and acknowledgement by the governor of the province" (Ministry of Forestry 1998: chap. 3, para. 6 [3]). This definition illustrates the asymmetrical position between forest-dependent communities and the state; an *adat* community exists only if it is recognized by the governor of the province. The *adat* community could not appeal on its own for legal acknowledgment by the judicial system. Hence, per state definition, an *adat* customary community is not an equal legal entity to the government under the law.

In response to the release of the ministerial decree, the *marga* leaders convened, reviewed the decree, and decided that it did not address their demands adequately. The political and economic crises that hit Indonesia in 1998 had delegitimized the state's power. Perceiving a coming of greater political space and opportunities for reform under a new government, the Krui community decided to maintain the status quo of the Krui *damar* gardens, to weather the crises, and to wait for an opportunity to again plead their case.

REVIVING THE *ADAT* COMMUNITY AND CONTESTATION
FROM THE FARMERS' ASSOCIATION

Despite nominating the *adat* community of Krui for Kalpataru, Team Krui members had concerns about using the term *adat* community in the ministerial decree (LATIN n.d.). They were concerned with potential misunderstandings about the meaning of *adat* community, *adat* law jurisdiction, and the land tenure regime. They pointed out that in the ministerial decree, *adat* law jurisdiction was equated with village jurisdiction, while in practice *adat* jurisdiction encompasses multiple villages. Further, the decree assumed *adat* land to be communally owned and managed,[26] while in practice the individual household is the locus of ownership and management. Team members also observed that the decree put the community at risk should the governor or the district chief decide that the *adat* community no longer existed. Team Krui members clearly recognized the vulnerability of using the *adat* argument, as one member noted: "Acknowledgment of the *adat* institution in the ministerial decree is worrisome and [puts the community into a] very vulnerable [position] because it depends on the goodwill of the bureaucrats. Revitalizing the *adat* institutions is still a big question for us" (LATIN n.d.: 2).

Despite doubts about the role of the *adat* community in the Ministry of Forestry decree, Team Krui continued to support the Marga Foundation. LATIN held a workshop for the sixteen *marga* leaders to discuss among themselves how to improve the livelihood of their *adat* community, develop a strategic plan to empower the *adat* community in Krui, and increase unity among *marga* leaders (LATIN n.d.). In practice, the workshop ended up focusing on the current debates surrounding the *adat* community, and Team Krui challenged the *marga* leaders to rethink prevailing wisdom on the topic. The workshop discussions appeared to have convinced the participants of the important role of *adat* institutions, such that when they were asked to formulate their personal objectives, the participants almost unanimously opted to revive the *adat* institutions in Krui.[27] Hardly any discussions touched on the management of the *damar* gardens, the thing that brought them the environmental award in the first place. Also, in spite of efforts to empower the *marga* leaders to be more independent, they insisted that the NGOS LATIN and WATALA continue to be involved in Krui and to support the local struggles.

That the workshop ended up discussing ways to resuscitate *adat* institutions raised concerns in the *damar* farmers' association. They were concerned about the speed of the Marga Foundation's development, which surpassed that of their own association. There was concern that they might never be recognized as the true inventors of the sophisticated *damar* agroforestry management system. They also were concerned about the lack of coordination between the farmers' association and the Marga Foundation and a perceived lack of openness in Team Krui's plans for the two organizations. One local farmer activist argued that the Marga Foundation was only an *adat* council (*penghulu adat*) that should only deal with *adat* issues and activities, while the farmers' association was the legitimate organization for the management of the *damar* gardens. However, realizing the asymmetrical power between the two organizations, farmers' association activists were reluctant to air their concerns openly. Instead, they directed their frustration at the local NGO WATALA, which they felt was trying to block their progress. The members of the farmers' association insisted that they needed to start moving forward with their programs,[28] to set up their own secretariat, and to plan their own budget. They wanted the association to be acknowledged as the legitimate representative of the farmers, and to be represented by their own nongovernmental organization in national and international fora. The farmers were acutely aware of the tenuousness of their current position in the shifting discourse of conservation and natural resource management. Unless they maintained their reputation as defenders of a sustainable natural resource management and as innovators of the *damar* agroforestry system, they felt they would lose legitimacy. With external pressure in the form of oil palm plantations looming, their sense of urgency was justified.

Discussion

Adat, long considered a legacy of a community's ancestors, is not purely a native construction. In many instances, colonial governments reshaped and redefined *adat* and custom to serve colonial political objectives (Zerner 1994a; Doolittle 1999; Kingston 1991). In Lampung, the *marga adat* system was shaped by its selective restoration in the 1920s by the Dutch, who used it for indirect tax collection and to retard communist influence on the locals (Kingston 1991). The people of Lampung fiercely rejected this revival of the *adat* system. They came to see the institution of *adat* as a colonial product that enhanced class differences between the commoners and the elite *marga* leaders.

In the 1990s, transnational symbolic politics paved the way for *adat* to gain a role on the stage in Indonesia's discourse of conservation and natural resource management. In the case of Krui, this developed into what I call the *adat* argument, the deployment of *adat*-related concepts in the discursive arena of conservation and natural resource management, referring especially to the invoking of *adat* land tenure rights and *adat* customary rights in legitimizing land ownership claims. Team Krui deployed the *adat* argument to aid the community in reclaiming ownership of *damar* forest gardens. *Adat* was represented as an indigenous institution that could serve as the basis for local resource governance, leaving aside the fact that it was also partly a colonial product. The advocacy strategy for the Krui community was so concentrated on reshaping and redefining *adat* institutions that it sidelined the farmers' achievements in inventing a sustainable forest management system and sophisticated silvicultural techniques, an achievement that, in contrast, owed nothing to the colonial government. If the community's success was framed as a modern achievement, it would be less powerful an argument, considering the ever-greater symbolic value granted to indigeneity. This framing has had the unfortunate consequence of portraying the farmers as inheritors of knowledge, not as dynamic inventors constantly modifying their technology in a changing landscape.

Discourse brings power relations into being because "it is in discourse that power and knowledge are joined together" (Michel Foucault, quoted in Lemert 1993: 522). As Foucault writes:

> We must conceive discourse as a series of discontinuous segments whose tactical function is neither uniform nor stable . . . as a multiplicity of discursive elements that can come into play in various strategies. It is this distribution that we must reconstruct, with the things said and those concealed, the enunciations required and those forbidden, that it comprises; with the variants and different effects—according to who is speaking, his position of power, the institutional context in which he happens to be situated—that it implies; and with the shifts and reutilizations of identical formulas for contrary objectives that it also includes" (quoted in Lemert 1993: 522).

The representation and reconstruction of *adat* institutions in Krui offer an example of the exercise of this sort of power. We need to be attentive to how power relations are affected by transforming the meaning of some terms (such as *adat*) and altering the meaning of others (such as the state's discursive domination over *adat* institutions), and how different actors use their roles in reconstructing and representing *adat* to alter power relations.

By reshaping and redefining *adat*, and by resuscitating the fading *adat* communities in Indonesia, Team Krui was trying to introduce new elements into the agrarian debate: a legal recognition for a customary community and a politics of law that respects the existence of traditional, marginalized communities in Indonesia. Team Krui's affirmation of the *adat* community's existence and its attempt to dominate the discourse of the *adat* community were aimed at creating a new, "normal" knowledge.[29]

It may take an essentializing argument to counter one (Dove, quoted in Brosius 1999b), but strategic essentializing benefits various segments of the community differently. The *adat* leaders adopted the strategy of strengthening the waning *adat* institutions and, consequently, enhancing their own influence, which had been eroded by various state interventions. Nevertheless, when encouraged to be more independent in planning for the revival of *adat* institutions, the overwhelming consensus was to retain the NGOs and other supporters as dominant players in the struggle. The *adat* leaders requested "to leave it to the NGOs . . . and the others to be involved in the arena and not to leave us alone."[30] This stood in stark contrast with the Repong Damar Farmers' Association's demands to become an NGO itself and for the local NGO to make way by lessening its own role.[31]

In this contest for power the farmers came to understand that, even though they also deployed the *adat* argument, the revival of *adat* institutions would obscure their own achievements in conservation. They understood that conservation and sustainable forest management were more highly valued than indigeneity in the national and international conservation debates. It was this logic that led the farmers to found a separate farmers' association, an organization that had nothing to do with *adat* but everything to do with *damar* gardens and forest management. The farmers did not favor supporting the *adat* argument, because this diminished their own role in the success of *damar* gardens and blurred their modern achievement with traditional culture.

The state, which continued to keep its grip on forests, was a key player in this discursive contest. Instead of succumbing to the "normalizing argument" waged by Team Krui, the Ministry of Forestry was successful in resisting the "new truth." The basic premises used in addressing *adat* communities in the ministerial decree remained orthodox ones. The premises of the legal decrees essentially retained the asymmetrical power structure between the state and its people. Despite the goodwill demonstrated by the minister, the bureaucracy essentially kept the Krui community in its place.

The *adat* argument as a point of entry in conservation and land struggles is both powerful and dangerous. In the case of Krui, it made possible winning the Kalpataru, an award that strengthened the local communities' political position. But in doing so, instead of winning recognition for their technological achievement, it gave undeserved credit to the *adat* institutions. It led to the *adat* institutions being revived, taking the stage, and diverting the focus from the more pressing issue: the prosperity and rights of the farmers to their *damar* gardens. Fighting back against this unintended marginalization, the farmers sought to introduce their own discourse to the arena: conservation and sustainable natural resource management.

Team Krui's search for local institutions, practices, and values to effect changes in the asymmetrical relationship between the state and the community are to be applauded. But attempting to revive *adat* institutions, forgetting that *adat* may well have been an undemocratic, class-based institution shaped to the needs of colonial rulers, can obscure the struggles of nonelite, marginalized farmers. In this case, the local farmers emerge as the savvy players, resisting the false allure of *adat* and instead presenting themselves as the stewards of conservation to win back their *damar* gardens.

Notes

The research for this essay was carried out during a two-year scholarship provided by Fulbright. I thank LATIN and SUWITO for access to reports and documentation of their work in Krui; I also thank the Krui community for the permission to observe and document their struggle. Michael Dove, Carol Carpenter, and Matthew Giancarlo extended generous support in the completion of this article.

1. As Krui was a high-priority issue for numerous institutions, a wealth of written documents was available for analysis. I was given access to various unpublished reports, informal notes, and transcripts of village meetings—which proved vital for my analysis—by individuals closely involved in the Krui case. These documents are otherwise inaccessible to the general public.

2. This shift is illustrated in the World Bank's policy: from the infamous transmigration schemes in the early 1980s that severely affected forest-dependent communities, the World Bank shifted its recognition of the local communities by sending a letter to the Indonesian government in June 1999, "suggesting that it listen to all stakeholders before rushing to pass the new forestry law" (Gautam et al. 2000).

3. Conklin and Graham (1995) argue that using a transnational image of indigeneity as a source of political power is dangerous because the power exists only so long as the political identity of the indigenous people echoes the current trend of transnational symbolic politics.

4. *Repong* and *tembawang* are local terms for forest gardens in Krui and West Kalimantan, respectively.

5. Bylaws of Aliansi Masyarakat Adat Nusantara (AMAN), chapter 5, article 19, note 2; http://dte.gn.apc.org/AMAN/kenal/ad.html (accessed July 7, 2010).

6. In Indonesian, Zakaria wrote this as "kedaulatan persekutuan hidup setempat atas wilayah yang menjadi wilayah ulayatnya itu." He defines sovereign rights as the right of self-government within the *adat* community territory. This encompasses the right to decide on the land tenure regime and *adat* regulations for the community. By definition, sovereign rights are broader than communal rights, which are almost always limited to land-tenure rights.

7. The Lampung *adat* community is divided into two main groups: Pepadun in the interior and Peminggir in the coastal areas. The Pesisir Krui community belongs to the latter and consists of sixteen *marga* (Hadikusuma 1988).

8. The local term for agroforest is *repong*. A *repong* is a piece of land planted with diverse productive perennial crops, such as *damar* (*Shorea javanica*), duku (*Lansium domesticatum*), durian (*Durio zibethinus*), nangka (*Artocarpus heterophyllus*), petai (*Parkia speciosa*), jengkol (*Phitecelebium jiringa*), manggis (*Garcinia mangostana*), and numerous timber species with economic value, mixed with wild species. A *repong* is called *repong damar* when *damar* trees dominate the plot (Lubis 1997).

9. This study is important in providing both scientific and political support to forest gardens in opposition to a plan to open oil palm plantations in Lampung Province.

10. See Michon et al. (2000) for a detailed historical documentation of *damar* activity in the area.

11. Three main factors explain this boom. First, the overtapping of existing wild *damar* had created a scarcity of product from the wild trees. Second, around 1920, pepper plantations were decimated by an unknown disease, which encouraged the villagers to substitute the cultivation of *damar*. And third, there was possible encouragement to cultivate *damar* trees by others involved in the *damar* trade, including, the local elites, Chinese traders, and the colonial administration (Michon et al. 2000).

12. This section is taken mainly from Kingston (1991).

13. According to Kusworo (1997), the Bukit Barisan Selatan National Park is composed of five different registered parcels of land based on the Bengkulen Resident's edicts No. 48, dated February 24, 1935, No. 80, dated March 8, 1935, and No. 117, dated March 19, 1935. The area totaled more than 190,000 hectares and had the status of game reserve.

14. This encroachment stopped when the state upgraded the status of the game reserve to national park, evicted the illegal squatters, and increased control over the park's boundaries.

15. A limited-production forest is a forest under the jurisdiction of the state, the dominant function of which is to maintain the forest's various ecological functions. However, limited and regulated exploitation is allowed, with taxes imposed on the exploiter. This means that villagers tapping the resin or taking out timber must pay taxes to the state, an ironic situation considering that the forest gardens were planted and managed by the villagers.

16. Consensus here only involved the state departments and the local governments. It was a top-down initiative in which the people's concerns were not taken into account.

17. See Torquebiau 1984 and Mary and Michon 1987 for examples.

18. Team Krui is an informal forum consisting of individuals and institutions that share similar concerns about Krui. The founding members are Orstom, the Ford Foundation, P3AEUI, WATALA, and LATIN. When additional institutions became involved in the Krui case, the team was expanded.

19. This meeting was held in Bandarlampung, the capital of Lampung Province, and was attended by seventy representatives from various institutions such as the University of Lampung, the Planning and Coordination Bureau of Lampung, the Agricultural Service, the Estate Crops Service, the Forestry Service, the Provincial Forestry Service, local government representatives, faculty from the University of Indonesia, and Team Krui members, including eleven *damar* farmers from Pesisir Krui.

20. Although the provincial forestry agency acknowledged that some *damar* trees were planted by the local community, they claimed that in Krui this amounted only to seven hundred and fifty hectares. They argued that the remaining forest in the area and within the national park was natural vegetation (Kanwil Kehutanan Propinsi Tingkat Lampung 1995).

21. The farmers' fluent use of conservation and scientific jargon is partially explained by their ongoing, and in some instances quite intimate, relationships with scientists, NGO community facilitators, and other researchers, who kept them abreast of cutting-edge debates on conservation and natural resource management.

22. This is the equivalent of a "town meeting," which every member of the territory attends to present his or her opinion on a particular issue.

23. A copy of this letter can be found in the offices of the Marga Malaya, North Pesisir subdistrict, Lampung.

24. I could not find documents that clearly state the purpose of the *adat* foundation Yayasan Penyimbang Marga. However, based on my interviews, it appears that the *marga* leaders felt they needed a forum to institutionalize their collective interests.

25. Out of the twelve *marga* leaders attending the workshop, six are no longer based in Krui. One is based in Jakarta, four in Bandarlampung, and one in Liwa, the capital of West Lampung District. All six work as either civil servants or in the private sector. Of the six locally based *marga* leaders, three work as farmers, one is a student, one a civil servant, and another is an entrepreneur.

26. See chapter 7, paragraph 12 (1) of the Ministry of Forestry's decree: "Extraction from managed *damar* forest gardens is carried out by the head of the *adat* community in line with regulations in this decree."

27. The *marga* leaders formulated three general objectives: to revitalize and revive the *adat* institutions in Pesisir Krui, i.e., the *marga* system, to strengthen the role of *marga* leaders and the foundation, and to improve internal unity among *marga* members.

28. At the time, the farmers' association had three urgent missions: (1) to follow up on the Ministerial Decree with action plans; (2) to conduct field investigations (there were reports of *repong damar* being cut down in the *desa* Bambang); and (3) to create a seed nursery to encourage more *damar* planting and cultivation.

29. The high-profile "Conference on Nusantara's Indigenous People" (Nusantara being a precolonial name for Indonesia) in 1998, the creation of the "Association of Nusantara's Indigenous Peoples," and the continuing debate on *adat* community–based forest management in Indonesia are some of the examples of this strategy.

30. This stance was the product of a workshop for *adat* leaders. The workshop, organized by Team Krui, was intended to help them prepare a strategic plan for the recently founded Yayasan Sai Batin, a foundation set up by the NGOs and donor institutions to support the local *adat* leaders in exercising their role as community leaders. In Indonesian, the statement reads, "terjadi konsensus sebagai berikut: 'Kami menyerahkan kepada LSM, seperti WATALA, LATIN, dan yang lain untuk ikut bermain, dan jangan dilepaskan sendiri.'" To this plea, the NGOs responded that their eventual phasing out was essential to their goal of allowing local communities to be independent and in full charge of their future. During the workshop, however, the *adat* leaders insisted on the NGOs' continuing involvement and refused to accept the argument that this undermined the prospects for their own independence.

31. One local community organizer in fact complained that the local NGO prevented them from taking opportunities to learn new skills such as participatory mapping and learning sophisticated technologies such as remote sensing.

REDEFINING NATIVE CUSTOMARY LAW
Struggles over Property Rights between Native Peoples
and Colonial Rulers in Sabah, Malaysia, 1950–1996

In this essay I explore the dynamic relationship between native customary law, colonial statutory law, and de facto land management practices in an agricultural community in Sabah, Malaysia. I seek to show how property rights and social relations formed around access to and ownership of resources can both constrain and increase the options available to community members. As farmers face inevitable changes in agricultural markets and the broader political economy, a property rights system that allows for dynamic changes is needed. Using a historical and site-specific perspective toward the ongoing transformation of property regimes, I am able to gain a window on several larger themes that are intertwined with changing property relations. First, this analysis enriches our understanding of state-society relations by looking beyond the notion of a colonial government that imposed its legal structures on a pliant and unresisting subject population (Berry 1993; S. Moore 1986a; Comaroff 1989). Instead I demonstrate the role of negotiation and cooperation between the colonial state and indigenous society, the diversity of discourses mobilized by various actors within both the community and the state (Agrawal 1999; Agrawal and Gibson 1999; D. Moore 1993, 1997), and the role of contingency in the expression of both state rule and local agency. I argue that the relationship between state authority and local people should not be conceptualized as an oppositional binary, based on modes of domination and resistance (Nugent 1994), but instead as mutually constitutive and deeply intertwined (Gupta 1998, 1995; Li 1999; Corrigan 1994; Nugent 1994). In brief, we cannot conclude that colonial law was simply a body of immutable rules imposed on the indigenous population. Nor can we conclude that if the indigenous people did not overtly resist colonial land settlement policies that they were passive subjects. Instead, we need to examine the nature of the negotiations between colonial administrators and local leaders and explore what creative forms of state control developed and what new forms of local autonomy emerged from these state-society interactions (see, e.g., S. Moore 1986a; Berry 1993; P. Peters 1994; Zerner 1994a).

Second, I highlight the notion that communities are internally diverse in terms of how individuals view property relations and the varying strategies that they employ to secure access to land (Li 1996; Agrawal 1999; Agrawal and Gibson 1999). Colonial governments tended to treat communities and their property regimes as homogenous and stable institutions. For instance, in North Borneo, G. C. Woolley, the commissioner of lands, published seven different reports focusing on the customary laws of six different ethnic groups (see Woolley 1953a, 1953b, 1953c, 1953d, 1953e, 1962a, 1962b). In five of his seven publications he failed to discuss the nuances in property rights regarding natural resources. In these five reports his discussion of land tenure focuses solely on the inheritance of land; he makes no mention of hunting rights or rights to forest products, fruit trees (wild or semidomesticated), or water. Static and simplified renderings of customary law, an institution that prior to colonial intervention was regarded as fluid and adaptable to local context, was a central strategy in colonial rule (Adewoye 1986; Berry 1993; Ranajit Guha 1963; Meek 1946; Merry 1988, 1991; S. Moore 1986a, 1986b; Okoth-Ogenda 1979; P. Peters 1994; Vandergeest and Peluso 1995; Zerner 1994a).

In contrast to the simplified colonial interpretation of native property rights this study shows how community members, in the face of external changes brought about under colonialism, were able to adapt both their rhetoric and their practices regarding property rights. Local villagers in a community I call Govuton did not necessarily present a single unified notion of traditional property rights to the colonial authorities. Rather, this case shows how villagers tried various ways and relied on distinct discourses to gain state-sanctioned ownership of the land that they had access to. Thus native customary laws may have been used in instrumental ways by local people for material and political gains, in a sense turning the tools of the elite to their advantage.[1]

Furthermore, I argue that since communities are deeply embedded in state regimes, property rights and other social relations are reconfigured through discursive practices that draw on both local and state-generated idioms in the context of changing political and economic conditions (Gupta 1995). Thus access to resources and the transformation of property rights is shaped by both internal village conflicts over cultural meaning, social identity, and power, as well as by the incorporation of rural areas into the colonial and national political economy (see, e.g., Berry 1993; Brosius 1997b; S. Moore 1986b; P. Peters 1994; Peluso 1992, 1996; Li 1996). Third, I explore notions of legal pluralism and native customary law and demonstrate that present-day understandings of native customary rights to natural resources took shape in the crucible of colonial rule through the formal codification of native customary law (S. Moore 1986b; Zerner 1994a). Thus current expressions of native customary law bear the imprint of regimes designed for rule. Furthermore, the colonial codification of customary law took a highly fluid set of practices and made them static, representing the nature of these laws in a certain period of time and a certain political-economic context (S. Moore 1986b; Zerner 1994a). In the village of Govuton the result is that some villagers find the colonial interpretation of native customary law to be restrictive, not enabling, forms of tenure. I conclude that in Govuton colonial interpretations of customary laws has locked villagers into a mode of agricultural production that is not empowering for all members of the community in the present political economy. This static and simplified rendering of native customary law has failed to adapt to significant changes in the regional political economy. Ultimately this frozen image of native customary law in the form of communal access to resources in the native reserve served to extend state control into rural livelihoods, rather than devolving decision making about community resources to the local level.[2] With this in mind, it is important to acknowledge and look for the diversity within and among households in a community. Seeking consensus among these differences inherent in social life suggests that compromises are necessary to achieve the goals of conservation and rural development.

In the following pages I explore the case study of the formation of the native reserve in Govuton to understand the ways in which local people responded to colonial policies of land settlement in the 1950s. In the first section of the essay I draw both on regional archival data from the Ranau District (see map 1) and on oral histories from the village of Govuton regarding the formation of the native reserve. In the second section I present ethnographic data examining contemporary land-use practices, property rights, and land disputes within in Govuton. The goal of this section is to trace the impact of the colonial codification of native customary laws on present-day society. While it is possible to understand many of the current land disputes as a legacy of the multiple legal systems implemented under colonial rule, it is also important to consider how land disputes are articulated in response to the values and social meanings attached to land and resource accumulation, and how these issues are shaped by changes in the

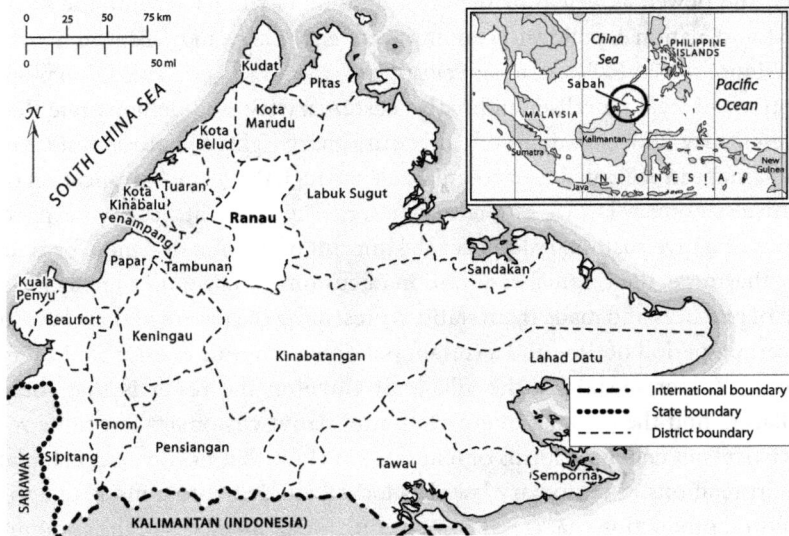

MAP 1 Location of Ranau District in Sabah

political economy. In brief, I argue that present-day land disputes are not simply an inevitable result of legal pluralism but that they are also the result of the continuously mobile nature of property relations.

I conclude that while the formal recognition of village-owned land under the concept of native reserve might have empowered local people to resist the outside appropriation of local lands in the 1950s, today the native reserve is not as empowering as it once might have been. In other words, the native reserve served to strengthen local autonomy in the late colonial period, effectively barring the state and other extralocals from appropriating village land that was communally owned. Yet in the contemporary period, the native reserve restricts local autonomy and decision making in terms of land. While the native reserve may still bar extralocals from encroaching on village land, it also keeps villagers in Govuton from the economic benefits associated with private property rights. As a result, some villagers express frustration at the restrictions associated with village ownership of land in the native reserve. Conversely, other villagers still benefit from the native reserve as they rely on the community-owned forested areas to support a particular lifestyle, one based either on subsistence needs or cultural values. In this essay I closely explore a specific land dispute to understand the multitudinous factors that influence peoples' present-day actions and discourses regarding the rights of access to land within the native reserve.

There are broader implications of this case study in terms of natural resource management. This analysis questions policies that propose to strengthen native customary law, in a sense returning to a time when communities had more autonomy in decision making, to achieve positive conservation and development outcomes. The impulse to support or "reinvigorate" native customary law in the search for sustainable natural resources management generally comes from two concerns. The first is a belief that governments and nongovernmental organizations have had little success with top-down approaches to conservation. Local communities are perceived to have a deeper long-term investment in their environment and a more nuanced knowledge of the ecosystem. Perhaps communities can suggest where other organizations have failed. The second concern is situated in a desire to link conservation goals with social justice. Is it argued that historically marginalized and exploited communities should be allowed an equal voice in the discussion about conservation and resource management (Brosius, Tsing, and Zerner 2005; Brechin et al. 2003). The details that this study highlights draw attention to the complexities of bolstering a particular imagining or rendition of native customary law without a thorough understanding of the context in which it was encoded and the benefits and restrictions that these laws may potentially bring to the entire community. In brief, there are many pitfalls for policy makers who uncritically adopt concepts of native customary rights as appropriate loci for local autonomy, social change, or the conservation of natural resources without examining where these rights came from and how they have changed over time.

Finally, this case demonstrates that conflicts, internal stratification, and competing individual needs make it difficult to view local communities as harmonious, idyllic, and commonly united in their resource needs (Agrawal 1999; Agrawal and Gibson 1999; Brosius, Tsing, and Zerner 2005; Li 1996; Zerner 1994a). This case illustrates that individuals in a community can have drastically varying notions about appropriate resource use, and that one individual's connections to external forms of power and authority can be used to suppress alternative discourses and practices regarding resource use and management.

The Native Reserve in Govuton

The village of Govuton lies in at the foot of Mount Kinabalu. The landscape is very steep and hilly and has been classified as lower montane rain forest (Kityama 1987). Govuton is a long-established village of swidden agriculturalists, with a current population of roughly two thousand people, which is divided into nine hamlets. Historically each of these hamlets would have represented a longhouse. Today people live in households of extended families. The contemporary economic base in the village is a complex, composite agrarian economy—families mix subsistence swidden agriculture with commercial agriculture of temperate vegetable crops. Some families also own wet rice fields and fruit orchards in distant villages. Additionally, most extended families have at least one wage earner. The majority of the villagers belongs to the Dusunic-speaking linguistic group and also refers to its ethnicity as Dusun.

In the late 1950s a group of local leaders in Govuton mobilized to protect their communally owned village lands from colonial appropriation. These lands included areas used for subsistence and commercial agriculture, as well as forested areas to which villagers traditionally had access for hunting and gathering. Used by villagers for centuries, these lands and forests not only held material importance in daily life but also symbolic importance in village folklore and ritual life (Doolittle 1999).

Rather than requesting private title to lands that each family could rightfully claim under the colonial land laws of 1953, local leaders in Govuton turned to a little-used section of the colonial land laws under which communal titles could be issued to villages, for "lands held for the common use and benefit of natives." Such communal titles were referred to in the land laws as "native reserves."

Both archival data and oral histories shed light on the course of events that led to the formation of the native reserve in Govuton, often providing contradictory accounts of who provided the motivating force behind the formation of the native reserve. In reports and letters produced by colonial administrators,[3] the paternalistic need to protect natives from foreign land speculators was emphasized. In letters from leaders in Govuton, as well as in oral histories, the desire to secure local access and traditional rights to lands in the face of state appropriation of village land (which would eventually open the door to land speculators) was emphasized. These accounts show how various actors employed colonial law in different ways to achieve the same goal—control over the land.

VIEWS FROM GOVUTON: PROTECTING TRADITIONAL LANDS

The impetus to form the native reserve in Govuton emerged as a result of local discontent at state polices that appeared to be aimed at appropriating village-owned lands. In 1957 the director of agriculture expressed an interest in the highlands surrounding Govuton. A road connecting Govuton to the coast was near completion, and it was hoped that temperate fruits and vegetables could be grown commercially in the cool climate. At the request of the director, the Department of Lands and Surveys began to survey land around Govuton. This was the first step required before land could be alienated to individuals for commercial agricultural development. In other words, the arrival of the surveyors in Govuton signified that the state considered the land "state land," available for alienation. To the villagers in Govuton this meant that outsiders would soon follow, staking claims to what local people perceived as village land.

Local leaders initially tried to resist the state's efforts to survey the land, claiming that the area the Lands and Surveys Department was surveying fell within the village's " 'reserve' for ratan [sic] and roof-making leaves" (Director of Agriculture 1958). Thus local leaders invoked notions of native customary rights of access, based on a tradition of communal usufruct rights to forest resources, to claim this area. These rights of access were acknowledged in the land laws of 1953, which recognized that villages should be allowed access to timber reserved for subsistence use. But these initial efforts to draw attention to native rights to the land proved unsuccessful. In April 1958, the West Coast resident asked the conservator of forests to find "an alternative source of domestic timber and housing material for the village of 'Govuton'" (West Coast Resident 1958). This letter indicated that the resident had no intention of allowing the villagers to claim the area in question based on native customary rights, as they were legally entitled to under the colonial land laws.

At the same time that some local leaders were trying to highlight the village's communal access to land, other villagers from Govuton attempted to take advantage of the colonial project of land settlement by filing applications for native titles (individual, private ownership of land issued under the 1957 land laws) to the land in question.[4] In other words, villagers from Govuton did not speak with one voice, nor did they act as a unified body in their attempts to seek state recognition of traditional rights of access to village land. These internal struggles over property rights within Govuton demonstrate how local people strategically invoke varying practices to gain access to land. In brief, villagers voiced different interpretations of local property relations and state laws in their attempts to get state-sanctioned ownership of the land they accessed. Some tried to assert the primacy of their customary rights; others tried to use the new colonial land laws as opportunities to secure private title to land.

The primary advocate of community ownership of the village land was a Govuton leader, Native Chief Gisil.[5] Gisil found an ally in the colonial administration with the arrival of John Dusing as the district officer for Ranau. In 1961, Dusing made the recommendation that the forested area around

Govuton be declared a native reserve, thereby officially initiating the process of state recognition of village ownership of the land.[6] Neither archival data nor oral histories state why Dusing supported the native reserve, but Dusing was the first district officer in Ranau who was a Dusun (the largest ethnic minority in Sabah and the ethnicity of the villagers in Govuton), and therefore we can speculate that he was sympathetic to native peoples' claims to land. There is some evidence to support this speculation. When Sabah achieved independence, the first native chief minister, Donald Stephens, appointed Dusing as his secretary of state. Tun Mustapha (the head of state) refused to support this nomination, claiming that Dusing would favor Dusun people in the distribution of government projects (Yahya Bin Mohamad 1990). The historical record therefore suggests that Dusing was noted for his pro-native sentiments. Regardless of the reason why the proposal for the native reserve was supported by the state, it was through political networks between the native chief from Govuton and the district officer from Ranau that the voices of leaders interested in village ownership of the land were privileged over other dissenting local voices. On November 1, 1966, Gisil and the *orang tua* (headman) in Govuton officially applied to the government for 3,120 acres of land surrounding Govuton to be proclaimed a native reserve. While the official signing of the gazette by the chief minister did not occur until 1983, some of the older men in Govuton remember helping the Lands and Surveys Department *tenom batu* (literally, "plant stones," meaning to set boundary markers) for the native reserve in 1968.[7] In the eyes of Govuton's residents, the native reserve and village ownership of the land and forests was recognized by the state in the late 1960s.

VIEWS FROM THE STATE: NATIVES AS UNSOPHISTICATED "WIGHTS"

During the period in which the North Borneo Chartered Company ruled Sabah, a dominant discourse surrounding native land rights and land-use systems emerged (Doolittle 1999). This discourse continued during crown rule, as evidenced in the colonial documents produced during this period. In one strand of this discourse, colonial administrators were concerned with the perceived inability of natives to manage their lands in a rapidly changing market economy.[8] Evidence of this concern is found in the restrictions against natives from selling their native titles to non-natives without governmental sanction. The notion behind this restriction was that natives did not understand commercial land transactions, and if they were not "protected from their own improvidence," they would sell all their land to foreign speculators and be left with no land of their own to cultivate ("Circular Notice to Officers" 1928).

Debates within the colonial administration over the possibilities of natives selling their lands to non-natives peaked in the late 1950s. On one side of the debate were colonial officers who believed that the "North Borneo native is a poor unsophisticated wight,[9] who is easy meat for a non-native land shark" (Director of Lands and Surveys 1957). These officers felt that it was their paternalistic duty to protect natives "against the cunningness of the sophisticated non-natives, who carrot-wise, dangle treasury notes or trade goods before their noses" (Director of Lands and Surveys 1957).[10] Yet other officers felt that the idea that natives were naive regarding the commercial value of land was pure fiction. "The average North Borneo native is just as acute a land shark as any non-native," stated one colonial official. "Every native knows the value of a scruffy bit of land . . . to which he has done nothing for years and years, but which suddenly appreciated in value because of the proximity to urban . . . development or because it is served by a new road" (Director of Lands and Surveys 1957).

Clearly colonial administrators held contradictory images of the ability of natives to manage their land without colonial intervention. The records in the Ranau District Office from the mid-1950s illustrate that this debate was resolved in a way that represented concessions to these competing colonial views: if natives wished to sell their lands to non-natives, they could do so with government sanction, but they would be charged a substantial premium for the transfer of their native titles. The colonial officers in Ranau hoped that the premiums would deter some natives, since they lowered the profits people would realize from the sale of their lands (West Coast Resident Chisholm 1957). Importantly, in terms of state strategies of rule, if natives did choose to sell their lands, it also would raise the profits of the state. Thus protection was offered to natives, yet natives could choose to forfeit state protection for a price.

This debate undoubtedly influenced colonial administrators' decisions regarding the formation of the native reserve in Govuton. While the impetus to form the native reserve came from leaders within Govuton, the final authority to proclaim the area a native reserve had to come from colonial authorities. Colonial land policies were dominated by a preoccupation with settling native claims only when such claims could be reduced to individual, private property. In other regions in North Borneo, native reserves had been abolished as the land was claimed by the state and made available for plantation cultivation (Doolittle 2005). Native rights to communally owned land, forest resources, and scattered fruit trees were generally ignored during the actual practice of land settlement. Colonial administrators considered this oversight necessary. It was the goal of the state to tax native lands and to systematically organize native property rights, thereby facilitating the alienation of so-called waste lands for plantations. Attention to all the details of native customary rights would only delay the profits that the state was eager to gain. The focus of the colonial administration on taxation over the settlement of all native rights was clearly stated by District Officer A. Francis, who wrote in 1902: "No native titles had been issued except for padi lands; orchards, house sites, grazing lands, timber reserves, sago, reserves for expansion have all been excluded from the settlement, mainly . . . because they were *not assessable with rents* or with only small rents."[11]

Why then did the colonial state support the native reserve in Govuton? One way to make sense of the state's support of a native reserve, a situation in which village communal property was protected and no tax revenue was derived by the state, is by placing the decision in the context of the administrative debate over protecting natives and their lands from non-native land speculators. In other words, colonial officers supported local communal rights to supposedly protect natives from losing their lands, despite the state's broader economic agendas. Viewed more critically, however, this form of protection also had the effect of intensifying state control over the native population.[12] Local traditional land rights became contingent on formal sanction and registration with the colonial government. Thus local customary rights, which previously flowed largely from local practices and changed through them, could only gain legitimacy through the final authority of the state and also could only be changed through it.[13] This form of territorial control, based on proscribing or prescribing specific activities with specific spatial boundaries, is a critical element of state power (Vandergeest and Peluso 1995). Moreover, the formation of the native reserve isolated the natives in Govuton from broader economic changes associated with the commercialization of land that the state and non-natives were enjoying throughout North Borneo, continuing the colonial trend of keeping natives in a marginal position in relation to the ruling elite. While "protecting" natives was the rhetoric used to justify certain forms of colonial property law, this rhetoric elides the important function of these laws in empowering the state.

The implementation of the colonial laws concerning the native reserve was brought about by the political and social networks extant between local society and state agents.[14] Without pressure from the local population, it is unlikely that the state would have sanctioned the formation of the native reserve. This evidence emphasizes that the relationship between state authority and local people is not always based on state domination and local resistance (cf. Nugent 1994). This data shows that at the local level there may exist conscious decision making and planning that result in local attempts to negotiate with state agents to influence state practices. This is not to suggest that the power held by the state and that held by society is equal, but rather I argue that state-society power relations are highly dynamic and contingent on the social, spatial, and temporal context.

Contemporary Property Rights and Land Disputes in the Native Reserve

According to the gazette notification for the native reserve, the headmen[15] of Govuton and the native chiefs of the district were appointed the "trustees" of the reserve.[16] The Gazette Notification reads "(1) The Reserve is for the sole use of the native at Kampung [village] Govuton for the construction of dwelling houses and other ancillary buildings; (2) The residents of the Kampung [village] may be permitted to practice traditional Kampung [village] industries, rearing of fowl, domestic animals and do cash crop farming within the Reserve."[17] Additionally, no individual may sell land in the native reserve since it is considered village communal property, not as individual private property. These rules are the extent of the statutory laws about land use in Govuton. They are the product of the colonial imaginings of how native communal tenure should be represented in statutory law. Like other aspects of the colonial project, this simplified reordering of property arrangements constituted an effort to make complex local society legible and manageable for centralized rule (cf. Scott 1998). In short, the written colonial law delineating land use in the native reserve does not capture the complexity of dynamic social relationships surrounding property rights in Govuton. In contemporary village life statutory laws are overlaid with *adat* (customary laws), creating competing sources of legitimacy for property claims in the native reserve. These differences, formed from the disjuncture between the panoptic view of the government and the complexities of local practices, form the starting point for the following discussion. Table 1, which summarizes the differences between statutory laws and local customary law governing land use in the native reserve, shows that the state categories of native property rights are much more restrictive than native categories.

Villagers in Govuton have a more elaborate land tenure system, following village *adat*, rather than the legal system extant in statutory law. Even though the gazette notification for the native reserve makes no distinction in how the land in the reserve is used, the villagers have divided it into two categories: *tanah kampong* (village land) and *hutan simpan* (forest reserve) (map 2). *Tanah kampong* is the area in the native reserve in which people may build houses, make vegetable and swidden gardens, graze their livestock, plant fruit trees, and build fishponds. The *hutan simpan* is an area of

TABLE 1 *Land Classification and Land Use in Govuton's Native Reserve*

State land classification	Approved uses under statutory law
Native Reserve	House construction, making commercial gardens, grazing of livestock, practicing traditional industries

Village land classification	Approved uses under customary law
Tanah kampong (village land)	House construction, making small home gardens, planting fruit trees, making fishponds, grazing livestock, making swidden gardens
Hutan simpan (reserved forest)	Cutting trees for household construction and fuel wood, hunting, collecting wild vegetables and other nontimber products

forest surrounding the village. According to village *adat*, people may collect forest resources and wood for personal consumption in the *hutan simpan*; it is also used for hunting. But even these two local categories of land, *tanah kampong* and *hutan simpan*, do not capture the complexity of daily practices or the de facto property relations formed through action. In the following sections I elaborate on the de facto practices of land use and property relations in Govuton and illustrate the ways in which contests over resource use play out in terms of divergent images of community and divergent interpretations of rights of access.[18] In sum, the colonial ideals of property rights do not even begin to encompass the complexity and fluidity of local rights to resources and land use systems.

Contemporary Property Regimes in *Tanah Kampong*

Farmers in Govuton combine swidden rice agriculture with the intensive cultivation of temperate vegetables for cash crops. As in many other communities in Malaysian and Indonesian Borneo, an indigenous labor theory of value forms the core of property relations in Govuton.[19] Thus a pioneer who clears the land from primary forest has permanent usufruct rights to that land. Although land in Govuton has been farmed for multiple generations and the original clearer of the forest has long since died, the notion that the first person to clear the forest can claim usufruct rights to the land is still embedded in local discourse surrounding property relations. On any plot of land in the native reserve, most adult village members can recount the

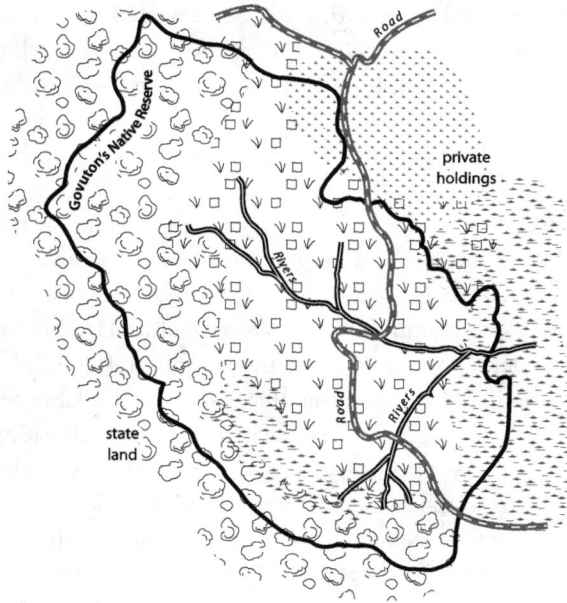

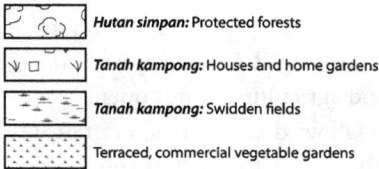

MAP 2 Land Classification and Use in the Native Reserve

Legend:

Hutan simpan: Protected forests

Tanah kampong: Houses and home gardens

Tanah kampong: Swidden fields

Terraced, commercial vegetable gardens

lineage of the person who cleared a specific plot of land and all the subsequent users. This notion, however, is intertwined with principles of inheritance, as the usufruct rights conferred on the first person to clear the forest are passed from parents to children for generations. When asked how they came to have access to a specific plot of land, most villagers in Govuton say that they have inherited it from their *nenek moyang* (ancestors), who first cleared the land.

Primary usufruct rights in Govuton also include the right to sell the use rights to another person if cash is needed. But in Govuton primary usufruct rights to agricultural land in the native reserve are limited in two ways. It is expected that kin and neighbors needing a place to farm can borrow any land left in fallow without having to pay compensation. And useful products growing naturally on fallow land (such as bamboo [*Gigantochlae* spp.] and wild vegetables), as well as abandoned or naturalized crops (such as taro [*Caladium* ssp.] and bananas [*Musa* spp.], which propagate independently as fallow fields return to secondary growth), can be harvested by anyone needing them for consumption. These rights of access to land in part delineate the boundaries of a community, loosely defined in terms of kinship, proximity, and need (for similar findings in Sulawesi, see Li 1996). The values of kinship and need become the parameters surrounding the exchange of rights of access to land.

Community access to fallow land is elaborated through what has generally been referred to in the literature as a "moral economy of peasants," in which reciprocity and exchange form the foundation of social relationships.[20] Such a notion of shared community access to agricultural land in the native reserve was possible in the past due to the agricultural cycle of a swidden field. Long periods of fallow meant that land was available for community use when the family with primary use rights was not cultivating it. The use of fallow land by other people is culturally elaborated as a valuable service in Govuton, since the secondary forest is kept back, and the owner is spared the hard labor of reopening the land in the future when he or she returns to it. This image of fallow land varies from other indigenous discourses about the function of the fallow period (cf. Dove 1983b). Rather than emphasizing the important ecological functions of regeneration for the swidden-fallow cycle, villagers in Govuton placed more emphasis on the value of keeping regeneration in check and agricultural lands open.

These community rights of access to fallowed village land are not well-articulated rights in present-day social life, and they are currently being subsumed as a result of two converging trends. First, as some villagers turn to wage labor instead of farming, their family's land lies in fallow for extended periods of time. Second, an increasing emphasis on intensive, commercial vegetable agriculture means that some land is cultivated on a permanent basis, in contrast to the cyclic nature of swidden. Therefore land planted in vegetables is removed from the swidden-fallow cycle and is no longer available to kin and other villagers in need. Thus, while there may be more land in fallow than in the past, once a farmer decides to use fallow land for permanent agriculture, he or she is effectively privatizing and enclosing land, making it inaccessible to others for an indefinite period of time. When a farmer uses previously fallowed land for permanent agriculture, he is changing the terms by which fallow land was previously accessible to the community. In other words, the shift from a reliance on swidden agriculture to intensive, permanent agriculture has rendered the property relations surrounding the community use of fallow land problematic.

In other communities in Indonesian Borneo, research has suggested that notions of community and identity can come together in powerful ways to support an "ethic of access" that tempers the inequitable accumulation of resources that can arise as a result of commercial intensification (Peluso 1996). In Indonesian Borneo, this ethic of access is most prominent with crops that have social and ritual meaning, such as the long-living durian trees, which have complex intergenerational inheritance patterns and imbue the landscape with a social history of the community. Thus kinship ties and community investment in the landscape link kin and villagers in shared ownership of the commercially valuable durian trees.[21] Conversely, Peluso shows that crops like rubber have different social meaning, and neither individuals nor communities form emotional attachments to rubber crops. Thus land cultivated with rubber lacks an ethic of access and as a result is considered private, individual property.

A similar, although less clear, trend can be seen in Govuton. The notion of community access to fallowed land can be seen as an ethic of access. Kinship and community ties and reciprocal relationships of exchange dictate that kin and neighbors in need should have access to fallow land. But presently, the more powerful emphasis lies on the commercial cultivation of vegetables, overriding the ethic of access. In areas in which people are cultivating plots of land for the commercial production of vegetables, there is an absence of locally agreed-on rules about how much land a single family can cultivate on a permanent basis. Wealthier families hire laborers to help them keep as much of their lands under cultivation as possible, reducing the amount of land that previously would have remained fallow. Also, sharecropping arrangements are arising, in which a family still allows neighbors or kin to use their land, but the primary owner demands a portion of the harvest or of the profit (usually one third) in return for the use of the land. Thus community rights of access to fallow lands, or a local ethic of access, is overridden by the growing commercial production of crops in which intensive cultivation is gradually replacing extensive cultivation.

Individual ambitions of wealth and resource accumulations are being emphasized over kin and neighbors' needs, and the changing property relations are disrupting other social institutions. Various village members depict this trend differently. Some villagers draw on national discourses of development and modernity to validate their hard work and initiative to participate in an economically based development of the land and resources in the village. One of the village native chiefs, for instance, believes that it is important for the village to do something more with their community forest reserves than "just letting trees grow." He has developed a master plan that would divide the village into different zones for vegetable cultivation, animal husbandry, a restaurant, and a hostel for tourists visiting Kinabalu Park. In discussion with political leaders in his region he has heard that there is interest from Japanese companies to establish tea, apple, mulberry, and mushroom agricultural schemes in Govuton. He believes that these developments would be good for the economy of the village. And in the process of rezoning the village he hopes that individuals could get more secure title to their land, putting them on par with other villages in the region where people have individual title to land.[22]

Other villagers draw on local discourses that validate indigenous technical knowledge, innovation, and farming skills allowing one farmer to be more successful than others. Gupak, for instance, is known by everyone in the village as the most successful commercial vegetable agriculturalist; he is also one of the older farmers in the village. Always eager to innovate in his garden, he experimented with planting coffee in the 1940s but found that it was too labor intensive. In 1946 he was the first person in Govuton to plant commercial vegetable crops, drawing on the resources of the Agricultural Department. He says he now makes a profit of about M$6,000 a year from his vegetables alone. With an understanding that farming is a long-term endeavor, he refused an offer of M$60,000 from an entrepreneur for his land. "I can make more than that from the land in ten years," he told me.[23]

Finally, other villagers see the trend toward resource accumulation as a disintegration of cultural values, which emphasize village equability in access to resources over individual accumulation. For instance, an anonymous letter was sent to the district officer complaining about one man's accumulation of land within Govuton. The letter, dated June 1993 and translated from Malay, says:

> The people of [Govuton] say with sadness and ask for sympathy from the District Officer since we have suffered abuse from Intang. He thinks he is rich and therefore can do whatever he wants. Those who have suffered under Intang need to work together to solve the problem that they are facing. If this issue is not defended by us we believe that our village will become the property of one person. . . . Nearly everyone, from the young children to old people, is involved in Intang's work as he buys our land either through sweet talk or force. . . . We are not more loved than dogs by Intang who extinguishes us as he wishes.

This letter gives us a clear image of the depth of local antipathy toward Intang, who has consolidated ownership over a large amount of village land for the extensive commercial cultivation of vegetables. Below I describe some of the mechanisms Intang has used to consolidate ownership of property.

New cultivation systems necessitate changes in village property relations

(Appell 1985, 1988). Since there have been few formal discussions about the distribution of rights under this trend toward commercial and intensive agriculture, the same plot of land can be claimed by different parties, each relying on various strategic discourses to substantiate their claims. For instance, a person may claim the right to cultivate fallow land through notions of communal rights of access based in values such as kinship and need. Yet with the new emphasis on commercial vegetable production, another person may claim the right to exclude community access based on his or her primary rights of access derived from inheritance principles when the use rights can be passed down through generations. Conflicting ways of legitimizing claims on land become particularly problematic when a person begins to cultivate permanent crops on land previously left fallow by another person, thereby indefinitely removing the land from community access.

Many people in Govuton do not anticipate a future as agriculturists. They are counting on profits from intensive vegetable production to educate their children and to launch them in nonagrarian careers. Enclosing land in the native reserve for permanent, commercial vegetable agriculture, effectively removing it from the swidden fallow cycle and from community access, is one of the ways through which profits can be realized. But as the following land dispute demonstrates, the ability to appropriate fallow village land for individual, permanent use depends, at least in part, on integration into the broader legal, administrative, and economic systems outside Govuton.

Intang, the man mentioned in the letter quoted above, was one of the wealthier and more influential men in Govuton in 1996. His power and wealth stemmed from his work of many years as a driver for the district officer. During this time he formed social networks with many state officials in the district office. These links to state agents and institutions allowed Intang to act as a broker of state-derived power at the local level. For instance, Intang had unusual success in getting native title to land that he owned in villages outside Govuton. Many people in Govuton owned land in neighboring villages and have filed applications for native title for these lands. Yet most of them have been waiting for over twenty-five years for their titles from the state. Without connections to individuals in the Lands and Surveys Department, connections that Intang had, applications for native title take decades to be processed. According to Intang, he got his titles quickly because he has close friends (*teman rapat*) in the department who fast-tracked his applications.

Title to land held outside Govuton has far-reaching implications. To get a bank loan, a person must provide a title to land as collateral. Thus Intang has been able to borrow money from the bank to launch several business enterprises, while most other people in Govuton are unable to do so. When people in Govuton need money, unable to turn to a bank, they turn to Intang. As a moneylender Intang provides a valuable service that is inaccessible to people without land title, yet at the same time he is often disliked and distrusted. Many people feel that he demands too high interest payments on his loans and that he takes advantage of less fortunate villagers. In short, Intang is overstepping culturally acceptable boundaries of exchange and reciprocity by making a profit by loaning money to villagers.

Intang's accumulation of wealth *outside* Govuton has facilitated his accumulation of land *within* the native reserve. He owns a large, permanent vegetable garden on village land in the native reserve. He sells the products from this garden in Kota Belud, a market center. This piece of land has several qualities that make it unusually valuable for the commercial production of vegetables. It is located near the road and near a source of water for irrigation. Furthermore, it is one of the flattest pieces of land in the hilly region, and therefore it is less work to cultivate. Intang bought the use rights to this particular plot of land from another villager, Momin, who claimed that although he had not cultivated the land for years, he inherited the land from his father and thus could sell the use rights. To gain access to this land, Intang thus strategically used the discourse of inheritance as the legitimate locus of ownership, buying the primary use rights from Momin, who had inherited them from his father. Once these rights were secured, Intang maintained his ownership of the land through permanent cultivation. By working the land, Intang's actions further legitimated his claims, since now the point of reference for these claims is not only inheritance but also a labor theory of ownership. But by planting permanent crops, Intang has effectively removed the land from community access, creating local tensions.

Other residents in Govuton questioned Momin's right to sell the land to Intang. Momin in fact no longer lived in Govuton. Villagers argue that since his father had essentially abandoned his land, leaving it uncultivated for decades, it had become available for communal access. Indeed, another family, headed by Kemburong, had built a house on a small portion of this land. When Intang bought the use rights to the land from Momin, he evicted Kemburong from the house she had lived in for more than ten years, claiming that he had legitimately bought the rights to the land.

Kemburong felt she had the primary rights, at least to the area of the land surrounding her house. She used the discourse of the rights of community access to abandoned land within the native reserve, based on need, to legitimate her claim. Intang felt that he had the primary rights to the land since he had purchased these rights from the person who claimed to have inherited the primary use rights. The dispute went to the Native Court at the district level. It was ruled that by abandoning the land and moving out of the village, Momin's father (and Momin himself) had relinquished their rights to the land. Thus the broader community had access to the land, and Kemburong the right to build a house there. The Native Court reasoned that abandoned village land had reverted to community land and that Momin no longer retained the primary use rights and could therefore not sell the rights to the land. The Native Court supported the customary law that allowed villagers in need to claim use rights to abandoned land in the native reserve.

Despite this ruling in the native court, Kemburong abandoned her house and Intang enclosed the land with a fence and established vegetable cultivation on the land. Local gossip suggested that Intang had offered Kemburong money to move off the valuable piece of land after the court ruling. Kemburong herself denied this, simply stating that she "felt bad" (*merasa sangat tidak senang*) about the land after the court battle and wanted to move.[24] This case demonstrates how cultural notions of access to property are adapted to meet new conditions and in turn structure daily land-use practices (see Li 1996: 509). By cultivating the land with permanent crops, by installing an irrigation system and by building a fence, Intang's land use strengthened his future claims to the land. Yet Kemburong's dilapidated house on one corner of Intang's garden provides a constant reminder that Intang acquired the land through contest and negotiation. That Intang should have failed in his appropriation of the land, since the Native Court ruled against him, complicates this analysis, but it also highlights the notion that Intang's externally derived wealth and power provided him with leverage in local disputes. Intang's power allowed him not only to override the court ruling through his actions but also to potentially alter the future trajectory of changes in property relations in the native reserve.

This dispute shows that land rights provided by rules and traditions are not guaranteed, but instead serve as points of leverage in ongoing negotiations over land rights (Li 1996). While the court attempted to enforce the rules associated with customary law, neither it nor other villagers were able to compel Intang to leave the land.[25] Instead, Intang successfully wielded his power and altered the ways in which customary rights of access to property are negotiated through his actions. Furthermore, villagers did not resort to social ostracism, since many depended on loans and work from Intang. Thus negotiations over land rights include references to power and wealth that are not necessarily encompassed in the rules and traditions of customary law but are still powerful enough to change rules of access. In this particular dispute, Intang's individual ambition to provide his children with finances to pursue livelihoods outside Govuton subsumed notions of community access to fallow land. At the moment it is too early to tell if this trend will continue. Perhaps other villagers will try to follow Intang's initiative and secure permanent rights to fallow land in the native reserve through the practice of permanent vegetable production.

It is instructive to focus our attention on land matters in the broader, regional context to understand the trend in Govuton toward the individual accumulation of land, a trend that is overriding older practices of community access to village land. Govuton's native reserve stands in marked contrast to areas governed by other forms of native property rights in contemporary Sabah. Many native claims to lands elsewhere in Sabah were recognized in colonial statutory law through the use of native titles. Today, farmers who have native titles may choose to sell or rent their land without much interference from the government (unlike the more restrictive period before 1950). They can also secure bank loans to buy additional lands or to improve current ones, using their native titles as collateral. Conversely, natives in Govuton's native reserve are restricted by the colonial static rendering of the once dynamic native customary laws that limit rights to land in the native reserve to use rights only. People in Govuton feel hampered by this; in today's political economy, where access to cash is equally important as access to land, the inability to sell or lease land or to secure bank loans is seen as a significant impediment to individual rights and economic mobility. For people to advance in contemporary Sabah, they need money for schools, transportation, medicine, and other material goods, expenses that were less important in the 1950s when access to good agricultural land and forest resources were the primary factors determining a family's quality of life.

Many people in Govuton express frustrations with the local property rights that govern the native reserve. Their concerns coalesce around the fact that usufruct rights to land in the native reserve are too limiting, particularly when compared with rights that indigenous people experience elsewhere in Sabah. For instance, some poor villagers complain that if they need money, they can only get a small sum from community members for the sale of their use rights to land; land without a defensible title has very little commercial value. If villagers in Govuton wish to sell their use rights to a plot of land they can usually get roughly M$500.[26] A similar plot of land held under native title outside the native reserve is worth between M$2,500 and M$3,000 because the title has value as collateral and is transferable without restrictions.

Other villagers complained that cultivating land in the native reserve is more work and more costly than cultivating land owned under native title, because goats, pigs, cows, and water buffalo all graze freely in the native reserve; to keep livestock out of their gardens farmers must build fences. Under native title the situation is reversed; people who own livestock must keep them tethered or fenced on their own land. Individual rights take precedence over community rights outside the native reserve, whereas community rights take precedence over individual rights inside it.

In this context, the actions and discursive strategies of people like Intang make sense—these farmers are trying to compete with farmers outside the native reserve who have private property. The formation of the native reserve produced a static colonial interpretation of native life and native property relations during a particular political and economic period when village ownership of the land, reciprocity, and barter were more valuable than individual title to land. While leaders in Govuton in the 1950s saw the native reserve as a way to assert local autonomy and protect traditional lands, today it provides villagers with less autonomy over land matters than owners of native land titles have. The regional political-economic context has changed, making the native reserve less appealing to some villagers, particularly those who desire to become more active in the commercialization of vegetable crops.

Conversely, other villagers find protection in and benefit from the native reserve, particularly when access to the *hutan simpan* (forest reserve) is considered. Table 2 represents some of the ways in which villagers utilize the forest reserve on a daily basis.[27] What is most interesting is that there does not appear to be any direct relationship between wage income and the use of the forest. For instance, Lumbow's wage is similar to those of Selimboi, Gitom, and Lehimboi (roughly M$1,200 per month), yet his family relies on forest resources much less than the other families do.

In a three-month period, Lumbow's household only spent only 4 percent of their days collecting forest resources. In contrast, Selimboi's, Gitom's, and Lehimboi's households each spent between 30 percent and 36 percent doing so. Cultural values may be as important as subsistence needs in determining the use of the forest reserve. For instance, according to Selimboi, "Sometimes you just prefer the taste of wild *lemiding* (forest ferns) over eggplant." And during special festival times, the older villagers enjoy roasted songbirds as an accompaniment to their *tapai* (rice wine). For these villagers, access to the forest reserve means that they can enjoy wild foods that they might not otherwise have access to. For other villagers, such as Gamid, whose average monthly wage is only M$201,[28] the forest reserve provides vital subsistence foods. He and his family rely heavily on access to the forest reserve to survive. In a three-month period, members of Gamid's household spent

TABLE 2 *Household Income and Dependence on Garden and Forest Resources in Govuton*

Name of head of household	Household size	Average monthly income	Percentage of income from sale of non-timber forest products and produce from garden	Percentage of days spent gathering non-timber forest products over three-month period	Types of resources collected
Lehimboi	9	$1,392	None	35%	Firewood, fish, squirrels, wild fruits, wild vegetables, fodder for pigs
Gitom	14	$1,164*	Garden, 14% Forest, none	34%	Firewood, fish, frogs, monkeys, birds, snakes, armadillos, wild fruits, wild vegetables, fodder for pigs
Lumbow	18	$1,078*	Garden, 75% Forest, none	4%	Fish, wild vegetables
Selimboi	15	$1,062*	Garden, 6% Forest, none	30%	Firewood, fish, wild fruits, wild vegetables, fodder for pigs
Siking	4	$1,025*	Garden, 30% Forest, none	10%	Firewood, fish, wild vegetables
Masin	10	$878*	Garden, 6% Forest, none	41%	Firewood, fish, wild fruits, wild vegetables, fodder for pigs
Gamid	5	$201*	Garden, 69% Forest, none	89%	Firewood, fish, wild fruits, wild vegetables, fodder for pigs

* This figure combines both wage labor and income from the sale of garden products in markets.

89 percent of their days in the forest gathering resources, a particularly high rate. They sold none of the resources, but rather used them for their own consumption. Thus villagers' opinions of the property rights and access to resources in the native reserve vary depending on both their material and cultural needs.

Many factors influence the transition from community rights to the individualization of rights. Some villagers, such as Intang, respond to material and political incentives to increase the trend toward the individualization of rights to land. Others express both material and cultural reasons for maintaining community access to land and natural resources. It is evident that the colonial codification of native customary law does not adequately capture the complexities of the contemporary resource use regimes in Govuton and that the village is becoming increasingly stratified internally, making it difficult for villagers to reach consensus regarding rights of access to resources. In the face of continuous mobile social and economic relations, the future of rights to land and resources in Govuton is by no means predictable. It is possible that the current trend toward the individualization of land will continue as certain individuals gain power and wealth, increasing village stratification. It is also possible that the villagers will once again mobilize to change the current statutory laws and reach a new agreement regarding the particularities of community ownership of the land and the natural resources in the native reserve. If this occurs, village stratification would be slowed and individuals seeking substantial profits from the use of natural resources would have to look outside the native reserve. What is certain is that the transformation of property relations will continue to reflect local and regional struggles over power, wealth, cultural meaning, and community identity. And these struggles will reveal the fractures and ambiguities experienced in a community that is economically stratified and diverse in its desire for change (see Ortner 1995).

The preceding case studies and analysis of property relations in Govuton from the 1950s to the late 1990s illustrate several points relating to the nature of both state-society relations and property relations. First, while states may in general attempt to systematize and organize the world according to principles that simplify local property relations (Scott 1998), specific local responses may pose obstacles or present alternatives that allow for creative and new forms of state control and local autonomy (Agrawal 1999; Agrawal and Gibson 1999). These interactions between state and society take place from positions of power that are asymmetrical. Thus while villagers in Govuton initially employed various techniques to resist state appropriation of village land in the 1950s, these actions took place within the new legal structures of colonial rule. In the end, a process of negotiation allowed for local control over community land, yet the realm of negotiation was restricted by the parameters of statutory law. State sanctioning of village community land as a native reserve stands out as an unusual practice in colonial law in Sabah. Elsewhere colonial administrators knowingly overlooked complex patterns of local property relations in their efforts to quickly settle native claims to land. In the case of Govuton, local leaders who refused to allow their communally owned village lands to be appropriated by the state influenced colonial administrators.

This leads to a second, parallel conclusion: any state intervention in local property relations, no matter how partial or incomplete, has the effect of consolidating state power over society. The native reserve emerged as an innovative form of state rule, created in response to local negotiation with state agents over property claims. In the case of the native reserve in Govuton, the state seems to have retreated in its ambitions of providing each person with individual title to his or her land. Yet at the same time, the colonial state asserted control over native lands by providing Govuton with a community-held title that was based on colonial imaginings of how native customary laws should be codified in a singular, static way, not in the plural, multiple, and diverse ways known for customary institutions. Colonial intervention in land matters was redefined by placing the emphasis on the paternalistic protection over natives rather than on economic growth. Therefore the native reserve also had the effect of intensifying state control, since recognition of native customary rights to land in statutory law was contingent on government sanction. While local people were able to alter the direction of state control over people and land, ultimately the broader trend toward rural incorporation into the state's project of land settlement continued.

Third, while colonial control was redefined through the formation of the native reserve, so were local property relations transformed. In the past, an ethic of access contributed to the notions of community access to fallow land in the native reserve. However, this ethic is increasingly eroded by individualized ownership of community land through the practice of permanent cultivation of commercial vegetables. The gradual enclosure of community land is motivated, in part, by farmers who are striving to make sufficient profits from agriculture so that their children can be educated and find employment outside the agricultural sector. But these individual goals are not compatible with an ethic of community access to fallow land. As a result, contests over access to land in Govuton are articulated in terms of competing representations of community and divergent images of rights of access. Notions of an ethic of access and rights to fallow land in the native reserve are placed in opposition to the principles associated with inheritance and the rights to transfer or sell inherited land. Thus the traditions associated with customary land rights in Govuton do not guarantee access to land, but instead serve as points of leverage in ongoing negotiations over access (see Li 1996). Furthermore, cultural notions of access to property are adapted to meet the growing market economy and in turn structure daily land-use practices.[29] Individual access to land is taking precedence over community access and is transforming property relations in the native reserve. Moreover, in struggles over access to land, negotiating points are not only based on internal village social relations but also rely on wealth and power that are based on external relationships. The most powerful people in Govuton, such as Intang, rely on power derived from outside the community in their contests over access to land at the local level (Scott 1976).

Fourth, by focusing on the temporal variations in local perceptions of the native reserve, this case demonstrates that legal definitions of customary land tenure may be empowering in one time and place and disempowering in another. The formation of the native reserve served to counter the colonial appropriation of local lands in the late 1950s, thereby strengthening local claims of access to traditional lands. Moreover, this move allowed the community of Govuton to claim collectively more land than they would have been able to claim individually. Under colonial land laws, natives could only claim access to land that they could prove they had cultivated continuously for at least three years. But by drawing on the colonial concept of a native reserve, the village of Govuton was able to secure community access to land that the village used not only for cultivation but also for hunting and the gathering of forest resources. Yet some villagers in Govuton feel that the native reserve is disempowering in today's political economy. Unable to gain access to credit or to sell their land for a profit, they express their frustrations at not having a permanent and transferable title to the land they use. For these villagers, the ability to be socially mobile in a growing market economy is limited by their inability to sell land. Other villagers, however, rely on the native reserve for access to forest resources for subsistence needs and cultural values and feel fewer pressures to convert their land to cash. Individual ownership of all the land in the native reserve would make these uses of the forest resources problematic. Villagers therefore continue to try various strategies to strengthen claims to land and to secure continued access to forest resources.

The case study of the native reserve in Govuton demonstrates how local leaders were able to use the colonial legal system to protect a large area of native traditional lands from alienation to outsiders. In the 1950s and 1960s local leaders were able to co-opt the colonial organizing symbols and legal practices to their advantage. Yet social relations and property relations are always in flux. While the formation of the native reserve may have frozen a particular vision of native customary law in the colonial legal system, it did not end local-level conflicts over cultural identity and power or over the ways in which these conflicts are used as negotiating points in struggles over access to land. The transformation of property relations in Govuton continues today, as these struggles are played out in daily practices that effectively overturn the statutory laws regarding the native reserve and that reflect other struggles in the social and historical contexts in which they are embedded (Peluso 1998).

Finally, these conclusions have significance for advocates of reinvigorating native customary law and community control over natural resources. Conservation groups in Sabah have argued that the legal entity of the native reserve may be able to promote both community control over natural resources and conservation (J. Payne, personal communication, December 5, 1995). Yet we have seen here that the notion of a native reserve relies on a static rendition of native customary law and on images of villagers in Sabah as homogenous and timeless in their cultural identity and their natural resource-management regimes (see Agrawal 1999; Agrawal and Gibson 1999; Brosius, Tsing, and Zerner 1998). This constitution of community and native customary law does not adequately represent local concerns and has even become the source of contemporary conflicts over natural resource use. As the case study of Govuton has illustrated, individuals in a community can have drastically varying notions about appropriate resource use, and one individual's connection to external forms of power and authority can be used to suppress alternative discourses and practices regarding resource use and management. External actors or institutions that attempt to form community-based conservation projects, or in any way alter a community's resource-use patterns, must seek to understand the nuances of community identity, the potential conflicts over resource use between various sectors of the community, which local voices are heard and which ones are suppressed, and the history of local changes in property regimes.

In closing, I want to return to one of the main points raised at the beginning of this analysis, namely, the replacement of a dynamic system of native customary law with a static system of colonial governance. While most agree that natural resource–management regimes need to be dynamic, adapting in response to environmental changes as well as changes in society and the political economy, it is rare to find equally dynamic governance regimes that can provide communities with the institutional support they need for local natural resource management without asserting external control and dominance that would undermine community self-determination and autonomy.

Notes

The research on which this chapter is based was made possible by the financial support of a Fulbright-Hays Doctoral Dissertation Research Abroad Award; a National Science Foundation, Law and Social Science Program, Dissertation Improvement Award (#SBR 9511470); and a Social Science Research Council/American Council of Learned Societies, Southeast Asia Program, International Doctoral Dissertation Award.

I thank many readers who helped me refine this analysis. They are Arun Agrawal, George and Laura Appell, Peter Brosius, Michael Doolittle, Michael Dove, Emily Harwell, Tania Li, Celia Lowe, Nancy Peluso, Hugh Raffles, James Scott, and Janet Sturgeon. The maps were made by Heather Salome at MetaGlyfix. An earlier version of this chapter appeared in Doolittle 2001.

1. For similar findings in the African context, see S. Moore 1986b. And see Brosius 1999a for a discussion of local people's strategic use of external labels and essentialism to assert local power.
2. See Schroeder 2005 for similar findings in the Gambia.
3. It is necessary to make a note on the term *colonial*. Sabah (known as North Borneo until independence in 1963) was ruled by the North Borneo Chartered Company from 1881 to 1941. After the Second World War, from 1946 until independence in 1963, North Borneo was a crown colony of England. This article begins with the late 1950s, the period during which North Borneo was a crown colony. These two periods of British rule, which I collectively refer to as colonial rule, differed in subtle ways. Nevertheless, many of the administrative policies and all the land laws instituted by the North Borneo Chartered Company remained in place under crown rule. In fact, many of the British colonial officers who came to North Borneo to serve under the North Borneo Chartered Company remained when North Borneo became a crown colony. For the most part, the laws and the men who implemented them remained constant between the two eras. Other authors have noted that very few changes occurred in Sabah during the fifteen years under crown rule (see, e.g., Ross-Larson, 1976; Ongkili 1972; and Tregonning 1965). Thus, for the purposes of this analysis, I focus on similarities rather than variations between company and crown rule.
4. Native title was the formal mechanism through which natives could get recognition of some of the land that they held under native customary rights in the colonial land laws. Natives could claim rights to the following types of land: (1) land under cultivation or land being used for housing; (2) land planted with fruit trees at the density of twenty or more per acre; (3) isolated fruit trees if enclosed by a fence; (4) grazing land stocked with animals; (5) wet and dry paddy land so long as it was cultivated for at least three years prior to registration;

(6) burial grounds; and (7) rights of way. While the land laws were relatively inclusive, for the most part natives were successful at gaining native title only when they could prove that the lands in question were used for permanent cultivation (Doolittle 1999).

5. The institution of the native chief was established under colonial rule, in which certain local leaders were given state recognition. Native chiefs were expected to serve as local representatives of the colonial government, collecting taxes and recording land transfers. They were also responsible for settling native disputes according to the local customary laws (see Kurus 1994 and Phelan 1988).

6. To protect the anonymity of the village I worked in and that of the people who live there I have chosen not to reveal the location of village-level archives.

7. While plans to proclaim the native reserve began in the early 1960s, for unknown reasons the area was not officially proclaimed by the chief minister until November 3, 1983.

8. The persistence of this concern between the eras of company rule and crown rule demonstrates that few if any changes occurred in how British rulers treated natives and their land rights between the two periods.

9. *Wight* is an old English word meaning "person . . . thing, creature of unknown origin" (*Oxford Encyclopedic English Dictionary* [Oxford: Clarendon, 1991]).

10. Harper (1997: 9) also comments on colonial polices of "official paternalism" in peninsular Malaya that were aimed at protecting the *orang asli* (literally, the original people) from "unseemly commercialization of Orang Asli life."

11. Memorandum by A. B. C. Francis, Colonial Office 874/985, 4; emphasis mine.

12. For more details on how the state management of local property rights can be seen as strategies aimed at intensifying state control, see Li 2005.

13. For documentation of similar findings elsewhere in Southeast Asia, see Peluso and Vandergeest 2001.

14. For a seminal study that focuses on the networks formed between state agents and local people, and the influences that these networks have on the transformation of property rights, see Berry 1993.

15. Govuton is a large village and is divided into several hamlets, since each hamlet has its own headman. Therefore, as a village, Govuton has many headmen.

16. "Declaration of the Kampung 'Govuton' Native Residential Reserve Ranau District," on file in the District Office in Ranau in the Land Office (n.d.).

17. Ibid.

18. Here I am building on the insights of Li, who argues that divergent and competing images of community become "culturally available points of leverage in ongoing processes of negotiation" over access to land (Li 1996: 509).

19. For more on "an indigenous labor theory of value," see Li (1996: 511–12). For other empirical literature on relationships among individuals, communities, and the formation of property relations in Borneo, see Peluso 1996; Appell, 1991, 1986; Dove 1983a, 1985b; D. Freeman 1970; and Tsing 1993.

20. For more on the "moral economy" of peasants and the "subsistence ethic," see E. Thompson 1975 and Scott 1976.

21. This relationship between kin and shared ownership of durian trees is so strong that according to one informant in Peluso's (1996: 538) study, "Selling your durian tree is like selling your own grandfather."

22. Field notes of interview with A. on June 23, 1996.

23. Field notes of interview with G. on January 17, 1996.

24. It is not an unusual cultural practice to move one's home after a tragic incident, although usually such incidents are associated with illness and/or death.

25. A common complaint about the Native Court in Sabah is that it lacks the power to enforce its decisions. The police are responsible for upholding civil law, not native law, and therefore there is no authoritative body to ensure compliance with Native Court decisions. Ultimately, the native chiefs and their courts are relatively powerless to resolve village matters.

26. At the time of research US$1.00 was equivalent to about M$2.50.

27. The data in this table were collected over a three-month period in which seven households in the hamlet (30 percent of the population) kept journals in which they recorded their daily wages from labor, income from the sale of forest products, income from the sale of cultivated products, and their consumption of forest products and cultivated products. The three-month period bridged the wet and dry seasons and occurred before the rice was planted. During my eleven months in the field, I observed more collection of forest resources during the dry season. In particular, firewood was collected heavily during this time to be set aside for the wet months. But the collection of vegetables and hunting for animals was consistent throughout the year. Most families have one or more persons who go into the forest several times a week. The sample size and time frame used to gather the data for table 2 provides only a snapshot of resources use, and therefore it is difficult to form definitive conclusions. Nevertheless, I believe the results do suggest certain trends in resource dependency, which is confirmed by the ethnographic data from interviews and by participant observation. To develop a more nuanced description of the variations in resource use and dependency it would be necessary to collect this type of data for the whole year and with a larger sample size.

28. By definition of the Malaysian government, the poverty line is the level of income necessary to obtain basic needs or the minimum necessities of life, including both food and nonfood items; a household is considered "hard-core poor" if its income is less than the food component of the poverty line. According to the permanent secretary of the Department of Rural Development, families of five people with an income below M$600 per month are living below the poverty line, and those families earning less than M$300 per month are hard-core poor (notes from interview July 31, 1996).

29. These conclusions support Li's (1996) findings regarding property relations in Sulawesi.

THE SOCIAL LIFE OF BOUNDARIES
Competing Territorial Claims and Conservation
Planning in the Danau Sentarum Wildlife Reserve,
West Kalimantan, Indonesia

It is now common for conservation planners to recommend that
resource tenure be clarified to rationalize resource management
and avoid conflict. The ubiquity of this policy prescription reflects
a convergence of interests in the importance of resource claims to
the management and physical condition of resources, as well as the
recognition that there are invariably multiple narratives of who holds
those resource rights. As other chapters in this volume attest, conservationists have for some time recognized that biological resources are
imbedded in social worlds, a reality that necessitates an expansion of
management planning beyond ecological boundaries to consider the
political and social contexts of local resource and territorial claims to
the habitats and species of conservation importance. Yet while the
contested nature of resource rights has emerged on the policy radar
screen, unqualified recommendations to simply "clarify tenure" suggest that the dynamic and social-political nature of boundaries remains underappreciated.

The aim of demarcating firm, stable boundaries reflects a notion
of property as a relation between owners and possessions that over-
looks the more salient relationships between people (Hohfeld 1913).
Other writers in this volume have demonstrated how power imbues
narratives of resource management. Here, too, the strength of the
narrative of "property as a thing" to be abstractly mapped in Cartesian space is no accident. In Carol Rose's (1994) analysis of the narratives underpinning different discourses of ownership, she argues
that the perception of property as a thing has proven to be so persistent precisely because it facilitates the prevailing capitalist trend
of the commodification, alienation, and consolidation of individual

property. However, she further argues that there can be no property rights over things without the recognition by nonrightsholders of the legitimacy of others' rights and their (sometimes coerced) agreement to the social obligation to respect their exclusivity. Rights therefore are more usefully viewed as inherently fused to a social contract between rightsholders and nonrightsholders. But what makes nonrightsholders either recognize or resist these boundaries? Who can enforce their claims and who cannot, and under what circumstances? Rights and obligations are forged and enforced in a political and social environment. The main argument of this essay is that boundaries are distilled representations of this environment, making them containers not only of geographic space and resources but also of identity, meaning, and power (P. Peters 1994; D. Moore 1998). As such, boundaries are by their very nature both multiple and fluid and therefore inherently resistant to uncomplicated clarification.

This essay analyzes the social environment around resource claims in the Danau Sentarum Wildlife Reserve (DSWR) in West Kalimantan, Indonesia (see map 1). I investigate the changes in meaning ascribed to the forest and fish resources of the DSWR region and the ways in which these meanings are influenced by history, markets, and relations with the state and neighbors. In turn, these meanings and the ideas of territory—the geographical spaces within which authority is exerted over people and resources—that they engender affect both the social and physical landscapes.

Following a site description, the first part of the essay describes how livelihoods and the identity of local resource users play leading roles in defining resource boundaries and coloring the interaction between different communities, traders, conservationists, and state officials. In addition, different forms of mobility and social networking are of central importance to structuring how rights and management are viewed by different local users. The rise and fall of markets for different products is closely associated with local perceptions of resource value and an interest in controlling that value through territories. Access to these changing markets depends on social networks, which are linked to ethnic, religious, political, and kinship identities. Yet there has been little attention to how various ideas of territory and livelihood articulate with different ethnic identities, histories, and relations with the state (Miles 1976; Hefner 1985). Seeking livelihood is never simply an economic endeavor but one that is deeply imbued with cultural and social politics, of which history plays a central role (Gudeman 1986; Shipton 1989; Harwell 2000).

Elsewhere in this volume authors have traced how national and global political economies, as well as finer-scale village and household histories

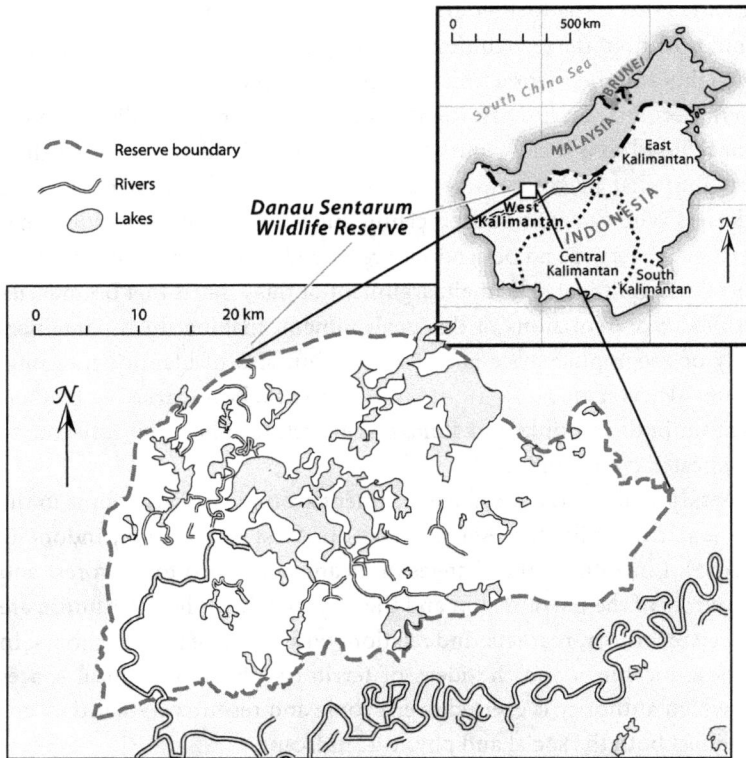

MAP 1 Location of Danau Sentarum Wildlife Reserve in West Kalimantan, Indonesia

have played out in local patterns of management and the physical attributes of resources. The second part of my essay positions local notions of identity, livelihood, and entitlement within particular histories that themselves serve as an important resource for those staking claims in the present. Having ancestors who farmed, fished, planted trees, or collected forest products in particular locations bestows complex inheritable rights on people to those resources, making it important for claimants to know their genealogy as well as the history of their ancestors' residence and management activity.

Further, as demonstrated by authors in this volume, the role of local communities' relations with the state and how conservation projects play into these relations variously for different people has often been underappreciated in conservation planning. As I detail below, many individuals demand access to resources in another group's claimed territory based on their own descent from a particular pioneer, ethnic group, or sultanate, and more recently, as citizens of sovereign nations. Likewise, states of various stripes and international communities also claim rights to territory and the resources contained therein based on history and legal instruments. Although informal local claims have a clear effect on management outcomes, states often ignore those claims and instead offer leases or titles to other parties based on their self-declared sovereignty over all resources within national boundaries. The military violence (or threats of it) with which states often enforce their version of boundaries plays an important role in establishing authority (Peluso 1993; Stonich and Vandergeest 2001). Indeed, as Vandergeest and Peluso (1995) detail for Thailand, this territorial sovereignty has been a characteristic strategy of states that consolidate power by establishing territorial authority over land, subjects, and resources. These assumptions of state ownership were the essential precursors to the establishment of legal codes ignoring local rights, criminalizing local practice, and bureaucratizing relations between people in the forest as well as between people and the forest. Drawing on Sack's (1986) and Soja's (1971, 1989) work on the sociology of space and territoriality, Thongchai's (1994) seminal work on the state and spatial boundaries (see also Anderson 1983), and the exploding field of inquiry into the politics of mapping (Harley 1989; Pickles 1995; Curry 1995), Peluso and Vandergeest (2001: 764) have examined the characteristic state strategy of consolidating power through the territorial and bureaucratic control of people and resources within national borders. "Sovereignty and discipline," they observe, "go hand in hand with government."

Nevertheless, this essay emphasizes that the power of state territoriality tactics is not absolute, as the DSWR residents' everyday spatial practices make territorial claims that resist state authority. Contradictory claims continue to be staked in spite of state efforts to insist on impermeable sovereignty. The third section of my essay argues that local territorial claims are in constant tension with other competing claims of neighbors and the state, underscoring the power dynamics involved in the recognition or disregard of other's boundaries.

In recent years there has been great interest in the evolution of particular visions of landscape (P. Peters 1994; Peet and Watts 1996; D. Moore 1998). Of special relevance to studies in this volume are analyses of the politics behind how "protected area" space becomes delineated and how authority is exerted over both people and resources within those spaces (Neumann 1992, 1997; Peluso 1993; Vandergeest and Peluso 1995; Brosius, Tsing, and Zerner 2005; Agrawal 2005; West 2006), as well as the convergence of environmental politics, territory, and local identity (Li 2000; Peluso and Harwell 2001). The present essay contributes to these debates by providing a watershed view of how different local residents resist or align themselves with state tactics of territoriality to their own benefit. The essay further illustrates how waxing and waning tides of state influence and changing economic markets, as well as efforts by a variety of local groups to stake claims, produce layers of territory and an interpretation of boundaries that will necessarily interface and interfere with conservationists' own attempts to produce clear boundaries and territories.

The essay closes with a call for a new ecology that does not abandon the idea of formally recognizing local resource boundaries, but instead pays explicit attention to the complexity, situational nature, and power dynamics associated with their production.

Fluid Landscapes and Strategies for Dealing with Uncertainty

The DSWR area is ideal for an investigation into diverse forms of boundary making because it is important to a variety of groups for different reasons. The physical landscape of the place lends itself to dynamism in territorial claims due to its dramatic seasonal flux in water levels and shifting availability of resources. Populated with two main ethnic groups, the Malays and the Iban, local villages have long defined their resource territories, which enclose forests and fishing grounds, although with varying degrees of precision and flexibility. Village legal institutions (*hukum adat*)[1] effective within these defined and defended village boundaries regulate local management practices.[2] These boundaries and regulations have changed in response to dynamic relations with states, markets, and neighbors, as well as with the mobility of people and resources.

The Kapuas lakes, the largest of which is called Danau Sentarum, are part of an intricate network of seasonally flooded lakes and blackwater swamp forests in the heart of Borneo, about twenty kilometers south of the Sarawak border. The area acts as an overflow valve from the great Kapuas River and is marked by a stunning difference between the wet and dry seasons. Under high rainfall, smaller rivers flow northward from the Kapuas River into the lakes, eventually flooding all the surrounding lowland forest. However, when rainfall drops below three hundred millimeters per month at the source of the Kapuas, the decreased water in the great river causes the smaller streams to reverse direction, draining the expansive flooded forest and lakes back into the Kapuas.

The variation in water levels between wet and dry seasons may reach twelve meters or more—a striking alteration in the landscape that creates a shift in both settlement and livelihoods. In the wet season, the expansive lakes and flooded forest are quiet—during a boat trip traversing the area one might only see two or three people fishing with hand lines, hoping for a bite. Entire villages sit vacant, as seasonal residents have returned to their homes on the Kapuas River. Even the area's diverse wildlife species rarely show themselves. With few fish caught and little extra cash to buy supplementary food, for some families this can be a long, lean season.

The dry season completely transforms the place. Travel across the lakes area (if the dry lakebeds can even be called that) is a daylong journey, and some villages are only accessible by a blistering, shadeless walk across the cracked lakebed. The traveler, whose boat in the wet season shortcuts easily among the trees in the flooded forest, in the dry season must pick her or his way carefully through a shallow, meandering channel, avoiding fishing nets splashing with fish concentrated from the wide floodplain. Scores of fisher-people, including small children, busily pull up nets and traps heavy with fish (which, if left too long unattended, are feasted on by otters, macaques, birds, snakes, turtles, crocodiles, and monitor lizards). Floating houses and *bandung* (small houseboats), many sporting satellite dishes, congregate into temporary villages in the middle of the dry lake beds—complete with soccer and volleyball fields on what was recently submerged lake bottom. Once-floating shops sit unevenly, grounded on their exposed floats. Late into the night, entire families squat on makeshift docks, cleaning and salting their catch in big pottery jars. In the morning, the salted fish are spread on every available surface to dry in the sun. Downstream, traders' massive houseboats stand by on the Kapuas, collecting the catch and selling gear and long-awaited luxuries like packaged food, brightly colored clothes and toys, pop music cassettes, gold jewelry, and televisions.

The unique seasonal flux of the lakes creates both great opportunities and great risks for people dependent on local resources. The rapid drop in water levels (sometimes, as in 1994, reaching almost a meter per day) causes fish migrations through small rivers leading to the Kapuas or strands them in seasonal ponds—conditions ideal for fishing.[3] Furthermore, the drop of water levels exposes upriver levees as fertile fields for rice agriculture.[4] Yet this seasonal water drop does not always materialize. In the four years during which I conducted fieldwork (1994–97), in two consecutive years the water level did not drop significantly (1995 and 1996). In wet years, fish remain dispersed throughout flooded forests and floodplains submerged or too poorly drained to be properly burned before planting (to release nutrients from slashed secondary growth). Worse yet, while rice varieties are locally selected to withstand some flooding, a newly planted field may be abruptly flooded by rising water that drowns the young plants. Local people must be prepared to deal with a flux in climate and resource availability, which demands great flexibility and a diversity of subsistence strategies, strategies that tend to differ along ethnic lines.

Social Networks and Livelihood Strategies

To understand the differing territorial claims of Malays and Iban to fish and forest resources, it is necessary to consider how livelihoods in the two groups differ and how they have changed in response to new markets, social conditions, and relations to the state. Both Iban Dayaks and Malays survive seasonal fluxes and hardships through mobility and social networking—but these practices vary in form and have different consequences.[5] Of course, there also exists great heterogeneity within the groups, as different individuals pursue or flee state association, and individuals cross ethnic boundaries through marriage, religious conversion, or adoption (Harwell 2000). Nevertheless, the assertion of bounded ethnic polarity—produced by states, international media, conservationists, regional ethnic and religious elites, and locals themselves—has greatly influenced discourses of entitlement in the region.[6]

There is a consequent difference between the two ethnic groups in the spatial and conceptual perceptions of resources and the cultural construction of rights around either waterways or forests. Iban, who are primarily farmers, see their territory (*pemakai menoa*) as anchored to forest and swidden agricultural resources, marked by vegetation types or hillsides. Malays, who are primarily fishers, perceive their village territory (*wilayah kampung*) first and foremost as fishing grounds, which are more easily marked by streams and lakes. Indeed, historically, Malays have had more or less unrestricted access to forest resources, although this has begun to change with increasing resource pressure as people begin to cut timber and rattan or to collect honey for sale.

Iban Dayaks have a diversified portfolio of subsistence and commercial activities. By far the most important activity for the Iban is a complex combination of rice agriculture and silviculture of rubber, fruit, and other perennials.[7] In addition to rice, Iban harvest various wild and cultivated nontimber forest products, hunt wild game, raise domestic animals, fish, and raise pond and cage-cultured fish. Most men engage in wage labor migration during some phase of their lives, if not regularly (Padoch 1982; Kedit 1993; Wadley 1997). A minority of Iban living close to the rivers supplement their subsistence activities with cash from the seasonal sale of freshwater turtles, honey, swidden vegetables, or rattan to passing traders' boats, nearby Malay communities, or timber camps. However, this income is both meager and sporadic, and as a general rule, Iban in the lakes area subsist on what they produce themselves, relying on fish, rubber, and wage labor to provide cash to pay for staples, gas for outboard motors and chainsaws, children's school fees, and other necessities.

In contrast to the Iban, most Malays in the lakes area use income from fish to buy their necessities—including the staple food, rice. Some Malays have secondary sources of income from harvesting honey, rattan, and timber, but most are completely dependent on fish to generate cash. Only a small minority of Malays in the area practice rice agriculture, and they have considerably less success than their neighbors—never enough for their own consumption, perhaps due to limited access to fertile farmland and shorter fallow cycles, poor rice varieties, less experience or expertise with farming, or reduced availability of labor for weeding (frequently limited by fishing activities; see Harwell 2000).

Rather than depending on a diversity of activities, many Malays survive seasonal hardship due to the flexibility that inheres in their mobility and wide social networks.[8] Until about fifty or sixty years ago, most Malays used the lakes area only seasonally for fishing, and returned home to towns along the Kapuas River during the high-water season when fish were scarce.[9] Today, many Malays have permanently settled in the lakes, but strong social ties continue to be maintained with the large Kapuas towns. In wet years when catches are low, many Malays fall back on social networks on the Kapuas River outside the lakes, relying on relatives or credit with riverboat traders whose floating shops frequently visit the area's villages. The reciprocal nature of these networks increases outsiders' access to village fish resources, as relatives from the Kapuas come in the dry season wanting to fish (and may decide to stay), and traders demand repayment of debts in fish rather than cash. Many traders may front the capital to buy fishing gear or salt in return for the fishers' promise of exclusive rights to their catch.[10] These factors, along with the dependence on fish rather than forest resources, contribute to a more permeable view of territory, as discussed below.

Iban have similar fallback strategies of mobility and social networks, but the patterns of resource access that result from employing these strategies are different. Like Malays, Iban rely on mobility for livelihood, but their journeys are most often throughout their rice fallows and orchards.[11] In addition, for Iban men, individual journeys (*bejalai*) stretch farther afield for wage labor (mainly to Sarawak), where they often remain for years (if not permanently) and where they establish new social ties yet periodically send remittances home (Kedit 1993; Wadley 1997). Like Malays, Iban also cultivate social networks, but these networks have very different effects on the notion of territory and on the resources themselves. Village cooperative work groups are often important for planting and harvesting fields. When harvests fail, people also depend on social ties to borrow rice. Harvest festivals (*gawa'*) are work to maintain ties with both the spirit and human spheres. Offerings act as a request for the continued favor of the spirit world, but harvest festivals are also a networking opportunity designed for socializing and strengthening ties with other households in the longhouse and with other villages who attend. While communal relations are an important aspect of Iban lives, Iban households are also very independent (Sutlive 1978) and tend to use these social networks as a fallback, rather than as a seasonal strategy (Harwell 2000). Likewise, Iban social networks make it possible for those from other longhouses to come and cut new farms or borrow farmland. However, such movements typically involve only single households, because the limited forest resource restricts the number of new residents that may be accommodated to make new farms (or borrow existing ones).[12]

Although movement has been an important part of Malay and Iban livelihoods, history has played a large role as both states and markets have greatly contributed to the reduced mobility of both groups. Historically, state pressure to sedentarize and subjugate warring populations during the late nineteenth century and the early twentieth contributed to the narrowing of Iban mobility from expansive migrations to localized movement across secondary fallows or individual (primarily) male journeys (Pringle 1968; Tarling 1978; Wadley 1997; Harwell 2000). Ironically, the increasing state involvement that suppressed Iban movement worked to encourage Malay movement, as it became safe to seasonally migrate to the lakes to fish without fear of Iban raiding. This contrast is part of a longer history of Iban seeking independence from state control, while Malays have sought refuge in state privilege, a difference reflected in the contrasting views of territorial entitlement and citizenship, as I will discuss below.

New markets, primarily for rubber and fresh fish, also encouraged people to alter their movements and have changed local notions of boundaries and territory. Iban migrations declined at the beginning of this century as more families wished to stay in one place to tend rubber and pepper trees (Dove 1994).[13] "Our feet have been stuck by rubber," one Iban pioneering family succinctly observed in Sutlive's (1992: 31) ethnography of Sarawak Iban. Iban on the Leboyan River, unlike their fellows in the Emperan and Empanang regions to the north and west of the lakes closer to the border, do not grow pepper because of the remoteness of lucrative Sarawak markets. Yet most families have access to rubber, which they tap during lulls in agricultural labor needs. The latex is sold to riverboat traders or in the Kapuas River markets. Likewise, new markets for fresh and live fish encouraged Malays (and some Iban)[14] to begin raising *toman* (*Channa micropeltes*) in cages for subsistence consumption and sold both live and salted to local, regional, and national markets. *Toman* require daily feeding and therefore generated more permanent settlement. Both these resources, rubber and *toman*, are frequently locally referred to by locals as *tabungan*, or "savings accounts," that can be drawn on in times of need or climatic extremes.

However, the effect of these two practices on local territory has not been identical, either for the resources themselves or for the social relations of those that manage them. Since *toman* are a carnivorous species, cage culture requires enormous amounts of smaller fish for food. The explosion of this aquaculture practice and the need for large quantities of baitfish (some fifty kilogram per cage per day) has led to ever more efficient fishing gear—primarily the *jermal*, a very large, staked funnel net. These nets are expensive (in 1997, about a million rupiah or US$430) and thus are frequently funded by traders in exchange for exclusive rights to the *toman* harvest. In this way, the expansion of cage culture has served to make Malays more sedentary and their territories ever more permeable to outsiders' access, and, through the demand for baitfish, has greatly increased pressure on the overall fishery.

In contrast, rubber trees are property held by the individual planter (or when she or he dies, equally held by descendants). The land immediately surrounding rubber trees also becomes de facto individual property, as it needs to be regularly slashed to clear surrounding vegetation for access. Therefore an increasing establishment of permanent tree crops such as rubber more firmly establish individual or descent group claims to land—a form of narrowing access to resources and the land around those resources (Peluso 1996).[15] Communal land, therefore, is gradually converted to individual land, and resource claims to territorial ones.

Thus Malay dependence on fish, with its specific technologies and social modalities of use, as well as Malay patterns of mobility and use of social networks means that access to Malay villages' resource territories are much less restricted than those of Iban. Rather than limit the community of allowable users, Malays instead tend to regulate types of gear and times and locations of allowable fishing. The mobile nature of the fish resource lends itself to a continued acceptance of new residents if this acceptance secures other social returns, such as the maintenance of social security nets and credit relations. Patches of common forest may be parceled out to a limited number of new immigrants until all available land is claimed, after which people wishing to immigrate must borrow, or in rare cases buy, fallowed land rather than clear their own. In contrast, mobile fish populations are a nonspatial, or congestible resource—one that can be shared up to a point with little noticeable impact until a threshold of extraction is reached, after which the resource population declines rapidly until it "crashes" (Rose 1991). Information about how many fish are left is always incomplete; that is, the crash point and where current extraction pressure is in relation to that point is unknown—a limitation that has consistently plagued efforts to determine a maximum sustainable yield concept for fisheries.

This incomplete knowledge allows fishing communities to continue to accept newcomers to access the benefits derived from maintaining social networks and to avoid the social friction that might result from closing the fishing grounds to outsiders. Weighing the benefits of maintaining social networks by accepting more new users against the costs of probable reduced overall catch requires a discourse of entitlement and value—that is, a cultural interpretation of rights and territories. Diverse discourses of rights are in constant competition, not only to secure access to valuable resources but also to define the meaning associated with histories, authority, and even communities and resources themselves.

These differences in perception of value have influenced the development of tenure regimes around resources. Charles Peters (1994) suggests that the increasing economic value and scarcity of forest resources due to increasing timber, honey, and rattan markets has caused access to those resources to be restricted, basically creating novel tenure regimes. This restriction of boundaries around increasingly marketable resources seems true for many Malay villages in their management of forest resources and is evidence of local ingenuity and the changing nature of territory. But it is important to remember that not everyone respects these new territories and barriers to use, especially those from villages on the Kapuas River, or the state itself and its commercial partners. It is by no means clear whether creative new discourses of entitlement are sufficient to control access to resources by those hoping to cash in on emerging commodity markets.

It is not just the perception of the local village residents that influences the openness of territories but also that of outsiders. Because there are no roads in the area, fishing territories are also compromised by the common view that rivers, as primary thoroughfares, are open to all. Furthermore, the idea that fisheries, as mobile resources that cannot be owned, should be available to everyone is reinforced by the government's ideas of national territory and the rights of citizenship (Harwell 2000). This ethic of access to waterways and fish resources is based on a perception of these resources as "God's gifts," which Malays frequently explain as a religious principle (*prinsip agama*).

This reasoning draws explicit attention to religious differences between the Muslim Malays and the Christian Dayaks.

In fact, one Malay village, in reference to its Iban neighbors who continue to restrict outsider access to forests, complained that the Iban want to establish a "monopoly" (their exact term), which goes against the Malays' perceived right of all Indonesian citizens to use resources in the country. Further, this national ethic of access stands in direct contrast to Iban claims to *tanah adat* (village customary lands) and individually held fallows and orchards.[16] Iban argue that the resource tenure and legal institutions, or *adat*, associated with these agroforestry activities reflect both their strong foundation in Iban culture and centuries of settlement in the area and long-established and defended Iban rights to resources.

These specificities illustrate that the practices of seeking one's livelihood in fish or farming are intimately associated with cultural idioms of meaning, value, and memory. The specific cultural interpretation of the value of a resource and one's historical entitlements to it have structured territorial claims in the lakes region. This is not to suggest, however, that territorial property regimes have simply evolved in response to unambiguous notions of scarcity and value.[17] To the contrary, I argue that the nature of value (and its relationship to scarcity) is culturally construed and that numerous claims and territories are simultaneously present—sometimes in direct conflict, sometimes simply operating behind each other's backs. These different perspectives on property and entitlement are in competition with each other, in constant negotiation, rather than one view obliterating all others.

It is also apparent, however, that the playing field on which different ideas of entitlement compete is not a level one, but one deeply imbued with power and uniquely situated in time and place. The view of space and resources as gifts from God and thus not subject to individual ownership is one that seems to contrast sharply with the Western (and capitalist) idea of property. Yet it is just this argument of the common good and the rights of all citizens to enjoy resources, employed under conditions of nationalism (and capitalism) in the New Order, that allowed for the state's appropriation of land claimed under customary title to be used for "national development" (Harwell 2000). This discourse of citizenship, so useful for the state's promotion of national development, has been creatively appropriated by the Malays in arguing against the monopoly of the Iban over God's gifts. In this instance, it is a powerful argument with the weight of the government and Islam behind it.[18] But these territorial claims also have rich histories, as I outline below.

Histories of Boundaries and Authority

The complex amalgam of claims and territories operating today in the lakes area is not only a product of various stakeholders competing for resources in the present but also evidence of multilayered power struggles taking place throughout history (see table 1 for a summary of the various property claims and supporting evidence). These contests for control occurred not only between warring local groups but also between the Dutch and British colonials who each sought to control territory as a means of controlling valuable natural resources, including swiftlet nests (for use in Chinese cuisine), *damar* resin (used in torches and for sealing/caulking boats), *gaharu* incense (from the fungus-infected wood of the *Aquilaria* tree), *bezoar* stones (traditional Chinese medicinal amulets from the gallbladders of bears and monkeys), native latexes, rattan, timber, coal, and fish (Enthoven 1903).[19]

Oral historical and Dutch archival data stretching back to the mid-1800s recount violent struggles for control over territory, resources, and subjects between the newly Islamic Malay sultanates on the Kapuas River and the Iban and Embaloh Dayaks (see Dutch residents' reports MvO/PVR/AVR 1860–86;[20] Veth 1854; Enthoven 1903; Kater 1883; V. King 1985). Complex networks of alliances and enmities created frequent tides of migration, in which groups fled the raids and attempts at economic and bureaucratic control of various states, only to have the vacant resources claimed by new settlers (Enthoven 1903; Pringle 1968; McKeown 1983; Wadley 1999, 2003; Harwell 2000). The Dutch began to intervene in the mid-1800s, as they had throughout the rest of Kalimantan, offering military support to the Malay sultans in return for loyalty—which included allowing the Dutch to control trade and to collect taxes (MvO 1860 and PVR 1861; Veth 1854; Enthoven 1903). Late in the 1870s, Dutch guard posts were set up in towns in and around the lakes—Badau, Pulau Majang, Kenelang, Ukit-Ukit, and various towns on the Kapuas, including Bunut and Semitau (AVR 1884, 1885; Enthoven 1903)—which temporarily reduced raiding and encouraged Malay use of the lakes.

Both Iban and Malay oral history recalls Dutch attempts at the beginning of the twentieth century to negotiate and map Malay, Iban, and Embaloh territories in an effort to reach peace and secure stability in the upper Kapuas. One such strategy for allocating territory was the presence of two valuable tree species. Malay interviews recount how land containing *belian* trees (or ironwood, *Eusideroxylon zwagerii*, an upland species that grows in land fertile for agriculture) was allocated to the Dayaks, and land with *kelansau* (*Dryobalanops abnormis*, a swamp species) to the Malays. This

TABLE 1 *Bases for Territorial Claims*

Type of claim	Supporting evidence
Tribal *adat* or sultanate territory	Oral history of settlement and conquest Kinship Physical landscape and markers (e.g., honey and rubber trees, farm fallows) Basic National Agrarian Law (1960)
Dutch state	Self-governing territorial contracts with sultanates (1850s–1916) Brokered peace treaties and territories (1860s–1920)
Independent state 1. National	National Constitution (1945) Basic National Forestry Law (1967, revised 1999) Basic National Law for Environmental Management (1982) National Law for Natural Resource and Ecosystem Conservation (1990)
2. *Kabupaten* (regency)	National Law for Administrative Decentralization (uu22/1999)
3. *Kecamatan* (district)	National Law for Regional Government (1974)
4. *Desa* (administrative village)	National Law for Village Government (1979)
International conservation community	Ramsar Conference on Wetlands of International Importance (1971) IUCN World Congress on National Parks and Protected Areas (1982)[1] IUCN Red List of Threatened Animals (1994) Convention on International Trade of Endangered Species (1973)[2]

[1]This congress established the agreement by signatories (of which Indonesia was one) to set aside 10 percent of land area as protected areas by 1990.
[2]While the CITES listed species are intended only to regulate the international *trade* of threatened and endangered species, the presence of CITES-listed species is commonly used as justification for an area's conservation value. See Lowe (1999) for a discussion on how species lists have increasingly dominated conservation debates.

serves as an example of how livelihood practices became spatially and rhetorically mapped onto ethnicities—the Iban were restricted to farming zones and Malays to fishing zones. Likewise, it is an illustration of the state's attempts to simplify complex claims into bounded territories.

These efforts were largely ineffective, complicated by continued raids by Sarawak Iban, many of whom were under orders from Charles Brooke, the second British rajah of Sarawak (Pringle 1968; Sandin 1967; Wadley 1999, 2003). In addition, to flee Iban raids as well as the demands for tribute and labor of the Malay sultanates, many Dayak groups continued to migrate throughout the officially designated "territories" (Enthoven 1903; Bouman 1952; McKeown 1983). Further, the Dutch territorial treaties also failed to resolve the existence of long-standing seasonal claims to fishing territory and honey trees in the lakes area.

The involvement of the state in delimiting areas of administrative control increased in independent Indonesia as government subdistricts, or *kecamatan*, were drawn up in 1974.[21] The boundaries of these *kecamatan* largely followed the boundaries of historical sultanates and Dayak territories, but they also deviated in many places. Indeed, even today, many of the official *kecamatan* boundaries are still disputed. Traditional claims of sultans or tribes are frequently used as evidence to support the present-day claims of a particular *kecamatan*. Documents from the *camat* (subdistrict head) of Batang Lupar make use of oral historical data regarding treaties between tribes and boundaries of sultanates to justify claims of the *kecamatan* territory.

In fact, many *kecamatan* claims are directly linked to historic natural resource claims. For example, the *kecamatan* of Selimbau bases its claim to the "right bank going upstream" of the Leboyan River on historic rights of the Sultanate of Selimbau to seven honey trees on that bank. These historic rights to trees, like the modern cultivation of rubber trees that have been used as evidence of contemporary land and territorial claims, have been dramatically expanded into wider claims to rights to land; the area now encompasses not only that surrounding the trees but also one entire side of the Leboyan River. The boundaries of tiny bureaucratic districts with no funds for administration nevertheless matter because they serve as rhetorical tools for staking claims to territory, resources, and power. For the *camats*, the extent of the territory under their control reflects their power and significance as officials and, in some cases, their status as descendants of sultans.

For the residents of the *kecamatan* (particularly those along the Kapuas River), territory is important because they view any areas within their own *kecamatan* as open for their settlement and use. Similarly, many individuals wishing to gain access to resources in another village's territory may claim rights based on their lineage from a particular sultanate. Among Malays who live in areas in which sultanate and *kecamatan* boundaries do not coincide, local rights to forest products are frequently determined by ancient sultanate boundaries (reflecting the importance of the physical presence of sultanate honey trees), while fishing rights are determined by *kecamatan* boundaries (Rona Dennis, personal communication, 1999). For example, residents of the Kapuas town of Suhaid claim rights to territory in the lakeside villages of Pulau Majang, Lubok Kelakati, and Kenelang because these are within the former sultanate of Suhaid. The village head of the larger village, Pulau Majang, claims that the residents of Lubok Kelakati originally came as seasonal fishers from Suhaid and asked permission to work in Majang's territory. "What could we do? We couldn't refuse them," the headman complained, "but then they stayed. . . . They built houses and a mosque, they brought their relatives, and now we will never be rid of them!"

Laid over this complexity was the independent Indonesian state's organization in the late 1970s of all communities (previously called *kampung*, now assigned the administrative label of *dusun*, or hamlet), into larger administrative villages, or *desa*, to be administered under one *kecamatan*.[22] The primary purpose of the Village Law of 1979 was the replacement of the vast diversity and complexity of village *adat* and institutions throughout the archipelago with uniformity, purportedly for better administration of development.[23] The Malay villages in the lakes area are all administratively grouped under a geographically remote *desa* on the Kapuas, which further complicates the village's ability to be self-governing and control access to its territory. For example, the village of Lubok Kelakati belongs to the *desa* of Suhaid, a large town on the Kapuas River that has many other constituent villages. Members of these other constituent villages may not, according to government officials, be denied access to resources within Lubok Kelakati territory, since, administratively, they are considered *desa* resources, the rights to which are shared by all *desa* members. Lubok Kelakati residents feel differently, even though they originally staked their claims to their territory within the village of Pulau Majang based on their historic rights as heirs to the Sultanate of Suhaid.

Other forces have further compromised village control of territory and the restriction of outsider access and have competed for the definition of "insider," redefining entitlements. One such change has been the classification of forests as state forests to be managed for the supposed greater good. During the 1970s, large-scale forest concessions in the area began to be assigned to outside parties based on the state's right to control national territory.[24] These concessionaires operate within the bounds of village territory, sometimes with and sometimes without local consent.

In one unique circumstance, one such concession was granted to three Iban elites in return for Iban assistance as trackers and intelligence gatherers during the 1964 Indonesian undeclared war, called the Confrontation, with Malaysia and the subsequent squelching of the communist guerilla resistance (Harwell 2000).[25] The "local" concessionaires then proceeded to secure permission to log forests in village territories within their concession by painting it as an Iban concession, which would use Iban forests to benefit Iban. Years later, villagers whose forests were logged with no return of benefits to them complained that the leaders had inappropriately used Iban ethnicity to expand their claims to resources that did not belong to them. "Those forests belong to my grandparents and to the grandparents of the others who first came here. We need those forests to build houses and canoes. If anyone is to benefit from them, it should be us, not those from the outside," complained one man.[26] These claims to resources based on ethnicity are revisited below in the discussion of decentralization.

The layers of territorial control and claims are deep, and they are maintained and reinvented rather than overwritten as people continue to make strategic use of them. In the same way that government officials refer to national legislation, local people refer to their history to support particular claims. In this way, different forms of authority—including history itself—are marshaled by various stakeholders when seeking to gain or maintain access to territory and natural resources.

Conservation and International Claims on Resources

In 1971, the Kapuas Lakes area was designated a Ramsar site,[27] that is, "a wetland of international importance," due to the unusual ecology, high species diversity, and number of endangered and endemic species. In 1982, a new layer was added to the already sedimented history of claims—the establishment of a wildlife reserve, backed up by legislation protecting resources under its control,[28] that again altered the meanings of the landscapes. The Indonesian government recognized the area's conservation value by delimiting 80,000 hectares of this freshwater-inundated forest as the Danau Sentarum Wildlife Reserve (DSWR). In 1996, the reserve was expanded to 132,000 hectares, encompassing more of the hills surrounding the lakes to protect the hydrological function and important wildlife habitats. A buffer zone bringing the total management area to 197,000 hectares was also later proposed. Table 2 summarizes the changes in reserve boundaries and the impact

TABLE 2 *Demographics of Different Conceptions of the* DSWR

Reserve boundaries and areas	1996 total population[1] (Wet season : Dry season)	Total number of settlements (Permanent : Seasonal	Ethnicity of settlements (Mixed : Dayak : Malay)
1982 boundaries (80,000 ha)	10,360 (4,415 : 5,945)	36 (29 : 7)	0 : 1 : 35
Proposed 1995 boundaries (132,000 ha)	15,850 (6,860 : 8,945)	55 (45 : 10)	1 : 2 : 52
Including proposed buffer zone (196,000 ha)	17,535 (7,560 : 9,975)	68 (53 : 15)	4 : 6 : 58
Including proposed buffer zone and villages on boundary of buffer zone (196,000 ha)	20,770 (9,130 : 11,640)	75 (59 : 16)	5 : 11 : 59

[1]Population estimates were calculated from project survey data on the number of families per settlement and then multiplied by an average figure of five persons per family (Aglionby 1995a).

these changes have on the number of people affected by the reserve. In addition to the international importance assigned to the wetland's diversity, its value as crucial habitat for the endangered *arowana* (*Scleropages formosus*) served as the primary basis for the establishment of the reserve. The *arowana* is an ornamental aquarium fish sold domestically and internationally throughout Asia (Dudley 2000). Believed by many Chinese and Japanese to bring good fortune, the *arowana* has been become a status symbol for the well-to-do in Indonesia and abroad.

However, the designation of the Kapuas Lakes as a reserve conflicts with the long-standing habitation and claims to the area as well as its importance for the livelihoods of local people. The proposed boundary extension and buffer zone of the reserve comprises seventy-six villages, an estimated dry season population of twelve thousand people (Aglionby 1995a). Although Indonesian law is interpreted as prohibiting the human habitation of protected areas, the project acknowledged that the resettlement of the many thousands of residents would be politically dangerous, prohibitively expensive, and ineffective (Giesen 1987).[29] Furthermore, budgetary constraints and a lack of state capacity make the strict enforcement of state-imposed regulations impossible. With this in mind, the Directorate of Forest Protection and Nature Conservation and Wetlands International, the international donor charged with the management of the reserve, were forced to recognize the need to provide reserve residents with a stake in the resource management planning.

The reserve team aimed to tap into local capacity by cataloging local institutions of resource management (*hukum adat*) and deploying some of them for the project's management goals (Harwell 1997). But the conservation team faced a difficult task in documenting *adat* and mapping jurisdictions, a topic explored elsewhere (Harwell 2000; Dennis 1997). What is important here is that the documentation worked to establish new forms of entitlement and authority, creating new meanings for landscapes and social relations of production. New territorial boundaries emerged and old ones solidified. Local communities developed rules for newly conceived violations, such as burning forest, fishing with electricity (which almost no one recalled occurring, and which fisheries experts say would be largely ineffective due to the water's low ion content; Richard Dudley personal communication, 1999), hunting "endangered species" or "animals protected by the government" (although very few could say which animals these were, and almost no one in the reserve hunted anyway since it is forbidden for Muslims to eat game other than venison). In the text of the village rules appeared new concerns with evidence (*bukti*), and previously flexible sanctions became standardized, echoing legalistic language. Rules that previously had sanctions in accordance with the offense changed to "punishment in accordance with national law, three months in prison and a fine of 10 million rupiah," a financial penalty unimaginable by local standards and a concept of incarceration that was without basis in local legal practice.

Clearly locals intended these rules as a message to the conservation project that the villagers were worthy resource stewards and deserving of project support (often at the expense of neighbors with whom they competed for resource control). An identifiable "conservation ethic" became a new means for communities to garner support for their claims by forming alliances with conservationists and their national and international networks.

Nowhere was the conflict of local interpretations more visible than in the project's mapping of territorial boundaries. Given the convoluted history of territories in the lakes area, mapping current village territory presented a dizzying task. In addition, the act of mapping itself served as a catalyst for some boundary disputes to be settled and new conflicts to arise. Ambiguity in practice is tolerable, but disagreement is bound to surface when someone wishes to record boundaries on an official map. Conflict was most common in areas of special economic value—timber, good fishing spots, and swiftlet caves in particular. Such conflict has intensified with the recent logging boom, as described below.

Conversely, many areas of ambiguity around the edges of village territory are more congenially described on project maps as *wilayah kerjasama*, or cooperation areas. This may be the result of close village ties, such as those derive from kinship or the fission of a large village into two. Such cooperative areas may also be indicative of the fact that as long as the resources involved are not too valuable, ambiguity is permitted. In any case, at least as they have been represented on project maps, *kerjasama* areas vastly outnumber the areas "not yet agreed upon" (*belum disepakati*), which is a sign of the great potential for cooperation among neighboring villages. It should also be emphasized that dispute constitutes a natural part of a continual process of renegotiation or reaffirmation of resource access between competing parties. In situations of such a complex layering of claims, it is the intensity and forms of dispute resolution that may be more accurate indicators of unworkable conflict, rather than the mere presence of disagreement.

Decentralization of Forest Control

In recent years, the disputes over boundaries and claims to territory have become more heated in the lakes area. Since the 1997 economic crash and subsequent International Monetary Fund (IMF) rescue loans, oil palm plantations have multiplied throughout the country. The area in the Kapuas Hulu regency is particularly targeted as an Integrated Economic Zone (KAPET), prioritizing oil palm as a way to generate capital. This designation has been further facilitated by an administrative decentralization that grants governors and *bupatis* (district heads) direct authority to issue permits and receive income from logging, mining, and plantations concessions. But the central government has not completely relinquished control of valuable resources, particularly in the arena of conservation, creating lingering uncertainty about which arm of the state has jurisdiction over which local resources.

Meanwhile, some local interests are also gaining ground in the territorial struggles. Small-scale logging in the lakes area (illegally smuggled across the Sarawak border into Malaysia, further complicating the notion of territory and national sovereignty) has boomed under decentralization, with Iban elites holding the local timber concession and acting as paid liaisons in setting up community logging and oil palm plantations after the valuable timber has been removed. Local villages are paid a commission for the timber, but this amounts to less than 1 percent of the export value of the wood (Wadley 2001).

Disputes have proliferated in villages over whether to participate in such high grading of village resources for such meager returns. Many see through the local logging elites' arguments that logging and oil palm constitute Iban development efforts, by and for Iban. Judging from the poor performance of the Iban logging concession in sharing benefits with local communities in the past, this latest scheme appears to be yet another business venture designed to enrich a few, while paying only the minimal amount needed to gain village consent, based on the latter's poor knowledge of timber values. Yet the scheme is cleverly couched in the language of ethnic empowerment and development and the local control of resources—especially salient during a time of decentralization and increasing ethnic awareness.

Additionally, bitter disputes between neighboring villages have emerged (or perhaps old wounds are reopening) over where the boundaries of village territories lie. Imprecise boundaries are commonplace around resources that are amicably shared or have little market value. But once resources begin to be sold, this lack of clarity quickly becomes problematic as different interpretations and sources of evidence are marshaled. In one example, involving a dispute between two neighboring Iban communities that each claim rights to upland forests, the rights to logging compensation and the ability to refuse logging altogether were at issue. Claimants in one village drew not only on oral history but also on community maps made by the conservation project as evidence of their boundaries and rights to resources. However, the project's maps were made by a Malay field assistant primarily for the purpose of mapping fishing boundaries that constitute the core of the reserve. Forest boundaries, in contrast to river courses and lakes, are hard to travel to, and the dense canopy made it difficult to get satellite readings, so these terrestrial boundaries were frequently estimated rather than specifically pinpointed using Global Positioning System (GPS) coordinates. Whatever the reason, in one particular instance the maps were made in consultation with the Malay communities living on the river rather than with the hillside Iban community who claimed the area. Therefore the maps could not be used as evidence in the boundary dispute because their veracity has been called into question. The project, therefore, missed a valuable opportunity to be more participatory and to engage different ethnic groups in the discussion of territory that might have avoided the dispute. Competing versions of oral history and the promise of financial gain, however slight, now make these disputes intractable. While this case illustrates that ethnicity matters in the perception of what is to be mapped, it also underscores the fact that village politics and personal affiliations continue to complicate consensus based on ethnic group alone and muddy the waters of boundary making.

These complexities illustrate that, rather than being instantiations of timeless tradition, diverse local ideas of territories are mutually constituted with local identities, histories, and livelihoods. Likewise, there are (and have been) multiple interpretations of these concepts of identity, livelihood, and territory—as well as of fundamental notions of value, authority, and entitlement. These competing views of lives, livelihoods, and landscapes, although unequal in power, do not obliterate one another but rather coexist, at times in conflict, at times in cooperation. Negotiations for access to resources, trade, and control over identity have been ongoing, but the terms change depending on the conditions. These negotiations, in which certain versions of history and legitimacy are used to stake claims, frequently occur through the daily practice of resource management.

Local Boundary Practice

Although it cannot be seen as static or isolated from social, political, or economic contexts, one convenient window on the multiple ways in which ideas of territory are manifest is local practice. If seen through the optics of history and multiple contests for control, local practice acts as one rhetorical device among many by which claims to territory are staked. The local boundary practices described below illustrate the multifaceted nature of territory—varying with time and place, as well as resource and actor. Different seasons and water levels, different ethnic perceptions of resources according to primary livelihoods, differing access to resources according to social networks and inheritance, different versions of history and authority all play into local people's everyday assertions of territory.

The following discussion of resource practice occurs in the framework of a village territory as the jurisdiction for rules and control of resources. Paramount to the effective delineation of territory is the ability to exclude outsiders. The permeability of village territory is a growing concern among local villages seeking to conserve resource stocks and to secure local livelihoods. The general understanding of rivers as common waterways presents a problem for villages whose territories are made up of fishing grounds. Still, it is common knowledge that those wishing to work in another village's territory—whether fishing, gathering rattan or other forest products—must first ask permission from the head fisherman (*ketua nelayan*) or village headman (*kepala dusun*) and agree to abide by village rules.

As discussed above, the inherited ownership of honey trees is important to the notion of territory in the lakes area. The ownership of honey rights, like that of rubber trees, has expanded rights to resources into rights to land. Honey is harvested both from wild nests on honey trees (*tapang* or *lalau*) or from traditional honeyboards (*tikung*) that encourage nesting and that are hung in trees in an individually owned area of honey trees called a *periau*. The income earned from honey is especially important to local residents because the harvest comes at a time when water levels are high and both fish and money are scarce.

As in the case of rubber trees, the area around honey trees is cleared of competing underbrush and therefore establishes a de facto right of control to the land around the tree. Iban refer to these areas as *pulau tapang*, literally, "honey tree islands." In Malay villages heavily involved in honey harvest, groups of hundreds of wild honey trees are inherited (frequently over several generations) as a *periau* area, rather than as individual trees. This inheritance establishes exclusive rights to all the honey from the trees in that area and the attendant right to protect those trees from damage through other practices.[30] It is likely that owned honey trees provide a powerful incentive to protect the forest from fire and logging. In fact, villages with a particularly high honey production do not allow the use of gill nets because people frequently burn forest to clear spots to hang these, which would damage the honey trees.

Although commercial honey harvesting has gone on for at least a century, recent market changes have also affected the idea of territory and its control around honey trees. As part of an effort to provide alternative income to residents, the conservation project offered training to increase the quality (and therefore price) of honey and acted as the intermediary trader of the product in domestic and international markets. In 1997, honey sold for Rp3,750/kg (from Rp1,500/kg in 1995 before the program), whether it is sold to the project or to local traders.[31] This increasing value of honey has led to increasingly defended territory around honey trees. During an initial investigation of the potential of honey markets, Charles Peters (1994) and Rouquette (1995) noted the existence of rules protecting tree ownership, limits on the number of people permitted to harvest honey, and fines for theft. Yet Rouquette also noted that as a result of increasing value, fines more than doubled in two years, reaching Rp250,000 per *tikung*.

It is important to consider the contexts within which these new claims are made—a new conservation project authority and increasing markets for fish and forest products—and the audiences to which they are directed, namely, powerful outsiders looking for a conservation ethic and competing users of the resource. As mentioned previously, not all those with conflicting views of entitlement to the resource will accept this new move to strengthen village territories. Much will depend on those who can be marshaled as allies in this claim.

TIMBER

Forest resources are important to both Malay and Iban communities, but their territorial control over them has historically differed, as explained earlier. Malay communities of the DSWR value timber for the construction of stilt houses and extensive networks of boardwalks (for use during the high-water season), as well as for canoes and houseboats. But the forest has not been a source of deep spiritual and historical significance for Malays, though it has been for Iban. Iban value forests to fulfill their subsistence wood needs, but mainly for long-fallow farming, silviculture, and an array of forests of local cultural and historical importance. Iban Dayak orchards are communally owned by longhouse communities, or if planted in a fallowed rice field by a single household, by all the descendents of that household. These orchards, or *tembawai*, in addition to producing useful fruits and fiber, also have great social significance. These forests are the sites of previous longhouse settlements and contain individual trees planted by longhouse members and passed to their descendants. As such, *tembawai* serve as cultural and historical markers. Frequently named after a significant event or pioneer ancestor who planted the trees, these stands are physical evidence of local history and kinship ties and, as such, imbue territory and boundaries with rich social meaning.

Likewise, Wadley, Colfer, and Hood (1996) explain in depth the diverse forms of Iban protected forest, called *pulau*, or "islands." These are believed to house spirits and are frequently explicitly recognized for their ecological function (as stream corridors, wildlife habitat, and seed sources for regeneration of fallows). There are complex and stringent regulations for the felling of trees in these protected groves. Indeed, the fines are required as ritual compensation to angered spirits, not as a commercial exchange of cash for damages. Not only do local people carefully monitor the local use of these forests but many local Iban villages have in fact successfully stopped cutting and levied *adat* fines against timber companies encroaching on these protected areas.

However, while Malays have not in the past defended their forest resources as forcefully as their fishing grounds, the advance of logging companies and the increasing value of timber and honey has caused Malays to also restrict outsider access to their forests. Many Malay villages have cooperative agreements with other villages to protect "community forests" (*hutan adat*) for local subsistence use only. A timber company that began felling trees in one such protected forest near the Kapuas town of Suhaid had its base camp burned and heavy equipment destroyed by angry locals. Operations were halted for some time, but the Department of Forestry granted the timber company special permission to log the area, even though it lies outside of its concession. This scenario offers a vivid example of the high stakes involved when outside interests compromise local management, as well as the costs of conflicting government goals for resource management even within the Ministry of Forestry itself, which both issues concessions and is charged with the conservation of forests. But the case also illustrates the diversity of sites and modalities of power that can challenge the success of nation-states in imposing boundaries that conflict with local claims.

BUBU RATTAN TRAPS

Even though fish are highly mobile in this environment, spaces in a village territory for setting fish traps are quite valuable and rights around them complex. One example are spots for setting *bubu,* staked cylindrical fish traps woven from rattan, ranging two to five meters in length and about one meter in diameter. *Bubu* are used only during periods of falling water levels, when the fish are migrating downstream, but they are very efficient, averaging catch rates of one kilogram per hour (Dudley 2000).

In many villages, *bubu* streams are hand dug and therefore inherited spaces, providing unique physical manifestations of the local management of waterways and fish populations. Both the labor invested and the historical claims based on long-term settlement firmly establish the priority of *bubu* owners' rights to fish in these areas over those of other gear users. The presence of such streams has become one of the prime sources of evidence in territorial disputes between neighboring communities in the lakes area.

LOTTERY OF FISHING SPOTS

Many villages have lotteries to distribute good fishing spots, for example, for *bubu* or the rights to use particular kinds of gear. These lotteries (if outsiders are not permitted to enter and only one ticket per person may be bought) in theory can help distribute fish catches equitably.[32] Some villages allow non-residents to enter the annual lotteries. The accessibility to outsiders provides an opportunity for fish traders to repeatedly win the raffle and to catch as many fish as they can without any concern for sustainability. In one example, wealthy residents from the *desa* of Suhaid (outside the reserve) repeatedly won the lottery of smaller hamlets (*dusun*) inside the reserve that are administratively considered part of that *desa*. Finding themselves excluded from the resources in their own village territory, reserve residents asked for the intervention of the conservation project to excluding these "outsiders" from their lottery. Aglionby (personal communication, 1999) reports that the Suhaid district head admonished the project, saying they had no right to restrict the access of Suhaid residents to resources located within their *desa* boundaries. These overlapping resource claims offer another illustration of the complex forms of territory at work around reserve resources.

JERMAL NETS

Jermal are probably the most controversial fishing gear used in the DSWR. They are large stationary funnel nets, typically about six meters in diameter and ten to twenty meters long, though they can be much larger. They are staked to block the stream flow and are left standing for several hours (although sometimes as briefly as forty-five minutes). *Jermal* come in various mesh sizes, but by far the most controversial is the use of the *jermal padat* with a 0.01 cm mesh (similar to mosquito netting) to catch huge amounts of very small fish, mostly for feeding caged *toman* but also for catching the valuable aquarium ornamental *ulang uli*, sold for export. Although Dudley's (2000) reported catch rates vary widely with season (from 1kg/hr to 70kg/hr), *jermal* are used throughout the year.

Dudley estimated in 1992 that there could be over 275 *jermal* in use in the DSWR, although the Fisheries Department has regulations restricting the total to 177. The fee for a permit is high (Rp30,000, about US$12), but the Fisheries Department does not have the resources to enforce its regulations, so not all *jermal* owners hold permits. Yet getting a permit does have some advantages. When locals complained to the conservation project and the Fisheries Department about another upstream village's use of *jermal*, which they believed was lowering catches in neighboring village territories, government officials declined to intervene since the offending *jermal* owners held permits.

No one is more aware of these problems than local fishers, although many are still reluctant to restrict the net's use due to the role *jermal* plays in the lucrative *toman* cage culture and *ulang uli* trade. Many villages restrict the mesh size allowed, others the place or time they may be used, or who may use them. This last form of restriction attempts to exclude nonresidents or absentee owners who extend credit or involve local residents in catch-share agreements.

The government has also recognized the *jermal*'s threat to the DSWR fishery, although like local fishers, they are ambivalent. The government has failed to act decisively to address the threat due to the conflicting goals of conservation under the subdirectorate of nature conservation (in the Ministry of Forestry) and of production under the subdirectorate of fisheries (in the Ministry of Agriculture). In 1995, an *adat* agreement crafted by the district head of Kapuas Hulu, signed by *adat* leaders and witnessed by the vice governor and various other government officials, outlawed the use of small-mesh gear, although the Fisheries Office continued to issue permits for it. This is a conflict of jurisdictions and claims to management territory, which I will discuss in more detail below.

FISH POISON

Tuba is a fish poison derived from *Derris spp.* root, which when mixed into the water paralyzes all aquatic life in a localized area. It is a very efficient and nonselective practice and can have a severe impact on downstream fish cages. In the past, the use of *tuba* by Dayaks was a common fishing practice before the increased settlement and explosion of caged *toman*. As Malays continue to move into the area, and their numbers of caged *toman* reach well over a million (Aglionby 1995c), the use of *tuba* has become objectionable, as witnessed by the presence of rules banning its use in many villages, including some Dayak villages.

Today, *tuba* (and its rhetorical role as a protest to *jermal* use) has become ethnically labeled, and it is the most politicized and highly contentious resource issue in the DSWR. The two practices have become emblematic of visions of territory and their ethnic associations. It is instructive to note the evolution of this situation and the implications of *tuba*'s high profile in resource-management discussions.

Several project sources report that *tuba* incidents occur "almost every dry season" (Aglionby 1995b; Dudley 2000; C. Peters 1994), but in 1994 extensive poisoning incidents used potassium cyanide (which is much more persistent and therefore causes much wider damage), not *tuba*. Nevertheless, the incident attracted the attention of the national media, the national parliament, the vice governor, the armed forces, and the police (Aglionby 1995b). By all accounts, the severity of the damage in the extremely low water levels of the drought of 1994 and the loss of income to those whose *toman* were wiped out were truly unprecedented.

The 1994 poisoning incident was presented by some locals as a resource claim—an act of protest, coming at a time when the presence of conservation staff and their involvement in local resource management was beginning to be felt. Although the project had not yet begun to make or enforce regulations, many local people were apprehensive about just what the project proposed to permit and what it would prohibit. Amid this uncertainty, the rhetoric of project extension and documents was and is still perceived as asymmetrically emphasizing the negative impact of occasional *tuba* use, while a blind eye is turned toward the far more widespread and damaging use of *jermal*. Locals (not just by those involved in the 1994 poisoning) widely see this both as an ethnic slur (since *tuba* was invariably presented by project staff as an Iban practice, although evidence showed that Malays as well as Javanese police were involved in the cyanide poisoning) and as a caving in to business interests outside the reserve.[33]

Yet it is also true that some of those directly involved with the poisoning and who benefited the most from the fish sales were members of the local Iban elite, the same ones who controlled the local timber concession and later began widespread logging and oil palm conversion in village territories. The Iban leadership made vociferous complaints about the encroachment of the government and "foreigners" (the internationally funded conservation project) into territory controlled by local ethnic groups, who they argued should have sovereignty over its management for local benefit. This was an argument that resonated with many local people, but it was also a convenient one for those elites who stood to gain from the unregulated resource extraction of the lakes area. Again, this scenario offers an interesting twist on the idea of territory and who the rightful community of users might be—as well as an illustration of how shifting alliances and political interests complicate these notions.

A different version of this narrative of rights and territory is illustrated by the poisoning event and the response it provoked from the state. The vice governor and regent organized a traditional agreement (*kesepakatan adat*) that required local leaders to swear not to use *tuba* or small-mesh *jermal*. The treaty proved to have little effect on fishing practice because there was no consultation with those actually involved in its daily practice and because the traditional leaders invited were in many cases not deemed to represent the local fishing populations in the reserve. Further, only local people were asked to swear to protect the fishery, while the Fisheries Department continued to issue permits for the *jermal*.

But what was accomplished was the reassertion of state authority over the reserve and its management. Although it had used (its own version of) local *adat* to soften the appearance of its claim, the state had essentially reasserted its territorial authority over what goes on inside the reserve. By taking a role in negotiating even a hopelessly flawed treaty, the state firmly staked its claim to leadership and sovereignty over resources and the implicit incapacity of local institutions to function without state guidance (*pembinaan*).

Conclusion

Colonial administrators, conservation staff, and community advocates interested in participatory mapping have frequently interpreted the contradictions in who has rights to what and why as historical inaccuracies—boundaries in need of "tidying up"—rather than as the very stuff of which claims and counterclaims are made and remade in the present. Local people refer to the past and the present to justify their rights, making history itself, in all its myriad reinterpretations, a malleable rhetorical resource for staking claims. Past experiences—ties of kinship, patronage, or social reciprocity, as well as bitter disputes and even violence—color the social relations at the heart of property relations, especially around mobile resources like fish or water. These relations are deeply permeated with power and struggle. Therefore this uncertainty about the "facts" of boundaries and "real" history cannot be banished with ever more analysis; instead it must serve as a window onto competing versions of meaning and authority (Harwell 2000). It is to be expected that conflicting claims, which reflect different perspectives and interests, cannot be completely channeled into uniformity and consensus.

But there are nevertheless many reasons to map boundaries. Counter-mapping tactics have been used by activists and local communities to resist state claims that the rural landscapes inhabited and managed by local people are empty or unowned (Jefferson Fox 1990; Peluso 1995). These strategies have proved useful in empowering local people by serving as public notice of local presence and jurisdiction, documenting their claims, and increasing their visibility to the state and their industry partners. In addition, the use of documented maps and rules as a means of making themselves "legible" to the state (Scott 1998), that is, to strategically distill the complexity of their local institutions for management and governance into forms recognizable to powerful actors, is not only a tactic of domination and control used by states but also a form of community resistance to state plans and narratives (Lynch and Harwell 2001; Jefferson Fox et al. 2008).

In the midst of such guerilla science, however, some critics worry about the costs of these tactics. In defining village territories, will the diverse suite of common, group, and individually held rights to resources within those territories become homogenized? Will mapping obliterate the complex rights that vary in time and space—such as seasonal rights to fishing holes or game or fruits? Will shared or contested boundaries that were fuzzily defined and flexible in the past become solidified? Who would win and lose in such shifts? These are valid and weighty concerns.

The field data presented here shows a trend beginning in the colonial period of the expansion of resource rights (to rubber or honey trees or fishing spots, for example) to territorial or land rights. But this was the case long before the conservation project began mapping village territories. In fact, colonial administrators attempted to map boundaries according to what they interpreted as ethnic territories. But these efforts proved largely ineffective (Harwell 2000; Wadley 1999, 2003), as people continued to assert claims and move across the landscape.

Further, conflict over boundaries has indeed increased around some resources, as in the recent logging dispute between neighboring villages, but this happened in response to changing political and market conditions, rather than in connection with any formal mapping activity. In fact, the formal maps made by the conservation team proved ineffective in either exacerbating or resolving this particular dispute, because the process by which they were made was called into question. Villages locked in conflict have instead turned to cockfights to resolve disputes, claiming the realm of spirit intervention, one therefore deemed incorruptible. The conservation project fumbled an opportunity to mediate such disputes by failing to grasp the contested and ethnically inflected nature of mapping and therefore producing maps dismissed as illegitimate by Iban villages. Villagers of both ethnicities have indeed turned their attention to the solidification of their village boundaries against outsiders (whether the state, the conservation project, neighboring villages, or entrepreneurial ethnic leaders). Yet the local boundary practices discussed earlier in the essay illustrate that the complexity of rights within those village boundaries has not been homogenized in the process.

This essay has argued that resource claims to territory are complex and dynamic and that translating them into clear boundaries makes for an enormously difficult task. But this is not to argue that such a task should never be attempted. Insisting on remaining "illegible" to the same interests who threaten to expropriate local resources seems, in most cases, an untenable strategy for maintaining control of territory. Some local villages, however, have clearly grasped the power that comes from allying themselves with the conservationists' language and tools (Jefferson Fox et al. 2008). Additionally, if deployed skillfully, mapped boundaries can still work to maintain flexibility, as they showed the potential to do in areas delineated as "cooperation areas" between villages. Such a designation reflects multiple claims and interests and encourages joint management among neighbors while helping to control access by river villages that reduces local fish catch.

Engaging mapping technologies does not mean the enforcement of stability or any clarity on boundaries. Rather, I argue for a new ecology that would embrace a dispute management approach rather than a consensus or dispute resolution approach to boundaries. Recent thinking on multistakeholder negotiations is useful in this distinction (Edmunds and Wollenberg 2001). This new approach acknowledges that power inequities and conflicting interests are unavoidable in negotiations and so-called participatory processes. This acknowledgment does not imply that such activities should be abandoned altogether, but rather approached with explicit attention paid to these differences and the degree to which they might be mediated. What is needed is a new ecology that does not unproblematically call for multistakeholder participation or community consultation but turns its intention specifically to revealing what kinds of obstacles constrain participation and collaboration. How might fractures between stakeholders be approached to encourage accountable, collective action for sound resource management, without requiring the resolution of all differences?

This distinction between dispute resolution and dispute management is significant because it affects not only policy objectives but also policy makers' approaches, namely, what methodologies, processes, and institutional arrangements are employed. If resolution is the goal, a tenuous consensus may be forced where it does not exist, allowing the mediators to consider the exercise completed. But achieving such fragile consensus inevitably silences some dissenting voices, which may emerge again to render any negotiated agreement ineffective. Instead, Edmunds and Wollenberg (2001) suggest that if inequities, disputes, and conflicting interests are *assumed*—that is, if the inherent politics of the process are directly confronted as part of the process—then the goal will be to manage such differences through mediation mechanisms and institutions, with explicit attention to the power differentials in these institutions and to improving accountability and access for the most disadvantaged stakeholders.

Such lessons are especially important in the current political climate in Indonesia, where increasing attention is paid to the need for community consultations—sometimes known as "socialization" (*sosialisasi*), a term that betrays the decidedly unparticipatory goal of such encounters. The rapidly changing political circumstances in present-day Indonesia since the resignation of Suharto and the onset of decentralization have altered the terms of the boundary struggles. While Suharto attempted to sever ethnic claims to territory to consolidate state control over national territory and the state forest, counterclaims by local governments and populist community movements and intensifying ethnic politics all battle (sometimes quite literally) with these blanket state territorial assertions. Local people are increasingly undertaking a "reclaiming" (the exact English phrase used) of land appropriated from village territories for plantations, industrial shrimp ponds, logging concessions, mines, golf courses, and resort facilities. Communities are successfully mobilizing to map their own boundaries and to submit them to local governments to have them removed from state-classified forest and placed under village management authority (Lynch and Harwell 2001), and they seem no longer content to simply be consulted, hungering for more meaningful participation in the management of their resources instead.

Although the Indonesian state administration may have become structurally altered under decentralization, the development imperatives pursued under the New Order go largely unchanged in the post-Suharto reform administrations. The issue now is to whom the spoils will go. Regional governments loom large in these new struggles for control, and it should not be assumed that decentralization has resulted in devolution of power to communities (Lynch and Harwell 2001). Unchecked, this context does not bode well for Indonesia's struggling democracy, pluralist management goals for its shrinking natural resource base, or its impoverished and disempowered populace. There is power in being able to enforce one's own ideas of territory. Given the increasing grip of international markets and state-sponsored extraction, these developments suggest that while there are great costs in attempting to define and officially recognize village territories, resource-dependent local people may have little other choice.

Notes

This essay is based on twenty-six months of field research during 1994–97, funded by the National Science Foundation (Grant No. SBR-9510422) and the Fulbright Foundation and sponsored by the Indonesian Institute of Science and Universitas Tanjungpura. I am grateful for the logistical assistance of the staff of the DSWR and for the field assistance of Valentinus Heri and Ade Jumhur. I am deeply indebted to the residents of the villages in Danau Sentarum for their hospitality, time, and insights, and to Michael Dove, Arthur Blundell, Eric Worby, Nancy Peluso, and Reed Wadley for their valuable input on earlier drafts of this essay. Any shortcomings that remain are my own.

1. *Hukum adat* is generally translated as "customary law," although this is a superficial definition. *Hukum adat* may be more accurately understood as the "moral ideal of social consonance and the behavioral imperative of propriety" (Geertz

1983: 207) and has roots in concepts of spirituality and connections between humans, ancestors, and their environment. In any case, it is a term with a much deeper cultural meaning and much broader implications than simply a "local resource management institution," as I will use the term in this chapter. However, as my essay will argue, these deep meanings are its greatest strength as a management institution.

2. Some of the data on local practices were collected by project staff Valentinus Heri (1996), Andi Erman and Rona Dennis (on village territory mapping, see Dennis 1997), and by KSDA (Konservasi Sumber Daya Alam, or Conservation of Natural Resources) staff Pangarimpunan Sinaga (1994a, 1994b, 1994c, 1996) and are supplemented by my interviews with field staff and government officials, as well as by my own ongoing independent fieldwork (from 1994 to 1998) and other project reports.

3. The lakes serve as spawning grounds for many Kapuas fish and have some of the highest freshwater fish diversity in the world, critical to the international conservation value of the area. The species list for fish of the lakes area currently comprises 315 species, many of them "new to science" (Widjnarti 1996).

4. However, such alluvial conditions are not present throughout the area, as many of the downstream areas are blackwater peat swamps or heavy acid clays unsuitable for agriculture. See Padoch 1982 for a description of this floodplain agriculture.

5. Other Dayak groups live around the edges of the reserve, but since they do not live inside its boundaries, they are not included in this analysis.

6. See also, "Horror Stories and Histories: The Cultural Politics of Ethnicity and Violence in West Kalimantan" in Harwell 2000: 196–226.

7. These systems are described elsewhere (D. Freeman 1955; Padoch 1982; Sather 1990; Sutlive 1978).

8. Berry (1993) noticed a similar attention to social networks in Africa, where because of the difficulty in gaining access to farmland, a proliferation of social networks that might increase such access took precedence over land management in people's time and capital investments.

9. Pulau Majang was established as a Dutch guard post in the late nineteenth century and according to local reports was quickly populated by Malays seeking access to fish resources and protection from Iban raiding. The seasonal use of the area stretches beyond the historical record—early reports from Veth (1854) and Enthoven (1903) report Malay and Dayak fishing in the area, although this use ebbed and flowed with waves of raiding.

10. I found little evidence, however, that the traders take advantage of credit relationships to buy fish at a low price. Rather, traders seemed to want to ensure that they had access to as much fish as possible (given the increasing number of traders competing for business).

11. See also Tsing 1993 for a description of Central Kalimantan Meratus Dayak mobility across farming landscapes.

12. Iban movement has been overemphasized as an "addiction to prodigal farming

methods" that forced them to constantly move in search of new farmland (D. Freeman 1955; Geertz 1983; Vayda 1961). However, Padoch (1982) showed that Iban also use alternatives to migration, including shifts to different swidden types such as swamp swiddens, the use of more labor-intensive cropping patterns, the management of secondary fallows, reduced fallow time, the borrowing of land, temporary wage labor migrations, and decreased fertility and population growth.

13. Male labor migrations, however, continue to be important sources of cash and social networks (Wadley 1997).

14. Although some Iban living on riverbanks have begun to raise *toman* in cages, far fewer have taken up the practice than Malays. This may be because of the high daily labor needed to catch large amounts of feeder fish for these carnivorous species, which need about fifty kilogram of food daily for a cage of about fifteen hundred fish. This labor may be needed for the agricultural pursuits that constitute the Iban primary form of subsistence. It is notable that among Malays, both men and women—usually jointly—keep *toman* and catch their food, whereas among Iban, it is exclusively men, because women are much more involved in rice agriculture.

15. Peluso (1996) recounts that among Salako Dayaks in the western part of the province, individuals have been planting rubber in descent-group gardens (*tembawang*), thus establishing circles of individual property within communal property.

16. These lands may encompass old house sites, cemeteries, or ritual groves, as well as communally held forest (Wadley, Colfer, and Hood 1996; Sather 1990). These rights are nominally recognized by national law (Basic Agrarian Law No. 5 1960, Article 5), although this legislation recognizes *adat* rights only insofar as they do not conflict with "national interests," an ambiguity exploited under the New Order to appropriate land and resources for the so-called common good.

17. Rose (1994) contrasts water rights in the western and eastern United States to show that property regimes are dependent on the nature and intended use of a resource, as well as the relations between users, and do not necessarily evolve in a unilinear manner from open access for all to common property to private property as a function of increasing scarcity. But see, in contrast, the economistic analysis of Demsetz 1967. Indeed, Rose argues that such a "scarcity" narrative is used as a rhetorical device to make private property seem the natural (and happy) outcome of all increasing scarcity scenarios, in spite of empirical evidence to the contrary.

18. Illustrating the situated nature of discourse, the same perspective that rejected individual ownership over God's gifts was not so successful for Native Americans, whose ideas of entitlement came into conflict with those of European settlers (Cronon 1983).

19. Trade in jungle produce among Borneo, China, and India dates to as early as the fifth century (Wolters 1967). According to Vaas (1952), probably no export market existed for fish until the early 1900s.

20. Located in the Arsip Nasional (National Archives), in Jakarta, Indonesia.

21. National Law No. 5 1974 on Concerning Basic Principles on Regional Administration.
22. National Law No. 5 1979 on Village Administration
23. See Lynch and Harwell 2001 on the intent and impact of the 1979 village administration law.
24. In accordance with the newly appointed president Suharto's goal of economic development through the extraction of natural resources, one of the first laws passed after the beginning of the New Order was the Basic Forestry Law No. 5 1967.
25. A debate continues among many Iban today over whether the concession was granted on behalf of the entire Iban community and then later co-opted by the three leaders. The leaders maintain this was not the case, but the dispute sheds some light on the knotty problem of who represents a community, or if such representation is locally accepted.
26. Granting this concession to local Iban leaders was extraordinary in a period when timber concessions were issued to military generals, members of Suharto's family, or their elite business partners. There are other unique aspects of tribal-state relations in this region. For example, in recognition of their participation in anti-Japanese resistance during the Second World War, some Dayak war leaders (including some from the lakes area) receive pensions from the government. Dayaks in West Kalimantan openly carry shotguns for hunting, which is illegal in all other parts of Indonesia. It is no small feat that local communities have been able to leverage such influence (even if in the case of the concession that influence has been co-opted by a few), and it is likely a reflection of the state's deep anxiety regarding the influx of foreign influence beyond its reach in this border area.
27. Conference on Wetlands of International Importance, Ramsar, Iran 1971 (see www.ramsar.org).
28. National Law No. 4 1982 on Principles of Environmental Management and National Law No. 5 1990 on Conservation of Natural Resources and Their Ecosystems.
29. Although relocation was a logistic and political impossibility, local people continued to fear displacement and enclosure of the reserve's resources. This was stimulated by the fact that Jakarta officials continually brought up the idea of relocation in project meetings, right up until the end of the contract. The fear of displacement and the idea that resources belong to the state proved powerful rhetorical tools for park guards and local officials.
30. The bees are migratory and so not every honey tree will bear a nest every year, although the bees do show preference for particular trees.
31. Wickham (1997a, 1997b) notes that the increased prices paid by the project induced traders also to increase their prices.
32. This method for ensuring local equity may be costly to sustainability, however. If winning a chance to fish in these locations comes only once, if at all, there is little incentive to limit fish catches. Also of concern is the practice in some villages of

selling more than one ticket per entrant. If unlimited tickets may be bought by a single entrant, obviously the lottery has become more of an auction than a protection strategy for distributive equity.

33. Further evidence of this resentment was illustrated by the public confrontation at a 1995 project seminar between the Iban *temenggung* (traditional leader) and project staff for what the former perceived as the smearing of the Iban name in a project document. The project was fined an *adat* settlement and required to make a public apology in local newspapers.

INTERPRETING "INDIGENOUS PEOPLES
AND SUSTAINABLE RESOURCE USE"
The Case of the T'boli in the Southern Philippines

A remarkable dimension of the current development discourse is its special focus on indigenous peoples, in particular their traditional knowledge systems and resource management strategies, as a source of insights into sustainability (Escobar 1995). Indigenous sociocultural systems are being plumbed for vital lessons and insights into conservation and environmental protection. Berkes (1999) suggests that this growth of interest in traditional knowledge has been spurred both by ecological insights from indigenous practices and by the need for a new ecological ethic in the face of natural resource depletion and environmental degradation.

Despite its appeal, the term *indigenous* is vulnerable to a certain romanticism, if not reification. As argued in this volume's introduction, the common portrayal of indigenous peoples as bearers of special knowledge of resource management can present them in a static, somewhat naively conflated category devoid of voice and agency. This "new traditionalist" discourse (Sinha, Gururani, and Greenberg 1997) essentializes indigenous peoples as actors following a predetermined script rather than as active and creative peoples in a dynamic, historical engagement with the rest of the world.

Whereas there is some validity to the argument that indigenous communities are often characterized by a culturally embedded conservation ethic supported by a traditional community-based political organization, references to indigeneity are sometimes trapped within claims of a past golden age. The concept of indigenous peoples living in homelands where they practice essentially sound and sustainable livelihoods is selectively invoking history to suit contemporary purposes (Cohen 1985). The premises behind this concept should be examined to see whether it is "only a case of imagining a past, (or) in an unintended way, also imagining the present" (Resurreccion 1999: 262). The conceptualization of indigeneity needs to be challenged in light of the contemporary reality of indigenous communities. In the Philippines, for example, most indigenous peoples are in a precarious state of recovering from extensive ecological destruction, displacement, and disempowerment. While there is significant temporal and spatial variation, the majority of these peoples have shared an experience during the past thirty years, in particular of internal colonialism and disenfranchisement, stemming from state-building activities, in-migration by settlers, logging, evangelizing, mining, agribusiness, and so on. Social stratification and polarization, resulting from increased relationships with the state and markets, have affected the values and goals of these communities. Power realignments and new ideological oppositions also have accompanied the increasing articulation between indigenous peoples, distant policy makers, and other stakeholders.

In the context of these changing relations among the natural environment, the state, and indigenous peoples, it will be argued here that indigeneity involves a continuous process of transaction and interchange. Caution is required when invoking notions of indigeneity, since the concept implies multiple contestations over power and identity (Peet and Watts 1996; Escobar 1995). It connotes politics of exclusivity and discrimination, and the achievement of the indigenous status is also relevant to the appropriation of rights and claims to particular territories and resources (Li 2000). In this essay, indigeneity is examined through the lens of political ecology, combined with a Gramscian analysis, to highlight various contestations, negotiations, compromises, and even cooperation among actors. Yet the analysis is not simply discursive, because beyond the "struggles over terminology" (Brosius, Tsing, and Zerner 2005: 7) also lies a struggle over territories and landscapes. In Southeast Asia, these territories usually involve forests and are, invariably, embroiled in resource politics (Bryant 1998; Bryant and Bailey 1997; Bryant, Rigg, and Stott 1993). As elsewhere in the third world, forests in the region constitute what Peluso and Watts (2001) refer to as violent environments, encompassing physical, symbolic, cultural, and emotional harm. The history of forest and resource use and management is politicized because rights are territorialized (Peluso and Harwell 2001; Vandergeest 1996). Indigeneity becomes more meaningful when it is seen as constitutive of forests as sites of competing interests. Indigeneity is actually a chameleon concept (Gray 1995) that has been deployed discursively and actively created, reshaped, and transmitted by different sets of actors at different times and places.

Indigenous Peoples, Land Rights, and Sustainable Development

The contemporary emergence of the concept of indigenous peoples needs to be viewed within two distinct social and political discourses. The first has been inspired by the politics of identity waged by indigenous peoples themselves, beginning in the 1970s, revolving around a sense of common concern and shared experiences in relation to states that disenfranchised them of their territories and culture. As in the cases discussed in chapters 4 and 5, the indigenous groups sought to capitalize on a history of neglect and persecution and used primordiality or the law of first possession (Brosius, Tsing, and Zerner 2005; Li 2001) as a political tool to garner political legitimacy. International meetings among indigenous leaders coalesced around an agenda that sought to project a cultural perspective to their territorial claims and assertion of their international sovereign rights (Wilmer 1993). This global reach strengthened the local articulation of indigenous rights even under conditions of tremendous duress from governments and encroaching capitalist interests.

The increasing visibility of indigenous peoples led to their accommodation in development discourses, and soon they became an object of transnational academic and development agendas. Indigenous peoples and indigeneity became important symbolic concepts in an era that increasingly subjected the notion of the nation-state to a critical examination. Interest in defining what is meant by indigeneity was heightened in the 1980s by the growing popularity of sustainable development and the attendant commitment to pluralistic resource governance and use. Given the desire for governance to not simply reflect power divisions or misrepresent disadvantaged groups (Jamison 1996), it was argued that new voices should be allowed into the arena of resource management. At the same time, the appropriation of the concept of indigeneity by indigenous peoples themselves has become a political project (Escobar 1998), and by employing resource use as a cultural emblem it also constitutes an ethnoterritorial movement (Grueso, Rosero, and Escobar 1996). The definition of indigeneity is a recursive process since the label is appropriated not only by the people in question but also by states and development agencies.

Indigenous peoples have been identified and labeled for multiple purposes, as is evidenced by a host of publications attempting to define and distinguish them from other sectors of a population (e.g., Kingsbury 1995; Gray 1995; Independent Commission on International Humanitarian Issues 1987; Burger 1987; United Nations 1982, 1983; Goodland 1982). The "rise of indigenous voice" (Friedman 1998: 568) led to a deepening of the understanding of their plight and conditions, yet as an empirical project of classification the result of the efforts has been problematic. There is still no uniform definition of indigeneity and interpretations are as varied as the number of publications that have come out. Nonetheless, from being a residual category, indigenous people have been able to achieve a more defined identity distinct from that of peasants on the basis of symbolic, place-, and culture-based assertions of their struggles (Rocheleau and Ross 1995). Notwithstanding the problems and lack of coherence in the application of the concept (Gray 1995), the label "indigenous peoples" has gained momentum in both local and international usage (Kingsbury 1995).

INDIGENOUS PEOPLES' MOVEMENTS

It is significant that the concept of indigeneity largely emerged out of the issue of land and resource disenfranchisement (Kingsbury 1995; Colchester 1995). This issue is linked to the upsurge of indigenous peoples' movements during the past half-century, originating in the Americas in the late 1960s and spreading internationally in the 1980s (Gray 1995). The Kayapo of Brazil, in their historic defense of their homelands, inspired similar struggles worldwide, and, alongside the Indian Chipko movement, became icons of environmentalism with cultural rights in the foreground. Environmental politics and activism permeated even beyond the boundaries of ethnoterritorial movements as demonstrated by the success of the Landless Workers' Movement (Wright and Wolford 2003) and the emergence of gendered grass-roots organizations (Rocheleau, Thomas-Slayter, and Wangari 1996).

The origins of indigeneity as a collective form of identity in political movements is reflected in Kingsbury's (1995: 33) list of requirements. These include, among others, self-identification and vulnerability, in addition to close cultural affinity with a particular area of land or territories. Many indigenous peoples in Southeast Asia derive much of their identity from their historically forest-based culture system (V. King 1998). Gray (1995: 57–58) urges us "to see the term [indigeneity] as a phenomenon to be understood and not as an analytic tool to aid understanding . . . as a political tool operating as an imperative term within a growing social movement." However, for indigenous peoples themselves, the culturally reifying label of "traditional environmentalist" poses a danger to their political advance as they negotiate their rights and the extent of their engagement with the larger society and economy (Laungaramsri 2001; Peluso and Harwell 2001; Ganjanapan 1998). Even as they have strategically deployed the politics of identity, their homelands remain contested territories (Gruner 2007) within the politics of community-based natural resource management (Brosius, Tsing, and Zerner 2005).

The visibility, intensity, and apparent strength of indigenous peoples' movements is reflected in the tremendous growth in the activities, publications, and advocacy of international and regional groups like Survival International, the International Working Group on Indigenous Affairs (IWGIA), Cultural Survival, Asia Indigenous Peoples' Pact, the Indian Council of South America, and many others. The establishment of an annual United Nations Working Group on Indigenous Populations (UNWGIP) in 1984 attested early on to the political impact of this movement. The UN declaration of the year 1993 as the International Year of Indigenous Peoples constituted another milestone in indigenous political activism.

What distinguishes indigenous peoples' struggles from other political movements is that their demand is not simply a human rights issue but an assertion of collective rights to land, culture, and, in many cases, self-determination. Grueso, Rosero, and Escobar (1996) refer to this as an ethnoterritorial movement, given the significance of territory, cultural rights, and self-representation in the face of externally-generated development. This movement emerged in the context of human rights violations, the plundering of resources and land incursions, transmigration and relocation, and even cultural genocide (Gray 1995). This is reflected in the fact that, as Colchester (1995: 61) writes, the claims made by indigenous peoples generally revolve around three central themes: the right to ownership and control of their territories, the right to self-determination, and the right to represent themselves through their own institutions. Invoking indigeneity became a "formula for encapsulating personal and cultural identity within a struggle for territorial decolonization" (Gray 1995: 56).

INDIGENOUS PEOPLES AND SUSTAINABLE RESOURCE USE

The rise of the indigenous peoples' movement was paralleled by the emergence of alternative models of development dedicated to greater social justice and environmental protection. Popularized by the United Nations' World Commission on Environment and Development (WCED) in the late 1980s, the phrase sustainable development quickly became a global catchword and the center of perhaps the most important discursive strategy for development. The phrase emerged from two significant concerns: the recognition of escalating environmental problems and an increased emphasis on community as the locus of development. Its appeal comes from its rhetoric, which encompasses an unstable amalgam of populist thinking (participatory and community-based approaches); a political agenda (equitable access to resources and decision making); increased attention to women and indigenous peoples; and an environmentalism deeply anchored in ethical and moral principles (Escobar 1995). In explicit opposition to an exploitative view of the society-environment relationship, sustainable development is dedicated to safeguarding the needs of future generations in the current management and use of natural resources.

The principles of sustainable development as laid out by the WCED articulated nicely with interest in the rights of indigenous peoples. Sustainable development, with its precepts of ecological capitalization and community-based resource-management strategies, meshes well with popular conceptions of indigenous peoples. It is widely thought that the production systems of indigenous peoples are based on a sustainable use of natural resources. This is attributed to the fact that their conservation practices are embedded in their sociocultural structure, whereas in modern industrial societies society and environment are supposedly separate and thus their relationship is extractive and unidirectional. The resource-management practices of indigenous peoples are seen as springing from a deep knowledge of the natural environment and its processes. Such knowledge, shared and conveyed through generations, is regarded as the key to their survival as communities. While admitting that traditional management systems are problematic in some respects, Berkes (1999: 161) nonetheless argues that they "inspire a new resource management science open to the participation of resource users in management, one that uses locally grounded alternatives to top-down centralized resource management."

Anthropologists' ethnographic accounts of indigenous peoples' management practices are drawn on to demonstrate the richness of their knowledge and to provide evidence of sound resource management (Berkes 1999; S. Stevens 1997; Colchester 1994). Even the World Bank has recognized that indigenous knowledge systems can provide lessons in sustainable development (Davis and Ebbe 1995). The manifest environmental crisis and the increasing vigilance of grass-roots civil society are obliging states to address resource governance even as they pursue the usual growth and development policies of state building. As a result, in the sustainable development discourse, indigenous peoples' society and knowledge have been reconstructed into an "eco-socialist" project (Pepper 1993). The state (and global) agenda of sustainable development has placed indigenous peoples at a central position in contemporary developmentalism, as recognized in the UN charter and other significant international documents, including the Draft Declaration on the Rights of Indigenous Peoples, ILO Convention 169, Agenda 21, and the Convention on Biological Diversity (CBD). States and international development agencies are promoting bottom-up models, like community-based resource management (CBNRM), ostensibly to strengthen local control and to accommodate community and tradition (Escobar 1995). In many instances, indigenous peoples have embraced principles of forest conservation and become willing and active agents of decentralized environmental regulation. This process epitomizes a kind of environmentality in which communities are transformed into environmental subjects requiring regulation and protection, following Agrawal (2005). Nonetheless, despite the rhetorics of alternative and more sustainable programs that claim to ameliorate the utilitarianism of early development regimes, indigenous peoples and their homelands continue to be objects of contest.

While these developments have brought recognition to indigenous peoples in some cases, they also have resulted in tensions and contradictions concerning the construction of indigenous peoples' identities (Brosius, Tsing, and Zerner 1998; Conklin and Graham 1995). The politics of translation and representation become more acute as opportunities and programs for indigenous peoples within the sustainable development discourse are defined, legally formulated, and implemented (Gatmaytan 2005). In particular, Li (2005) argues that programs like the CBNRM tend to oversimplify the meaning of communities, resulting in the misrepresentation and even marginalization of indigenous communities, an observation shared by a number of scholars (cf. Kingsbury 1998; Ganjanapan 1998; Ghimire and Pimbert 1997). This is reflected in a critique by the Indigenous People's Biodiversity Network (IPBN) of the shortcomings and ambiguities of international documents and policies concerning the definition of indigenous peoples and indigenous knowledge and the terms of the Convention on Biological Diversity (IPBN 2000). The process of cross-cultural translation alone has become an intrinsic political problem in the face of increasing differentiation, complexity, and the contingent character of indigenous representation (West 2005; cf. Gatmaytan 2005). Even as the debate goes on, its terms are being played out in specific local contexts, through peoples' daily activities and their interactions with their resource base.

Indigenous Peoples in the Philippines and Cultural Marginalization

The Philippines is characterized by not just biological but also cultural diversity. Throughout the Philippines there are at least one hundred indigenous peoples with distinct cultural characteristics. The terms *tribal Filipinos* and *ethnic or cultural minorities* are used to distinguish these indigenous peoples, who have historically retained more of their traditions, from those who were more acculturated during the colonial rule of both the Spanish and the American regimes. Collectively, the indigenous peoples account for just 20 percent of the country's population (amounting to between 4.5 and 7.5 million) but for a much greater percentage of the country's cultural diversity. They are also largely associated with the upland or mountain environment (or hinterlands) of the Philippines, the lowlands being where the mainstream ethnic groups dwell. Centuries of colonialism are responsible for this cultural and geographic dichotomy and the marginalization of the indigenous peoples from national, political citizenship.

THE ANCESTRAL DOMAIN AND STATE POLICIES

Colonial rule dismembered or outright eliminated ancestral domains—the term used by advocacy groups to refer to the traditional territories of indigenous peoples. The Regalian doctrine—which placed all lands in the Philippines under the Spanish crown—established the theoretical basis for the current national land laws. When Spain sold the Philippines to the United States, the latter respected Spanish land grants and church lands but ignored customary and indigenous rights to lands and resources. The indigenous peoples' loss of their ancestral domain was further exacerbated by the adoption of the American Torrens system, which legitimized individual property rights over lands. Within this system, rights to lands are recognized only if there is documentary evidence of tenurial claims. The Land Registration Act of 1902 and the Public Lands Act of 1905 declared indigenous peoples' land to be public land available for commercial development, as did the successive Public Land Law Acts of 1913, 1914, and 1919. Throughout the islands as a result, ancestral territories were absorbed into state territory as public domain. The development of land laws since then has given no recognition to the communal ancestral domains. State laws and policies have consistently discriminated against the land rights of indigenous peoples. As a result of colonial and postcolonial policies, as well as the politics of state building, ancestral lands have become public lands and the indigenous peoples have been turned into virtual squatters in their homelands (Leonen 1993; Lynch 1982, 1984, 1986; Fernandez 1980).

Following the global trends toward recognition of the sovereign status of indigenous peoples (Wilmer 1993), recent constitutional reform in the Philippines has included provisions that finally legitimize the status of indigenous communities. After the overthrow of the Marcos dictatorship, provisions pertaining to the special autonomy and rights of indigenous peoples were incorporated into the new constitution. However, the constitution still reflects contradictions in the government stance regarding the rights of indigenous peoples in the Philippines:

> All lands of the public domain, waters, minerals, coal, petroleum, and other mineral oils, all forces of potential energy, fisheries, forests or timber, wildlife, flora and fauna, and other natural resources are owned by the State. With the exception of agricultural lands, all other natural resources shall not be alienated. (art. 12, sec. 2.)

> The State, subject to the provisions of the Constitution and the national development policies and programs, shall protect the rights of indigenous cultural communities to their ancestral lands to ensure their economic, social and cultural well-being. The Congress may provide for the applicability of customary laws governing property rights or relations in determining the ownership and extent of ancestral domain. (art. 12, sec. 5.)

These passages reveal a continued state bias against ancestral territories, which are still "subject to the provisions of the Constitution and national development policies and programs." The primacy accorded to national policies and programs, such as hydroelectric power and infrastructure for export-oriented agribusiness, continues to result in the dispossession and relocation of indigenous peoples. Furthermore, the constitution leaves it completely to the discretion of Congress to recognize customary laws with respect to the determination of ancestral domain.

In response to mounting pressure from alliances of indigenous peoples and advocacy groups, the Philippine Congress finally enacted the Indigenous Peoples' Rights Act (IPRA) in 1998. Yet a lawyer challenged the constitutionality of the IPRA shortly after it was passed; and its implementation was stalled by transitions between successive government administrations. In 2001, the Supreme Court of the Philippines finally ruled that the IPRA was constitutional, and soon after the government implemented it. Yet the process for the recognition of ancestral domain that is laid out in the act involves a complex and lengthy procedure in the field and in the government bureaucracy. This includes a petition by the local community; proofs of claims through written accounts that specify customs, traditions, and other cultural markers; maps from and endorsement by the National Commission on Indigenous Peoples (NCIP); and the issuance and registration of a certificate of ancestral domain. This documentation involves lengthy transactions and negotiations with a variety of government agencies, a process for which many indigenous peoples in the Philippines do not have the resources or the expertise.

The Spaniards and the Americans never succeeded in colonizing the indigenous peoples in Mindanao. At most, the U.S. colonial government was able to set up agricultural colonies in Davao, Bukidnon, and a few other provinces in 1919 with landless peasants from Luzon and Visayas. Mindanao was regarded as a "wilderness," which continued to inspire the postcolonial government's transmigration programs in the 1950s. This period witnessed a massive exodus to Mindanao by Visayans from the central islands of the Philippines and Ilocanos from northern Luzon, followed by the arrival of agribusiness corporations (Tadem 1992: 7). This process of internal colonialism was accompanied by the development of paternalistic social relations between the migrant settlers and the indigenous peoples of Mindanao, resulting in an ethnic hierarchy that represents the latter as lazy and unproductive and needing sustained supervision from their "more advanced and civilized" fellow Filipinos (Duhaylungsod 1993: 140). This historic process of marginalization has been the catalyst for the development of cultural identity and political cohesion among the indigenous peoples in Mindanao.

Compared with the indigenous peoples of the northern Philippines, the non-Muslim indigenous peoples of Mindanao have garnered little academic interest. Early ethnographic work centered on the Manobo (Garvan 1931; Manuel 1973) and the Subanon (Frake 1955); the Higaonon of Bukidnon also have received some attention (Cole 1956; Biernatzki 1973; Cullen 1973; Arcilla 1979; Edgerton 1982). Since the late 1970s, however, interest in the other indigenous peoples of Mindanao has increased (e.g., Schlegel 1979; Duhaylungsod and Hyndman 1993; and, recently, Gaspar 1997; Nazarea 1998; and Manuta 2001), as has the political visibility of its Muslims. Martial law under the Marcos dictatorship and the common experience of disenfranchisement has contributed much to the evolution of a collective consciousness among Mindanao's indigenous peoples. This is signified by the appropriation by political advocates for these peoples of the Visayan term *Lumad* over the past twenty years. This term is used to refer to the non-Muslim indigenous peoples (Rodil 1990: 5) and literally means "indigenous" or "grown from the place" (Agbayani 1990: 25). It is also an acronym for Lumadnong Alyansa Alang sa Demokrasya—Indigenous Alliance for Democracy.

Lumad identity as a political community dates to 1983 and the activities of a group of advocates from the religious sector of Philippine society. In 1986, the common interests of different indigenous communities were forged into what is now Lumad-Mindanao—a coalition of local and regional organizations determined to defend their ancestral domains. Both Protestant and

Catholic organizations have helped facilitate this wider alliance. Since then, they have begun to join in national political struggles on behalf of ancestral domain and self-determination by the Philippines' indigenous peoples (Duhaylungsod and Hyndman 1993: 141–47; Hyndman and Duhaylungsod 1993; Rodil 1990; Bennagen 1985: 8–17).

Like indigenous peoples' organizations in the northern Philippines, Lumad-Mindanao built its political platform on the argument that its members have retained many distinguishing cultural characteristics that set them apart from the mainstream Filipinos. They argue that for them land is not something that can be expropriated and that their claims to land do not follow the Western concept of ownership. Even where there is a cultural notion of private property, they claim that there are indigenous rules on rights to dispose of the land.

The appropriation of the term *Lumad* in an attempt to unify the non-Muslim indigenous peoples of Mindanao represents a strategic form of negotiation within the state. However, the struggle of indigenous peoples in Mindanao has taken many forms apart from political mobilization. Claims to secure community territories are not only articulated through national political advocacy. The indigenous peoples of Mindanao are employing a variety of localized ways of representing their cultural identity, especially those groups that are involved both in negotiation with the state and in linkages with the wider economy. One such group is the T'boli.

The T'boli

The T'boli have not yet been reached by the Lumad-Mindanao coalition, but they are similarly engaged in asserting their cultural identity and defending their remaining ancestral domain. For example, a number of T'boli communities have reoccupied expropriated lands, which has become one of their best strategies for reclaiming ancestral domain (Duhaylungsod and Hyndman 1993: 150–60). Other strategies include strengthened local leadership and the revival, if not revitalization, of some of their cultural institutions.

THE PEOPLE AND THEIR ANCESTRAL DOMAIN

The T'boli inhabit South Cotabato and the newly formed Sarangani province in Mindanao. Their ancestral homeland encompasses a two thousand square kilometer heartland forming a triangle between the towns of Polomolok, Surallah, and Kiamba (map 1). The Cotabato Cordillera that extends along the coast for more than 190 kilometers rises abruptly from the sea in the T'boli homeland, extends inland for over 55 kilometers, and averages some twelve hundred meters above sea level in elevation. The T'boli

MAP 1 Location of Upo in the T'boli Homeland in Southern Minanao, Philippines

distinguish between coastal (Mohin) and mountain (S'bu) communities. Their total population was estimated to be about 150,000 in 1992 (Duhaylungsod and Hyndman 1993).

The government transmigration policy of the 1950s resulted in the invasion of T'boli lands by Ilocanos on the coast and Visayans in the interior lowlands—who eventually gained land titles. The T'boli have also been seriously threatened by a variety of other groups, including loggers, evangelists, gold miners, agribusinesses, and cattle ranchers (see Duhaylungsod and Hyndman 1992a, 1992b, 1993; Hyndman and Duhaylungsod 1990, 1992, 1993). The advance of this frontier into T'boli territories was made possible in part by state legal institutions that denied recognition of the T'boli ancestral domain claims.

Although many T'boli learned to speak Visayan and Ilocano languages as the migrant population increased, they continued to identify themselves as T'boli. Retreat to interior lands remained a cultural safety valve for many T'boli in the early years of colonization, which allowed them to retain their cultural autonomy. Presently, however, they find themselves directly confronted with and enmeshed in an antagonistic cultural system and logic of economic production.

ECONOMIC AND SOCIAL STRUCTURE

As is the case for most indigenous peoples in the Philippines and elsewhere in Southeast Asia, the T'boli livelihood was traditionally characterized by slash-and-burn cultivation (*tniba*) or swidden, supplemented by hunting, fishing, and the collecting and gathering of nondomesticated resources. In contemporary times, however, they are increasingly forced to engage in sedentary agriculture as a result of limited land resources, declining forest cover, and population growth.

The T'boli are polygynous, and extended families usually reside in scattered homesteads of three to four houses, where they claim and exercise rights to an ancestral domain under a local leader (*datu*). Land tenure for swidden agriculture and hunting follows ancestral rights. Each family inherits rights to land for cultivation and to their "spirit owner of rice" (*sfu halay*) (Forsberg 1988: 52) through the line of male descent. As one T'boli put it, "We are not aware of [land] titles. Our plants are our titles, for example the bananas, the bamboos, the big trees, and the waters. So even if we do not have titles, there was never any trouble among us, because we have our own law since our grandfathers were still living."[1]

The *datu* plays a key role in the social structure of T'boli society. A *datu's* following of mostly kin and affines usually constitutes a small-scale village. A *datu* is a man endowed with the attributes of *s'basa*. *S'basa* is the general T'boli term for all forms of reciprocal sharing (Duhaylungsod 1993; Duhaylungsod and Hyndman 1993: 62), which permeates the group's economic and sociocultural structure. T'boli women can achieve distinction in their community as *boi*, the female equivalent of a *datu*.

UPO: A RECLAIMED ANCESTRAL DOMAIN

The construction and representation of ethnic identity among the contemporary T'boli is achieved through symbolic inversion and boundary maintenance. Claims to both territory and cultural identity are asserted by deploying a sense of history and homeland, as in the case of the T'boli in Upo.

Upo is a T'boli Mohin village in Sarangani Province about seven kilometers from the nearest town of Maitum (map 1) and covering about forty-nine hectares. In 1998, Upo had a population of 1,195 consisting of 259 households. Until recently, accessibility to the community was through hiking the seven-kilometer distance between the village center and the municipality of Maitum. Today, mobility has been enhanced by road improvements and the consequent expansion of transportation facilities, making Upo accessible via a dirt road usable by three- and four-wheeled vehicles.

The community is dominated and led by the Kusin family. Perido Kusin, an educated, outspoken member of the family, argues that Visayan settlers and other non-T'boli people misrepresent the T'boli by referring to them as Tagabili, a pejorative term roughly translated as "peddlers" (Duhaylungsod and Hyndman 1993: 59). T'boli is a self-ascribed label recognized and widely used throughout the T'boli region. The use of the label *Tagabili* versus *T'boli* is symbolic of the ongoing divide in social relations between the migrants and the T'boli. Perido further explains that the T'boli themselves distinguish the Tao Mohin (T'boli of the sea), the Tao S'bu (T'boli of the lake), and the Tao B'lai (T'boli of the western mountains near the Manobo) (Duhaylungsod and Hyndman 1993: 159), which reflects the fact that identity representation is a relational matter (Resurreccion 1999: 37).

The ancestral domain of Upo forms part of the vast mountain ranges of the Cotabato Cordillera. The total land area of Upo is 2,450 hectares, 175.5 hectares of which, with some dipterocarps, have been declared a state forest reserve. Patches of secondary growth forest are still abundant, interspersed with the T'boli swidden farms. Important cultural landmarks such as sacred sites and tabooed and ceremonial areas, all of which are intimately known to the local T'boli and imprinted in their collective memory, cover the entire landscape.

In the 1940s, portions of Upo's ancestral lands were lost to an American named James Strong, who established a coconut plantation, and to a Muslim land speculator. The latter, with political power and access to legal resources, obtained title to and occupied some lands, although his actions were consistently contested by the Kusins. These actions exemplify the way in which the T'boli have been pushed to the interior as a consequence of encroaching outsiders. In 1993, however, realizing that interior lands were no longer available for retreat, the T'boli of Upo took the initiative to reclaim ancestral lands that had been lost to speculators. Kubli, a thirty-six-year-old who eventually became the *barangay kapitan* of Upo,[2] encouraged his covillagers, mostly kin, to stay put in the lands they were cultivating. Whereas this formed part of an effort to start protecting and projecting their indigenous identity, it necessitated some adjustment in their traditional livelihood of shifting cultivation, which was itself a culture marker. They began to engage

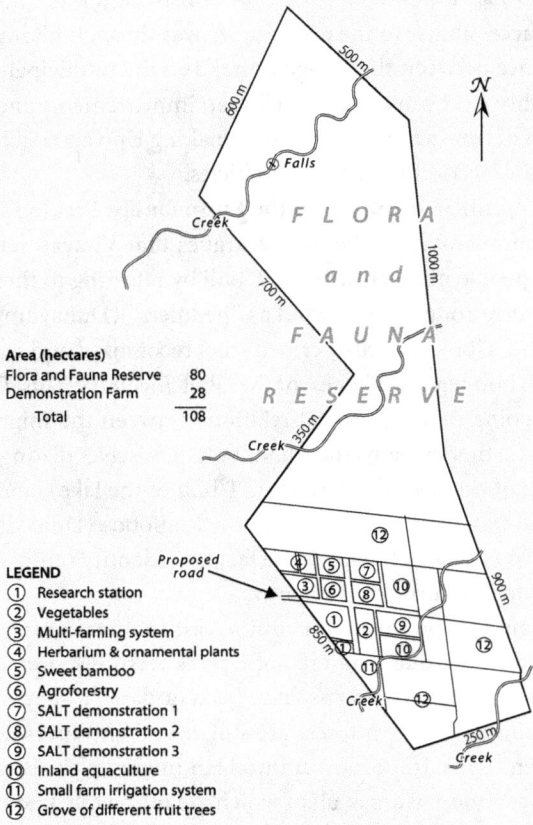

Area (hectares)

Flora and Fauna Reserve	80
Demonstration Farm	28
Total	108

LEGEND

Proposed road

① Research station
② Vegetables
③ Multi-farming system
④ Herbarium & ornamental plants
⑤ Sweet bamboo
⑥ Agroforestry
⑦ SALT demonstration 1
⑧ SALT demonstration 2
⑨ SALT demonstration 3
⑩ Inland aquaculture
⑪ Small farm irrigation system
⑫ Grove of different fruit trees

MAP 2 Map of Proposed Agricultural Station and Land-Use Zoning for Upo Lands

in more sedentary forms of agriculture, although they still maintained some of their swidden gardens in the forest.

The reoccupation of Upo's ancestral lands was based on an assertion of cultural identity, buttressed by the reassertion of historical ties with their forbearers. The accounts of early migrant settlers in Maitum, their differences with the T'boli notwithstanding, actually corroborated the Kusins' territorial claims. Their initiative was also supported by Kubli's personal connections with the local Municipal Environment and Natural Resources Office (MENRO), which provided a connection to the provincial Department of Agriculture (DA) for development programs.

The establishment of these links with state agencies during the subsequent decade led to the entry into Upo of a variety of development programs and to a marked increase in the political activities of the local government in the village—all occurring during a period when the development gaze of the state came to first include indigenous peoples. During the Ramos administration (1995–98), Upo was chosen as a pilot site for the Social Reform Agenda (SRA) program, which facilitated the extension to the village of various development initiatives, including the development of a spring, of roads, primary educational facilities, and various livelihood schemes. In addition, the new ties to state agencies resulted in a land-use zoning plan of the Upo ancestral domain (map 2). All these interventions led to dramatic transformations in the sociocultural and material conditions of the Upo T'boli.

FOREST PROTECTION AND MANAGEMENT SCHEMES

The spring development project involved the Davao Medical School Foundation, a medical nongovernmental organization (NGO), and the provincial and local governments. Many development projects come as a complete package, but in this case the Upo leadership simply requested the materials required to set up the water delivery system and then mobilized work groups of local T'boli men to carry it out. The project led to the formation of the Barangay Upo Water Sanitation Association (BUWSA), which directed the villagers to plant rubber and coffee trees around the spring, among other activities.[3] The provincial government provided water buffalo and goats for distribution to Upo households as part of a project aimed at livelihood improvement through an improved farming system. Sedentary agriculture is a relatively recent practice among the Upo T'boli and the use of draft animals and the plough is still quite new to them. In addition, an afforestation project was initiated by MENRO to replace the trees washed away by a flash flood in June 1999.

Another related development involves the organization of male villagers into a forest management group called *bantay gubat/ilog* (forest/river guardians). This group was formed even prior to the issuance of Sarangani Provincial Ordinance No. 11, which is the principal provincial legislation for the protection and conservation of the forest. Upo still has much to conserve: of the 2,450 hectares in the ancestral domain of Upo, about 1,307 hectare (53 percent) still lie under primary forest while, secondary growth covers over 1,000 hectares. The primary goal of the village forest management group is to deter outside timber poachers, and their initial performance looks promising. If this group can be sustained, and if the provincial ordinance is also enforced, the quality of the remaining forest cover of Upo, which has been declared a protected area by the *barangay* council, can likely be maintained.

The forest cover of the Upo ancestral domain is both a cultural marker and a resource base. In national as well as global development agendas, forests are seen as rich reservoirs of biodiversity that must be protected. This is the basis for the new role of indigenous people like the T'boli as "stewards" of this new form of capital (Escobar 1996; Schroeder and Neumann 1995). Village initiatives like the forest management group and the reforestation scheme reflect the potential for collaboration with external actors in this stewardship, although the underlying premises and motives are not the same.

HUMAN LIVELIHOOD INTERVENTIONS

Upo also has been the recent recipient of a number of development projects aimed at improving human livelihood. In 1998, the Mindanao Baptist Rural Life Center and the provincial government distributed water buffalos and goats to Upo households and sponsored a farming technology seminar for cultivating steeply sloping lands. The national Integrated Social Forestry (ISF) project also has been extended to Upo, giving each of the 98 T'boli households three hectares for agroforestry cultivation. A second batch of applications for this project from another 143 households is still currently pending.

Following discussions with the provincial and the municipal government, an Agricultural Research Center was established in Upo in 1998, complete with a ten-hectare nursery and agricultural research station (see map 2). The motivation for establishing the center was primarily to support future commercial forestry development on T'boli lands and in neighboring areas. Rubber seedlings have already been planted in the area.

The Barangay head of Upo, Kubli, lobbied the Fiber Industry Development Authority for a technical assistance scheme to develop T'boli's traditional *abaca* fiber craft. This scheme has supported massive and ongoing *abaca* planting in three *sitios* in Upo.[4] The local leadership hopes to use this to revitalize the traditional *t'nalak* weaving of *abaca* craft goods and to transform it into a commercial enterprise.[5]

The Women in Development of Sarangani, a program created in 1995 by the provincial government and now active throughout all rural municipalities, has also reached Upo. The goal of this program is to support women's empowerment through income generation. Under its auspices, the women of Upo have been organized into a task force and have been given livestock and sewing machines to improve their livelihoods. In addition, the Upo chapter is playing a role in forest management by accepting weeding contracts from farm owners in the nearby communities. They are also in charge of the Upo day care center and have established an herbal garden beside it.

Social relations embedded in the traditional subsistence production of indigenous peoples are often very different from the social relations associated with the logic of commodity production and market economies (Duhaylungsod 1998; Duhaylungsod and Hyndman 1995). An increased linkage to cash economies inevitably reshapes social and ecological relations. Engaging T'boli women in development and increasing their articulation with the market, both measures of sustainable development, may radically transform the way in which the Upo T'boli eventually see themselves.

Continuities and Transformations

The greater inclusion of the Upo T'boli in the orbit of development has increased their access to the benefits of modernization such as social services and education. The most significant benefit has been the reclamation of much of their ancestral territory. Together with their other resources, this will greatly facilitate the material and cultural revitalization of the Upo T'boli. The social and political space created by Upo's increased access to state agencies has resulted in a number of opportunities for sustainable resource management. Further degradation of the forest has been halted, and there are visible signs of regeneration in the quality of the ecosystem. Collaboration with the state has also resulted in adjustments and changes in the social and political organization of Upo. It remains to be seen, first, whether these efforts will be long-lived and, second, whether they will in fact ensure the community's survival as a people with a distinct culture.

Upo has become a development showplace for both the provincial and the municipal governments, which holds pitfalls as well as promises. The government's discourse of "green growth" and sustainable development may lead to intensified resource use and to the greater commodification of Upo's traditional resources. There are already indications of this in the *abaca* fiber cultivation development project, among others. The Upo T'boli will have to grapple with these and other sociocultural consequences of increasing commodification and the attendant threats to an indigenous economy. As Peet and Watts (1996) observe, commodification destabilizes not only control over resources but the entire indigenous society. Traditional concepts of social capital, which permeate the T'boli sociocultural system, may be especially vulnerable to the massive opening of the T'boli to these external forces. The logic of the market economy, with its ethic of individualism, generally works against the norms of reciprocity and community solidarity that underlie the traditional subsistence economy.

The shift from a largely subsistence- to a more market-oriented agroforestry system by itself entails important changes in traditional production relations. In particular, the encouragement to engage in more intensive *abaca* production will have an impact on gender relations. *Abaca* development is leading to an increased involvement by women in marketing (which has been further enhanced by the road improvement). Previously, men and women more or less equally shared all livelihood activities. But today, men are focusing on farm and forest activities whereas women are focusing on market-related activities. In the past, men were more involved in occasional market transactions and had more control over market-based income, which tended to be used for group reproduction activities (like drinking). Today, with women responsible for most market sales, the income is used more for household reproduction. This shift has implications for women's empowerment, but its precise impact still needs to be further analyzed.

Notwithstanding some recent empowerment of indigenous peoples, their rights and resources are still at the mercy of state prerogatives, especially in terms of overall national development. Tensions arise when both local customary and national statutory laws come into play, because they usually pursue opposite goals. An example is the Integrated Social Forestry (ISF) project. The Upo headman and the local MENRO office believe that the community is not able to exercise adequate control of the project areas because of the terms imposed on them by the national government through the Department of Environment and Natural Resources. The local leadership hopes for a revised or alternate scheme that will free them of such constraints. In addition, the prospects of regaining the lands appropriated by the American James Strong, which are strategically located in the center of the ancestral domain of Upo, are bleak because they have already been titled under national law. Gatmaytan, a Filipino lawyer and anthropologist, has noted that legal advocacy for indigenous peoples may contribute to their legal protection, but it also "results in many of the nuances and meanings inscribed in the indigenous languages of space being lost in translation" (Gatmaytan 2005: 467).

Upo is also threatened by new forms of land appropriation in the context of sustainable development and environmentalist discourses (similar to those noted by Schroeder and Neumann 1995). For example, the land-use zoning plan has set aside lands for a "flora and fauna zone" to promote natural capital. A substantial parcel of 108 hectares also has been appropriated as an agricultural research area, to be managed by the provincial government. The planned agricultural research center will provide services not just to Upo but to the entire province of Sarangani, and since the Upo T'boli do not have educational qualifications as agronomists or horticulturists, it is unlikely that they will play any role in the center except as laborers. Since it is a provincial government project, the government will likely eventually appropriate the area. In addition, the research station may destabilize local rights to Upo's social forestry project lands that it adjoins and that are but weakly controlled by the community at present

These developments among the Upo T'boli illustrate the complex transactions and negotiations that indigenous peoples face today in their interaction with the state and other external actors. From indigenous peoples, they have been turned into what Agrawal (2005) described as willing and active environmental subjects and agents of decentralized environmental regulation and conservation. The issues at stake concern sustainable livelihoods and cultural survival. The spaces opened up by these new external relations are spaces of hope, but they are also spaces of further contestation.

Conclusion

It would be anachronistic and misleading to invoke the conservation and resource management of indigenous peoples as simply a rich and traditional example of the integrity of the relationship between nature and culture. On the contrary, their environmental relations are an arena of constant negotiation, as communities constantly reconstruct and transform their social and ecological realities according to new social conditions and events. As shown in the T'boli case (and similarly in Harwell's West Kalimatan case; see chapter 6), the forces of colonization and the postcolonial activities of state and development agencies have profoundly altered the landscape of most indigenous peoples. At the same time, and in spite of the continued precariousness of their claims to traditional territories, many indigenous peoples are struggling to sustain their communities.

Notwithstanding a history of dispossession and marginalization, many indigenous peoples are now using political activism to call attention to the problems they face. This struggle has enabled some of them to win inclusion in both global and national political fora. Their political capital has been strengthened by the global environmental crisis, which has led to a development reorientation that valorizes the resource-management practices of indigenous peoples. These converging trends have led to an identification of indigeneity with the promise of sustainable development.

In the face of increasing articulation with the larger society, however, a reconfiguration of power relations and resources within communities is inevitable. As Cernea (2000) argues, risks come with all development. Social, economic, and cultural transformations are simultaneous processes that will continually impinge on traditions in communities and their resource management practices. To buffer these changes, the challenge facing indigenous peoples' communities is to build resilience and adaptive capacity, which implies "learning to live with change and uncertainty; nurturing diversity for re-organization and renewal; combining different types of knowledge for learning; and creating opportunity for self-organization" (Folke, Colding, and Berkes 2003: 383).

From engaging in the politics of identity that celebrated indigeneity, many indigenous peoples are now challenged to explore social and livelihood alternatives and to create new possibilities for themselves under the discourse of sustainable development. Their traditional resource-management domains, although they are the basis of their cultural identity, are not unchanging. Indigenous societies are continually evolving as the result of a dialectical relationship between themselves, their natural environment, and the state regime under which they live (Hyndman and Duhaylungsod 1994: 105). Indigeneity is a constructed concept that indigenous peoples have to continually articulate. The concept of indigeneity has been essential to their recent political struggles and has had practical consequences for them. But its ultimate test lies in how it will help or hinder indigenous peoples to define their choices, actions, and goals as they continue to engage in a struggle for material and cultural survival.

Notes

I am grateful to the Center for International Forestry Research (CIFOR) and the Gender Program of the University of the Philippines, Los Baños (UPLB), for some of the data on the T'boli Upo, gathered under the research project "Creating Space for Local Forest Management."

1. Perido Kusin, T'boli. Field notes of Levita Duhaylungsod and David Hyndman in Upo, April 1991.
2. A *barangay* is the smallest political unit in the villages in the Philippines. The *kapitan* is the *barangay* head. Kubli Kusin eventually rose to becoming a municipal councilor.
3. In 1990 the national government, through the Department of Environment and Natural Resources, issued the Master Plan for Forestry Development that included the Integrated Social Forestry project, participation in which is one of the central sites of negotiation over ancestral lands.
4. *Sitios* are household clusters that comprise a *barangay*.
5. *Abaca*, the fiber plant source for *t'nalak* weaving, grows wild in the forests of the region.

III

**RECONSTRUCTING AND REPRESENTING
INDIGENOUS ENVIRONMENTAL KNOWLEDGE**

EIGHT

Endah Sulistyawati

THE HISTORICAL DEMOGRAPHY OF RESOURCE USE
IN A SWIDDEN COMMUNITY IN WEST KALIMANTAN

Tropical forests are major reserves of biodiversity and play a crucial role in maintaining ecosystem services critical to the biosphere as a whole (Noble and Dirzo 1997), as well as providing resources for human beings. Human actions in utilizing the forest resources represent a form of disturbance that potentially alters the forest structure and composition and, in its extreme cases, converts the forest into other land uses such as agricultural fields. The type and frequency of such disturbance eventually affect the long-term dynamics of the landscape (Forman and Godron 1986). Therefore the composition of the landscape is, in part, a consequence of historical events (Dale et al. 2002: 649) associated with land-use decisions carried out by various agents.

In the forest margins in many parts of the world, the farm household is the major agent of land-use decisions (e.g., Dove 1983a, 1983b; Vosti and Witcover 1996; Walker and Homma 1996; An et al. 2001; Perz 2002; Roy Chowdhury 2007). Vosti and Witcover (1996) argue that the availability of resources (natural, human, and capital) that can be brought to achieve the household's primary objectives, and constraints imposed by external conditions (e.g., policies, institutional arrangements), are the key factors determining land-use decisions. In this context, household demography affects land-use decisions through its role in determining the availability of resources, particularly household labor for farming and the level of domestic food demand. Household demographic characteristics include the size and composition of the household, which changes dynamically as the household goes through its developmental or life cycle (Fortes, 1971) due to the demographic processes of aging, birth, marriage, death, and migration.

In many traditional communities, the formation, proliferation, and dissolution of households are regulated by the community's sociocultural norms related to marriage, such as marriage preference, prohibitions, and postmarital residence (e.g., Lee 1988). Generally, those norms also encompass matters like access to land resources and inheritance (e.g., J. Hansen et al. 2005; Ferrera 2007; Takane 2007). While in some traditional communities land resources constitute a common community property in which member households only receive user rights (e.g., Takane 2007), in others, the household is the main unit for land (property) acquisition. The latter is exemplified by the case of the Kantu' and Iban of Borneo (Dove 1985b, 1985c; D. Freeman 1992). In those two tribal groups, the demographic history of households affects the pattern of property accumulation. The presence of a productive male in particular is critical to determining the ability of a household to accumulate title to land, which is established by clearing primary forest for swiddens. The community's social rules regulate how the household's lands thus acquired will be inherited by each member as they start to establish their own households. In the Kantu' and Iban cases, it is clear that while the dynamics of household demography determine how many resources can be accumulated, the social rules of inheritance determine how these resources will be distributed through the generations.

Swidden cultivation is an interesting case for exploring the interactions between the household economy, household demography, and socioculture because many aspects of the swidden communities' sociocultural life, such as social structure, rituals, and rules, revolve around their system of swidden cultivation (Dove 1983b; Dressler 2005). In most swidden communities, land and labor are the main resources in the agricultural system, and the dynamics of these resources are regulated by the prevailing social system. Community rules regarding tenure regulate access to land resources. Similarly, the social system affects the organization of production and consumption, in which the household is generally the basic unit. The household is the main source of labor in most swidden cultivation communities; therefore the extent of cultivation by the household correlates with its structure and composition (e.g., Dove 1984).

Despite its prominent occurrence, especially in the outer islands of Indonesia (Sunderlin et al. 2000), the merits of the swidden cultivation system have long been debated. Opponents regard swidden cultivation as a wasteful land-use system because it requires much land and has low productivity. Proponents, however, maintain that swidden cultivation is a logical choice for farmers in many parts of the tropics who face the problem of inherently poor soils and limited access to chemical inputs. Natural processes occurring during the fallow efficiently restore soil fertility and control weeds (e.g.,

Kleiman, Bryant, and Pimentel 1996; Ramakrishnan 1992: 370), and this enables poor farmers to grow food crops with relatively minimal labor inputs. Dove (1983b) has argued that much of the condemnation stems from a willful ignorance toward the culture of swidden cultivation communities. Many aspects of their life and agriculture are misunderstood, and therefore the economic benefits of this system are not fully appreciated (Cramb 1989a). An understanding of the cultivation system, including of its sociocultural aspects, is thus crucial to identifying the system's strengths and weaknesses, so that the positive can be enhanced and the negative mitigated.

This essay explores the connection between the dynamics of the household and the dynamics of the landscape, as mediated by land-use decisions among households, using a computer simulation model of the swidden cultivation system. The model provides a framework to study the interactions between the economic, sociocultural, and environmental factors of the swidden cultivation system over the longer term. The case study presented is the swidden cultivation system of the Kantu', an indigenous group living in West Kalimantan, Indonesia. The model is developed based on data mostly derived from Dove's reports on the Kantu' (mainly Dove 1981, 1984, 1985b, 1985d, and 1993c). They are supplemented by data from other similar ethnic groups in Kalimantan and elsewhere in Borneo (Drake 1982; D. Freeman 1992; Padoch 1982) and my brief observations of Dove's study site carried out in November 1999.

The main strength of this simulation model, as compared to other simulation studies of the swidden system (e.g., Wilkie and Finn 1988; Gilruth, Marsh, and Itami 1995), is its extensive representation of the sociocultural aspects of the swidden community. This is the basis for this study's contribution to the central themes of this volume, particularly the issue of indigenous environmental knowledge in resource use. The approach taken here also allows an examination of how indigenous knowledge, together with other factors of resource use and culture (e.g., labor, access to land) and under conditions of increasing population pressure, affects the pattern of landscape transformation and the sustainability of the swidden cultivation system itself.

The goal of the analysis is to use this modeling approach to help us understand the strengths and limits of the swidden cultivation system. This essay will first present the ethnographic background of the Kantu', followed by a description of the model. In the second part of the essay, the model will be used to assess the future of a hypothetical Kantu' community living under increased population pressure. The third part will present simulation experiments in which some core assumptions of the model (i.e., agricultural strategy and constant commodity prices) are modified. Finally, the limitations of the model and possibilities for further modifications will be discussed.

Kantu'

The Kantu' ethnically belong to the Ibanic group of peoples and inhabit the region along the Empanang River (see map 1 in chapter 3), in addition to the headwaters of the Kapuas River.[1] Dove studied one subgrouping of the Kantu', the Melaban Kantu' living at Tikul Batu longhouse,[2] located near the intersection between the Empanang and Kantu' Rivers. The longhouse was established in 1957 by nine households and occupied a territory of about 10 km². By 1976, there were sixteen households, with a total population of 123 persons. At the time of Dove's study, the area was isolated and only accessible by river.

The Kantu' lived in a largely subsistence economy in which the swidden cultivation of rice (typically mixed with maize, cassava, and vegetables) was the main activity. They also engaged in several nonsubsistence activities to earn cash for buying basic trade goods including salt, tobacco, kerosene, and cloth. Cash-earning activities included working for wages across the border in Sarawak, growing pepper, and trading; but tapping rubber was the primary source of cash for the Kantu'. This combination of subsistence and cash cropping has been widely found among swidden cultivators in Indonesia (e.g. Angelsen 1995; Gouyon, de Forest, and Levang 1993; Lawrence, Peart, and Leighton 1998). Usually rubber seedlings are mixed with crops in the swidden and left to grow along with some other species when the swidden is fallowed. Income from rubber provides a crucial cushion in the event of economic misfortune, especially harvest failure, when farmers have to buy additional rice. Rubber cultivation and other market-oriented activities, as Dove (1981: 85) suggested, remained subsidiary to swidden cultivation, their central economic activity.

The Kantu' have two major social groupings, the longhouse, which is equivalent to a village, and the household. The longhouse consists of multiple apartments, each of which is occupied by a household (*bilek*). The household is the primary unit of production, consumption, and land appropriation. Swiddens are cultivated largely by household labor. Extra household labor is only sought when needed. Rice is produced mainly for household consumption, but it is occasionally given to others as a loan or wage payment. Marriage marks an important phase in the household developmental cycle. Newlywed couples usually practice a period of postmarital residence with the parents of either the wife (uxorilocal) or the husband (virilocal). The departure of an out-marrying child means that he or she will lose rights to the property of the natal household but will enter into the rights of the new household. An out-marrying child will thus not inherit any land from the parent. After a period of postmarital residence, married couples leave the parents' household to establish their own through a process of household partition. An exception to this involves the youngest sibling, who usually will stay permanently with the parents and thus perpetuate the household as a unit.

Households acquire ownership to land by felling primary forest to make a swidden. The feller has the foremost right to reuse the secondary growth (fallow) plot; nevertheless, under certain circumstances others can borrow the plot for a season. In household partition, rubber groves and other property are divided between the parents and the child's newly formed household, but this procedure does not apply to fallow swidden plots. In this case, the rights to reuse the secondary growth plots are shared between the parents (first generation) and the households of the children (second generation) created through partition (but not with out-marrying children). However, these rights cannot be shared with additional households created by the partition of the children's households (i.e., the third generation, from the perspective of the feller's household). In other words, the rights to secondary growth plots can only be shared between two successive generations (Dove 1985d: 18–19). Dove (ibid.) asserted that when third-generation households are formed, the first and second generations will divide their rights to secondary growth. Households of the second generation could then share their rights with the households of their children without violating the "two successive generations" rule.

The role of the indigenous environmental knowledge of the Kantu' in resource management can be seen in the way they select sites for swiddens. The Kantu' tend to make multiple swiddens in different locales with respect to vegetation type, drainage characteristics, and altitude. The aim is to maximize the benefit of different microenvironments and at the same time to minimize risks. In selecting the swidden sites, the Kantu' employ several different criteria, including vegetation type, soil, ownership, and proximity to the previous year's swidden, other households' swiddens, and the household's other current swiddens. In terms of vegetation type, they assess the differences between swiddens made in primary and secondary forests in terms of labor requirements and productivity. Dove's data (1985d: 377) suggested that the labor input for swiddens made in primary forest is much less than that for secondary forest, that is, 88 and 168 man-days per hectare for primary and secondary forest, respectively. The difference mainly derives from the absence of weeding in the former. However, secondary forest generally yields higher productivity due to the difficulty of achieving a thorough burn in primary forest (it is not a matter of soil fertility). The Kantu' generally want to balance the benefits of the two types of swiddens. Therefore each household ideally makes at least one primary forest and one secondary forest swidden each year (Dove 1985d: 78).

The Model

The model is designed to simulate the land-use decisions made by individual households and to examine their consequences for the landscape and the welfare of the community. The model simulates the development of a landscape from the beginning of territorial occupation by pioneering households. In this model, the history of each person (e.g., age, sex, birth, death, and marriages) and each household (e.g., number of members, wealth) is followed through time. Land-use decisions to cultivate swiddens and tap rubber are simulated as a function of the household's consumption needs and the availability of land and labor. The model also takes into account the land tenure system, which regulates land appropriation as well as inheritance. The model does not treat households as homogenous units. Household differentiation can arise in many ways, and this modeling exercise helps to show how demographic events bring it about.

The model uses a rule-based approach as opposed to a mathematical or statistical one. The dynamics of the system are simulated according to a set of rules that are technically represented as a series of IF-THEN-ELSE statements. These rules constitute a necessary simplification of the complexities of real life. Some of them are rules that already exist in the community regarding such matters as incest, inheritance, or postmarital residence; but others are assumed based on the information available. The model is organized in four modules: population dynamics, land-use decision making, vegetation dynamics, and production evaluation. In general, the simulation is run on a yearly basis, except in part of the land-use decision-making module, where a 365-day calendar is used to simulate the execution of swidden cultivation tasks (e.g., slashing, felling, and harvesting). The following section provides a brief description of the model. More detailed accounts of the model can be found in Sulistyawati (2001).

THE MODEL'S COMPONENTS

The model encompasses several discrete entities, which can be classified into human entities (persons and households) and physical entities (blocks, plots, and landscapes). Each entity has a set of attributes, many of which are more or less analogous to those we find in a real life (table 1). The landscape

TABLE 1 *The Main Components of the Model*

Person	Household	Block
• Identity number	• Identity number	• Identity number
• Household affiliation	• Head	• Coordinate location
• Age	• Heir	• Plot affiliation
• Sex	• Senior household	• Vegetation type
• Biological father	• List of junior households	• Years since disturbance
• Biological mother	• List of members	
• Adoptive father	• List of owned plots	**Plot**
• Adoptive mother	• List of cultivated plots	• Identity number
• Marital status	• List of borrowed plots	• List of blocks
• Spouse	• Rice stock	• Owner
• List of children	• Cash stock	• Tenant
• Years since last birth	• Rice debt	• Year acquired
• Residence type	• Cash debt	• Method of acquisition

is represented spatially as grids of blocks, each of which has an area of 0.5 hectare. For the current application, the landscape initially consists of free-access primary forest blocks (figure 1a). A plot is a management unit that consists of a group of blocks. Plot formation occurs when primary forest is cleared for swiddens, with the result that a group of blocks is taken out from the free-access land pool and placed under the ownership of the clearing household.

THE POPULATION DYNAMICS MODULE

The population module simulates five demographic events: birth, adoption, marriage, death, and household partition. Birth, marriage, and death are simulated randomly for each eligible person based on a set of probabilities derived from demographic rates such as age-specific fertility and mortality rates. These probabilities were used to determine whether an eligible person in a given sex and/or age would give birth, get married, or die in any one year. The model simulates a closed community in which the population size is entirely determined by birth and death. Each person belongs to a household through birth, marriage, or adoption. An adoption is triggered for a

(a) Initial vegetation land-use

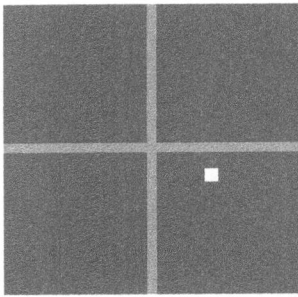

(b) Topography

▨ Primary forest	☐ Class 1 **Low altitude**
▨ River/transport route	▨ Class 2
☐ Longhouse	■ Class 3
	☐ Class 4 **High altitude**

FIGURE 1 Initial landscape configuration

childless couple when the wife is older than forty years, with a sibling's child being given the priority for adoptee (D. Freeman 1992: 19). The model simulates marriages among single or widowed persons of marriageable ages (fifteen to forty-nine for females and fifteen to sixty-four for males), except among those prohibited by incest taboos. Incest taboos prohibit marriage between persons living in the same household, siblings (including half siblings), first cousins, uncle and niece, and aunt and nephew. Spouse selection is based on the preferred age difference, which is simulated randomly with a mean age difference between husband and wife of five years and a standard deviation of three years. The model assumes that a newlywed couple practices a period of residence with the parents of either the wife or the husband. The decision on the location of the postmarital residence depends on the size of the landholdings of the two households involved and the membership status (see below) of the marrying persons.

The model assumes a hierarchical structure within the household in which each person is assigned one of the following membership statuses: head of household, heir, or member. In deciding the location of the postmarital residence, a head or heir will not change residence, instead always bringing the spouse into his or her household. If the statuses of both marrying persons are the same in this respect, the person from the household with the smaller number of plots moves to the household with the larger number. The subsequent departure of a married person from the parents' household generally is triggered by the presence of two married siblings in the same household at the same time. In this case, the earlier married sibling who has stayed more than five years will move out from the parents' household (senior household) to establish a separate one (junior household), by means of household partition. Household partitions can happen several times within the same household. On each occasion, the junior household is given some of the household's rubber plots, rice, and cash but none of the secondary forest plots. The secondary forest plots are shared between the senior and the junior household(s) until land inheritance takes place, which is triggered by the partition of one of the junior households. On this occasion, all junior households acquire their own portion of secondary forest plots, so that they no longer share land with the senior household.

THE LAND-USE DECISION-MAKING MODULE

The model simulates a community living in a subsistence economy in which the level of agricultural production is geared toward fulfilling the household's customary day-to-day requirements (Sahlins 1972: 77) rather than maximizing profits. Household labor constitutes the main means to achieve this goal. The demographic structure and composition of the household will influence the level of agricultural production, a relationship that was noted by Chayanov (1966). The model simulates two agricultural activities, namely, the swidden cultivation of rice and cash production by tapping rubber. Given the subsistence-oriented nature of the simulated community, it is assumed that the size of the household (the number of consumers) determines the household's demand for products. Furthermore, the model assumes that households have a strong preference for growing their own foods. Accordingly, cultivating swidden always takes precedence over tapping rubber, which means that tapping rubber is done only when there is no work in the swiddens. This assumption reflects a characteristic of many swidden cultivation communities including the Kantu' (Dove 1993c: 142–43), the Mualang (Drake 1982: 142–46), and the Iban (D. Freeman 1992: 269–70), in which rubber is not seen as an alternative to rice production. Cramb (1989a: 108) summarizes the reasons for the strong preference toward the swidden cultivation of rice, including a taste preference for traditional varieties, the benefits of producing intercropped vegetables at low opportunity cost, and the security it provides against fluctuation of market prices for rice and rubber.

Based on these assumptions, the household's rice production in this model is driven by the demand for eating rice, seed rice, and rice debt. The magnitude of the demand changes dynamically throughout the simulation. The demand for eating rice is the sum of the annual rice requirement of each

household member, which is calculated from the age-sex daily rice requirements provided by Dove (1984: 128). The magnitude of the rice debt is
determined by the debt history of each household. The ability of the household to fulfill these demands depends on its labor force, that is, the number
of its workers (persons aged fifteen to sixty-five). The model also specifies
that at least two workers are required to cultivate each plot. In other words,
the model imposes a maximum number of plots that a household with a
given number of workers can cultivate.

The model distinguishes two types of swidden cultivation: swiddens
planted with rice only and swiddens in which rice is intercropped with
rubber. Intercropping rubber with rice is the only method for rubber planting in this model. When such a swidden is fallowed after one year of rice
cultivation, rubber will eventually dominate the stand. In any one year
households may cultivate one or more swiddens depending on the level of
rice demand. However, rice-rubber swiddens are not made every year. The
model assumes that when land is available, each household attempts to
ensure the continuity of rubber production by having at least one mature
(tappable) and one immature rubber garden. Rubber planting is therefore
triggered soon after the most immature plot has come into production. The
model assumes that rubber is planted only when the household has at least
two fallow plots. This implies a relatively low rubber-planting frequency.
The size of the rubber plot is set at one hectare. Thus if rubber is planted in a
two-hectare swidden, only one hectare of it becomes the rubber garden. The
extent of swidden cultivation for each household is simulated by repeatedly
adding one more swidden to the household's cultivation list until the expected rice yield has exceeded the rice demand *or* until the number of
swiddens has attained the maximum.

During the process of swidden selection, each household is assumed to
select the best sites from the free-access primary forest blocks or existing
(secondary forest) fallow plots. A site index, as described in table 2, is used to
summarize the relative attractiveness of the potential sites (plots or blocks).
The factors to be considered in the swidden selection (F) are based on Dove's
report (1985d: 55–99). The weights (W) in this formulation reflect the relative importance of each factor. As there are no empirical data on this matter,
the relative importance of these factors is approximated, but the weight
scoring follows the procedure described in Eastman (1997: 9–10).

The complete steps for scoring site selection factors will not be described
here (see Sulistyawati 2001), but the basic principles are as follows. The state
of a free-access block or plot with respect to each site selection factor is rated
on a one-to-five scale, with the higher numbers corresponding to increasing

TABLE 2 *Site Index*

Site index $= F_1W_1 + F_2W_2 + F_3W_3 + \ldots + F_8W_8$

Factor	Description	Weight
F_1	Distance to longhouse	0.0533
F_2	Closest distance to the river	0.0533
F_3	Closest distance to previous year's swidden	0.1487
F_4	Farthest distance to same-year swidden	0.0533
F_5	Closest distance to other households' swiddens	0.1487
F_6	Vegetation type preference	0.3404
F_7	Ownership preference	0.1487
F_8	Topographical preference	0.0533

preference. With regard to topography, sites at lower altitude are preferable, and thus given higher scores, than those at higher altitude (figure 1b). Distances are classified into five classes and scored based on the following assumptions. The magnitude of preference increases as the distance decreases relative to the longhouse, the river, the previous year's swidden(s), and other households' swiddens. These assumptions reflect a desire to minimize travel time. The preference to make swiddens close to the previous year's swidden(s) in particular is related to the need to harvest nonrice cultigens from the previous year's swidden and to reuse the swidden hut (Dove 1985d: 62–69). The preference to locate swiddens close to other households' swiddens reflects the need to facilitate labor exchanges between households. Having several swiddens adjacent to each other also reduces the severity of pest attacks since the attacks are thereby spread over a larger area of cultivation (Dove 1985d: 70–75). These strategies are not unique to the Kantu'; other swidden communities have been reported to pursue similar strategies (e.g., Colfer 1997; Mackie 1986). On the other hand, the site preference decreases with decreasing distance from the household's other current swiddens. This reflects the desire to diversify the environmental characteristics of the household's swiddens made during the same year (Dove 1985d: 69).

As far as vegetation preference is concerned in the model, the score assigned to each vegetation type depends on the state of the household, particularly with respect to labor availability and the type of swidden intended. When making rice-rubber swiddens, "old" (and less productive) rubber gardens are the preferred choice. When making rice swiddens, the model assumes that each household wants to balance primary forest and secondary forest. Therefore each household attempts to clear at least one primary forest plot per year,[3] but this attempt is made only if the household contains at least one male adult, otherwise they clear secondary forest plots. If the household plans to clear primary forest, then primary forest blocks are assigned a higher score than other vegetation types. When the plan is to clear a secondary forest plot, potential sites are evaluated according to the duration of fallow. Fallow duration is classified into long, medium, and short fallow, with the corresponding length of ten, five to nine, and less than five years, respectively. Generally, long fallow plots are given a higher score than other vegetation types, except when there is no adult male in the household. In that case, the medium fallow plot becomes the preferred one. In all cases, short fallow plots are the least preferred ones. In this model, fallow duration is the only factor determining the productivity of the land. In practice, rice yields decrease as the fallow is shortened. The rice yields of swiddens made in primary forest, long-fallow secondary forest, medium-fallow secondary forest, and short-fallow secondary forest are set at 500, 750, 500, and 300 kilograms per hectare, respectively.[4]

With regards to ownership, the model takes into account the possibility of borrowing land, but it imposes restrictions in terms of who can lend and who can borrow land. Only households that have five plots or less are eligible to borrow in the model; and only those who have more than five plots for themselves as well as for each of the junior households with whom they share land are classified as eligible to lend. The best plots in terms of fallow length are reserved for the owner's use and are not lent.[5] The model's restrictions on the lending side are designed to ensure that owners reserve a sufficient amount of land to avoid having to cultivate short-fallow plots. In the model's selection process, noneligible plots are eliminated from the list of potential sites. Among the eligible sites, free-access lands and owned or shared plots are given higher scores than plots owned by others. These rules apply only for selecting sites for rice swiddens. Because of tenurial implications, rice-rubber swiddens can only be established in free-access primary forest or on secondary forest owned by the household in question.

Once the site index is calculated for all potential plots in the model, the one with the highest site index score is selected for the swidden. If the selected site is a fallow plot, the entire plot is recleared. If it is a primary forest block, the model attempts to delineate a two-hectare plot (consisting of four connected one-half hectare blocks). However, as more and more of the land is allocated to households in the model, this goal may not be achievable, in which case the model attempts to delineate a one-hectare plot. There are only two possible sizes of plots—one or two hectares.

Within the site index, considerations of vegetation type are given more weight than ownership and spatial considerations. Recall that long-fallow secondary forest is the most desirable plot and the short fallow is the least desirable one. This means that, for example, households would likely choose a long-fallow plot located on a less desirable spot over a short-fallow plot located on a more desirable spot. Such an over-weighting of vegetation criteria might not always be the case in reality. Mackie (1986: 52) reported that in her study area farmers often ended up choosing an inferior site to farm simply because it was near an established farming group. The assumption in the current model thus implies a conservative strategy in which the selection of short-fallow plots would be avoided except in extreme cases, such as when the household has limited landholdings. Therefore the selection of short-fallow plots in the simulation always reflects a situation in which the pressure on land is high.

Whereas the model as a whole operates on a yearly basis, the execution of cultivation tasks (i.e., slashing, felling, burning, planting, weeding, and harvesting) is simulated on a daily basis. The timing of cultivation tasks is based on the labor schedule provided by Dove (1985d: 377–81). The timing of the tasks for each plot is simulated randomly within the reported calendar range. In the case of households with multiple swiddens, the household's labor is divided among the plots (with a minimum of two workers per plot). The model allows labor assigned to one plot to be used in the other one when it would otherwise be idle. If a task is constrained by time and the household's labor does not suffice to finish it within the specified time frame, the household makes a demand for external labor. This demand is only triggered by a labor shortage. Likewise, unused household labor is placed in the pool of labor available for hiring. After all tasks are completed for each household, the model simulates the labor-hiring process by matching the daily labor demand and supply. When demand is greater than supply, it is assumed that labor from outside the village is hired. In this way, each household has opportunities to both hire and perform wage labor. Wages are paid in rice at the rate of 3.5 kg (unhusked rice) per worker per day (Dove 1984: 115). In reality, a number of labor-exchange arrangements exist, such as gift labor, cooperative labor, and reciprocal labor (see Dove 1985d: 24), none of which involve rice payments. Yet hiring labor is the only method of recruiting external labor in this model. Once labor expenditures for swidden cultivation have been calculated, it becomes possible to calculate the use of the remaining labor available for tapping rubber, which is simulated according to the module described next.

THE PRODUCTION EVALUATION MODULE

This module first evaluates whether rice currently available (rice stored, rice harvested, and rice earned from wages) can meet the household's demand. Second, it simulates cash production from rubber tapping. The production of rubber is driven by the annual cash demand for buying basic trade items (sugar, cooking oil, kerosene, textiles, and salt),[6] for repaying cash debts, and for buying market rice in households whose own production fails to meet demands. In other words, rice shortfalls will intensify rubber production. In this model, tapping is simulated per worker per half hectare rubber block on a daily basis when there is no swidden work, though the exact timing is not set. The model assumes low rubber productivity, with production from mature and old rubber plots set at two and one kilograms per day, per worker, per half hectare, respectively. Tapping can be carried out on a maximum of 180 days a year.[7] Shared tapping arrangements can provide access to other households' tappable plots, with the proceeds being divided (50:50) between the tapper and the owner (Drake 1982: 146; D. Freeman 1992: 269). If the proceeds from tapping rubber fail to cover the deficits in rice and cash, a loan-seeking procedure is triggered for needy households.

THE VEGETATION DYNAMICS MODULE

Vegetation is classified as swidden, shrub, young secondary forest, old secondary forest, primary forest, immature rubber, mature rubber, and old rubber. This classification to some extent corresponds to the way in which a number of ethnic groups in Kalimantan classify vegetation based on the sequential changes following swidden cultivation (e.g., Colfer 1997: 94–113; Dove 1985d: 50–51; Mackie 1986: 76; Wadley 1997: 95–96). In this model, the timing of transition from one state to the next succeeding state is determined simply by the number of years elapsed since the swidden was fallowed. The model specifies two paths of vegetative changes depending on the type of swidden. Following the abandonment of a rice swidden, the succession proceeds in the following order: shrub (one to five years old), young secondary forest (six to twenty years old), and old secondary forest (at least twenty-one years old). For an abandoned swidden previously planted with rice and rubber, the succession is immature rubber (one to ten years old), mature rubber (eleven to forty years old), and old rubber (forty-one to fifty years old).

The model simulates a hypothetical Kantu' community living in a territory of 12.5 km^2 (1250 ha), divided into a landscape of 50 x 50 blocks of one-half hectare each. This simulated area is close to the estimated territory of the Kantu' at Tikul Batu, that is, 10 km^2. The initial community consists of nine households (the same as the number of households that established the longhouse of Tikul Batu), each of them having seven members (arbitrarily assigned). The landscape is initially covered by primary forest. The configuration of the initial landscape is shown in figure 1b. The topography is an arbitrary construction representing a longhouse located in a valley. The duration of all simulations in this analysis is seventy-five years, which is set to correspond to the period 1957 to 2032. The timing of Dove's initial study would correspond approximately to year twenty of the simulation. Since the model is a stochastic (random) model in which each run yields a nonidentical result, multiple runs are required to capture the variation in the dynamics of the system. The seventy-five-year simulation was therefore repeated twenty times and the results are presented as averages of the twenty replicates.

Simulation Scenarios

The dynamics of the swidden system will be investigated by running the model under a number of simulation scenarios (modes). In the first scenario, the model is run using all the rules and assumptions described in the previous section, and this simulation set is called the default mode. The results from the default mode will be presented to provide an overview of the dynamics of the system. Although numerous aspects of the system can be investigated using this model, the present analysis focuses only on select aspects, namely, the consequences of population pressure; the pattern of agricultural expansion and its implications for the landscape; and the landholding distribution among households and its implications for fallowing practices as well as the economic welfare of the community in terms of rice and cash sufficiency. In two other scenarios, some of the rules and assumptions of the default mode are changed to address specific questions. One core assumption in the default mode is that swidden cultivation is always given priority over tapping rubber. In the second scenario, this assumption is relaxed by allowing households to make a choice between growing rice in swiddens or focusing on tapping rubber to earn cash that is later used to buy rice. In the third scenario, an exogenous driving force in the form of price fluctuations is introduced into the system.

Analyzing the Dynamics of the Swidden System of the Kantu':
Results from the Default Mode

POPULATION DYNAMICS

During the seventy-five-year run of the model, the simulated population increases from 63 to 421 persons, with the number of households increasing from nine to eighty. The increase follows an exponential growth curve at an annual rate of 2.6 percent. This magnitude of population pressure is very close to the estimated rate of population growth in the Nanga Kantu' village, to which the study site belongs administratively, for the years 1984 to 1999, which was 2.9 percent. During the model runs, the population density in this model increases from around five to thirty-six persons per square kilometer, which overlaps with the ranges reported for traditional forms of swidden cultivation in Kalimantan (figure 2a). But the simulation ends at a higher density than these reported figures, which implies that the situation in the final year represents very high pressure on the land.

More insights on households' dynamics can be obtained by examining the distribution of household size (figure 2b). The distribution of household sizes approximates a bell shape, which is to some extent similar to that reported for a number of Borneo communities (D. Freeman 1992: 10; Drake 1982: 80; Wadley 1997: 66; Dove, personal communication 1998). The wide range of household sizes indicates the presence of households at different developmental stages. The small households are mostly newly formed ones, while the larger ones are likely to be households that contain married children and grandchildren and that have not yet undergone household partition. This bell-shape pattern remains more or less constant over time in that a few small and a few very large households are always present. Significant differentiation in terms of the demography of households thus always exists at any given time.

PATTERN OF THE AGRICULTURAL EXPANSION
AND ITS IMPLICATIONS ON THE LANDSCAPE

Apart from population growth, the extent of swidden cultivation of each household and the preferences of each household concerning the best location for swiddens are the other main factors that determine the pattern of agricultural expansion. The simulation results show that the average number of swiddens per household ranges from 1.4 to 2.6 over the entire simulation period, with the corresponding cultivated area ranging from 2.5 to 5.2 hectares per household. This compares with Dove's figures for the Kantu', which show each household each year making on average 2.3 swiddens with a total

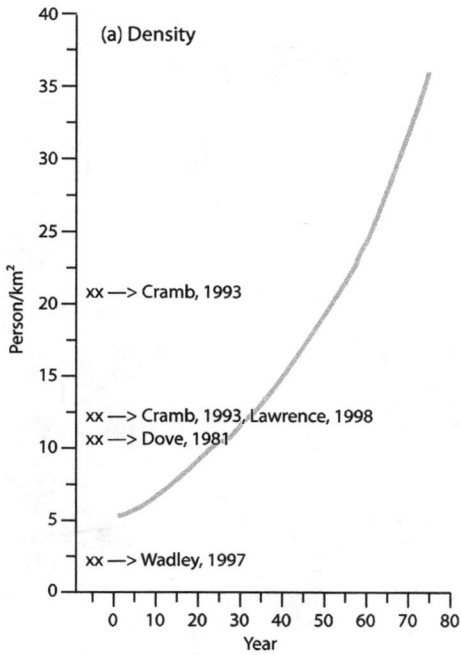

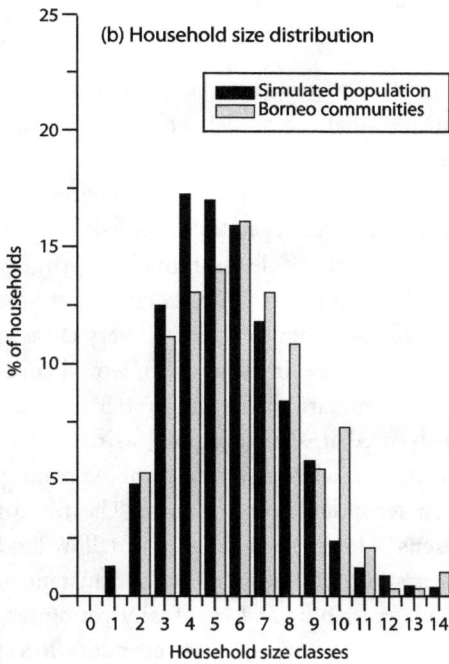

FIGURES 2A AND 2B
Population density and household size distribution

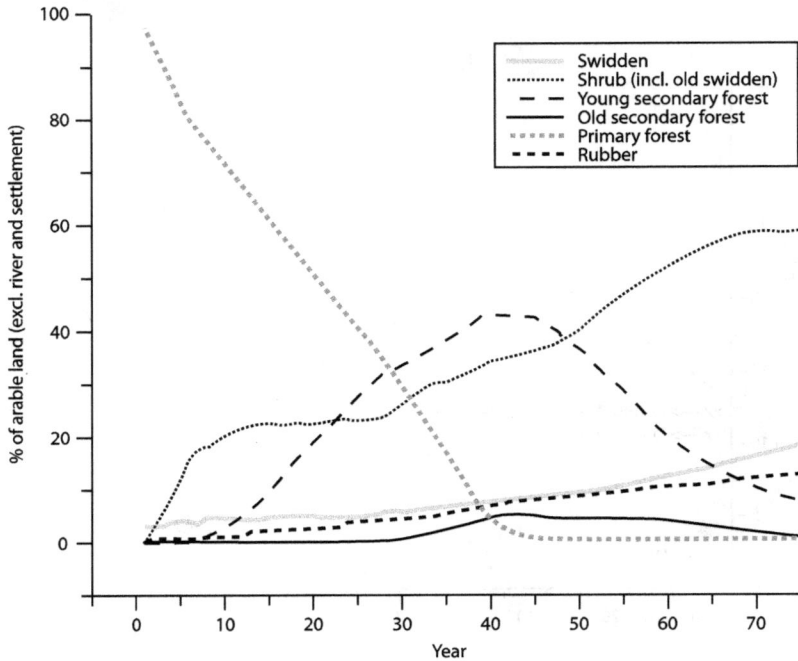

FIGURE 3 Vegetative composition of the landscape

area of 4.6 hectares (Dove 1981: 81, 88). The fact that the simulated values encompass the empirical ones indicates that the model is able to present a more or less realistic rate of agriculture expansion.

The extent of each vegetation type in any given year in the simulation is a function of the rates of both land clearing and vegetation succession. With the model's current assumptions about land-use decision making, primary forest decreases at an average rate of around twenty-five hectares per year and disappears entirely after forty-five years (figure 3). This is very close to what happened in reality. During my field visit in 1999, which would compare to year forty to forty-five of the simulation, the Kantu' told me that primary forest had been completely used up. Furthermore, as secondary forest fallow plots begin to accumulate, households start to combine making swiddens in primary forest and in secondary forest fallows. The rate of fallow land conversion into swiddens is lower than the rate of fallow land formation because primary forest is still preferred in some circumstances. Consequently, the shrub (one to five years old) and eventually young secondary forest (six to twenty years old) accumulates. Old secondary forest starts to appear after about twenty-five years from the initial clearing of primary forest yet the extent of old secondary forest is generally small. After all primary forest is cleared the stock of young secondary forest rapidly diminishes, since all swiddens now have to be made in fallow land. In addition, the plots of young secondary forest with a fallow period of more than ten years are cleared first since this is the preferred vegetation. As population grows, the area of swiddens continues to increase and consequently so does the shrub land, which is formed following the fallowing of swiddens. At the end of the simulation, shrub land dominates the landscape.

Access to Land and Its Implications for Fallowing Practices

LANDHOLDING DISTRIBUTION

There are two mechanisms by which households acquire ownership to land: first, by clearing and making swiddens in primary forest and, second, by inheriting land from parents. We can distinguish two critical periods in the development of a community that determine the distribution of secondary forest fallow holdings. The first period occurs before the disappearance of primary forest. Although land inheritance may also take place in some households, clearing primary forest is the main means to acquire land during this period. Since clearing primary forest requires the presence of at least one male worker, the distribution of landholdings is largely a function of the age of the household and the continuing presence of male workers. Households that are established earlier and contain male workers simply have more chances to accumulate plots. The rate of land accumulation is especially high in the nine initial households; the average landholding among these initial households after twenty years is twenty-six plots. This simulated figure is slightly less than the average rate of land accumulation among pioneering Kantu' households twenty years following initial settlement, which was thirty-one plots (Dove 1985d: 79). The second period occurs after all primary forest has been cleared, so that land inheritance constitutes the only mechanism to acquire land. During this period, the amount of land acquired through inheritance depends on the size of the parents' landholdings, the number of eligible children, and, from the perspective of receiving children, the pattern of postmarital residence in the parents' household. Since out-marrying children are not eligible to inherit land, households with a high frequency of in-marrying spouses will have a larger number of eligible children and thus a smaller land inheritance for each child.

The longitudinal trend in the distribution of landholdings is presented in figure 4a. Before discussing the pattern further, it should be borne in mind that due to the time lag between household formation and land inheritance, there are always groups of senior and junior households sharing the land.

Therefore the size of actual landholdings only partially reflects the access to land. Access to land by senior households is always less than what they formally hold since some of the lands may be used by junior households. Similarly, junior households may formally hold no land at all. Yet this does not mean that these households have no access to land since they are entitled to use their senior household's lands. During the simulation, the proportion of households sharing land is high, amounting on average to 70 percent each year. Therefore a better assessment of a household's access to land would need to take into account the number of junior households with which its holdings are shared. For this purpose, a proxy measure of household access to land is calculated as the adjusted number of plots that each household would acquire if land inheritance was carried out that year. This calculation takes into account the number of eligible siblings and the duration of their stay in the parents' household.

Interhousehold variation in landholdings over the course of the simulation is presented in figure 4b. The number of households with adjusted landholdings of five plots or less is particularly important since these households are likely to practice short-fallow rotation. As expected, the data show that the size of this group increases dramatically during the last twenty-five years of the simulation. It is interesting to note that when the population pressure is at its highest level (at the end of the simulation), there are a few households (6 percent) that are well endowed with land (more than fifteen plots). The presence of these very large landholding households, albeit small in number, presumably indicates that even under increasing population pressure, random chance alone will result in the persistence of a few large landholding households.

The emergence of land inequality has been reported for many swidden-cultivation communities including the Kantu' (e.g., Coomes, Grimard, and Burt 2000; Cramb 1989b; Dove 1985d; Padoch 1982). During my own field visit to the Kantu', I found wide variation in landholdings, ranging from as few as three to as many as twenty plots per household. While inheritance constitutes the only mechanism to transfer land ownership in this model, selling and buying land happens quite often in reality. Note that there is less inequity in rubber holdings than in swidden land. No land-sharing arrangements exist for rubber plots. The low intensity of rubber planting assumed by the model results in relatively uniform rubber holdings, with most households owning one to five plots (figure 4c). In comparison, Dove (1993c: 138) reported that each Kantu' household owned an average of five rubber plots in 1976.

(a) Distribution of households according to the 'formal' fallow land holding

(b) Distribution of households according to the de facto fallow land holding

(c) Distribution of households according to the rubber holding

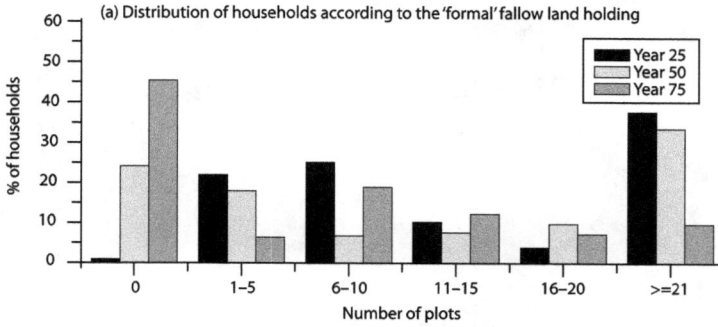

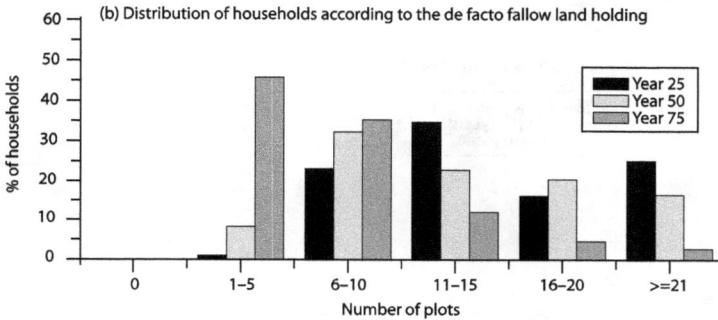

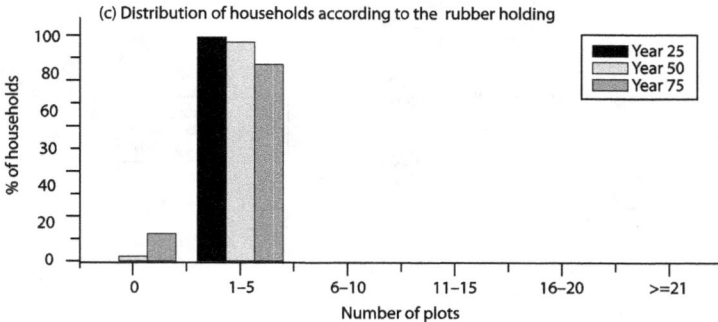

FIGURES 4A, 4B, AND 4C Distribution of fallow landholdings and rubber landholdings

FALLOWING PRACTICE

The most immediate implication of the differential access to land concerns the length of the fallow. Figure 5a presents the average age of forest (i.e., fallow length) cut for all nonprimary forest swiddens. In the simulation, the initially increasing fallow length corresponds to the period during which the fallow plots accumulate as a result of continuing primary-forest plot clear-

(a) Average fallow length of non primary forest plots

(b) Distribution of plots by fallow classes

Short fallow (1–4 years)
Medium fallow (5–9 years)
Long fallow (>=10 years)

(c) Average land holding by fallow classes

Short fallow
Medium fallow
Long fallow

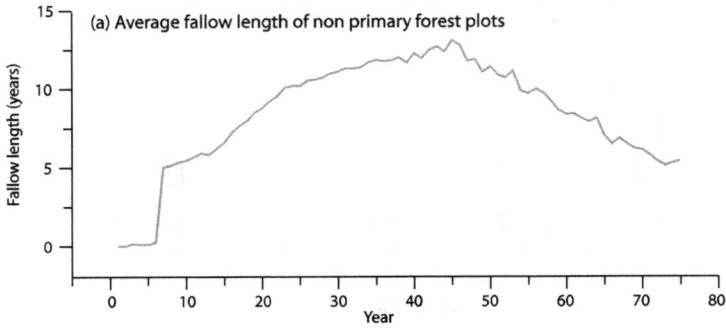

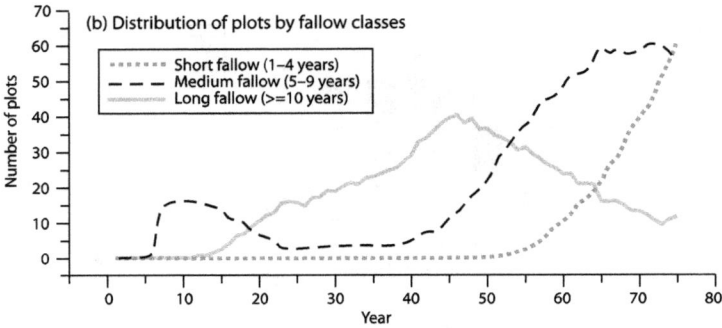

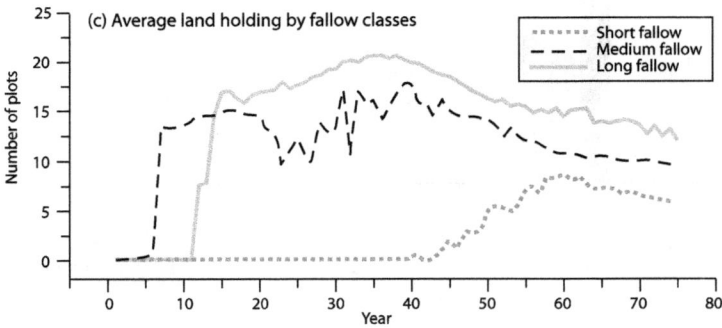

FIGURES 5A, 5B, AND 5C Fallowing practices

ing. The average fallow length declines following the disappearance of the primary forest. By comparing the distribution of swiddens based on their fallow length (figure 5b) and the landholdings of the households in question (figure 5c), it is clear that differences in landholdings affect fallowing practices. Those with more access to land can observe the desired long fallow, whereas others are forced to shorten their fallow periods. Almost half of all swiddens are made in short-fallow plots (one to four years old) by the end of the simulation, with most of the rest cut from medium-length fallow plots.

Rice Sufficiency

My analysis of rice production focuses not on the absolute level of production but the extent to which overall production can meet demands. At the end of each simulation, an evaluation is carried out to determine whether the rice currently available—which consists of the previous year's swidden harvest, rice stores from earlier years, and rice earned from wage labor—can meet the annual rice demand—which consists of eating rice, seed rice, and rice for paying wages. Hence a variable called rice sufficiency is constructed, which is defined as the proportion of the rice currently available to the annual rice demand. To facilitate an assessment of the interhousehold differentiation in sufficiency, households are divided into three classes as defined in table 3.

The results of the simulation (figure 6a) show that during years twenty through fifty, an average of 54 percent of households attain the high sufficiency level, which is similar to the empirical data reported for the Kantu'. According to Dove (1985d: 294), in 1975 and 1976, 50 percent of households were able to meet their consumption and seed rice requirements from a combination of harvested and stored rice. As the simulation continues, however, a general decline in rice-sufficiency levels manifests. By the end of the simulation at year seventy-five, around 80 percent of the households failed to meet their rice demands. While in reality the success of burning, the extent of pest attacks, and other such environmental factors also affect household rice production, in this model it is determined by access to land and labor only.

The length of fallow periods solely determines the level of rice yield in this model. This means that the shorter the fallow the household is obliged to accept, the more land it needs to cultivate to fulfill its rice demands. However, this is eventually limited by the size of the household labor force, because the model imposes a maximum number of plots that a given number of workers can cultivate. Labor is thus the ultimate constraint on rice production, although it is mediated by a lack of access to more productive land.

The simulations show that as the population increases, the proportion of households with small landholdings increases (figure 4b) and the average fallow length decreases (figure 5a). The reduction in rice yield due to

TABLE 3 *Rice Sufficiency Classes*

Class	Degree of rice sufficiency
High sufficiency	Rice currently available can meet the annual demand
Medium sufficiency	Rice currently available is 50–99 percent of the annual demand
Low sufficiency	Rice currently available is less than 50 percent of the annual demand

shortening fallows is obviously a factor in the general decline in the rice sufficiency level. The simulations also show that the length of fallow correlates with the size of landholdings (figure 5c), such that households with smaller landholdings tend to practice shorter fallows. It is tempting to attribute the occurrence of low to medium rice sufficiency entirely to the problem of decreasing fallows. However, households with low and medium rice sufficiency levels appear even in the early years of the simulation, years during which access to land is not yet a problem. This should prompt us to examine the role of other production factors that pertain not to the land but to the demographic structure of households.

Rice demand is a function of the demographic structure of the household, which determines both the size of the household (number of consumers) and the size of the labor force (number of workers). Holding other factors constant, the larger the household, the greater the rice demand and consequently the larger the area of swidden needing to be cultivated. But the household labor force imposes a maximum limit on this area: a household with two workers, for example, can only cultivate one plot, regardless of the number of its consumers. The actual pressures on production involve not merely the number of workers but also the relative ratio of consumers to workers.

The ratio of consumers to workers changes dynamically as the household goes through its developmental cycle. This ratio is likely to be high early on during household formation, when children do not yet contribute to the household's labor force; and it will gradually decrease as the children reach working age. We can theorize, therefore, that the labor shortages resulting from unfavorable consumer-to-worker ratios may be an important factor contributing to the low and medium rice sufficiency levels. An examination of the mean consumer-to-worker ratio of each rice sufficiency class as presented in figure 6b supports this contention. The overall trend suggests that

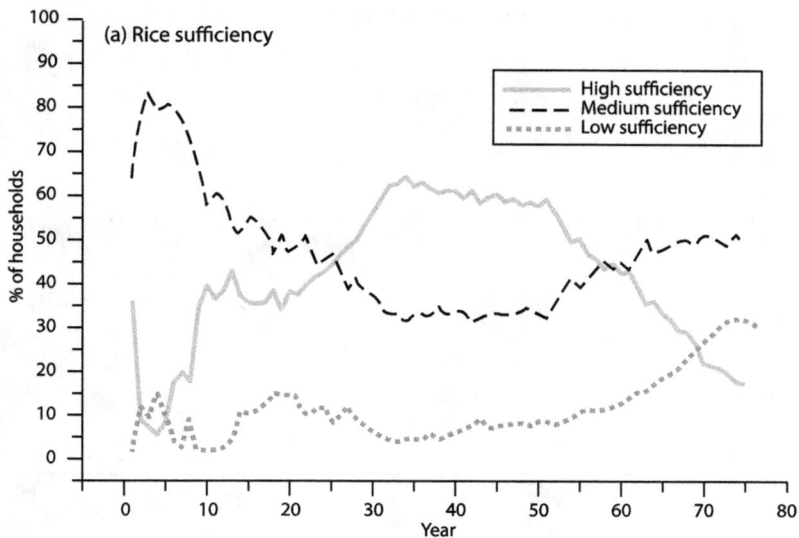

(a) Rice sufficiency

% of households

High sufficiency
Medium sufficiency
Low sufficiency

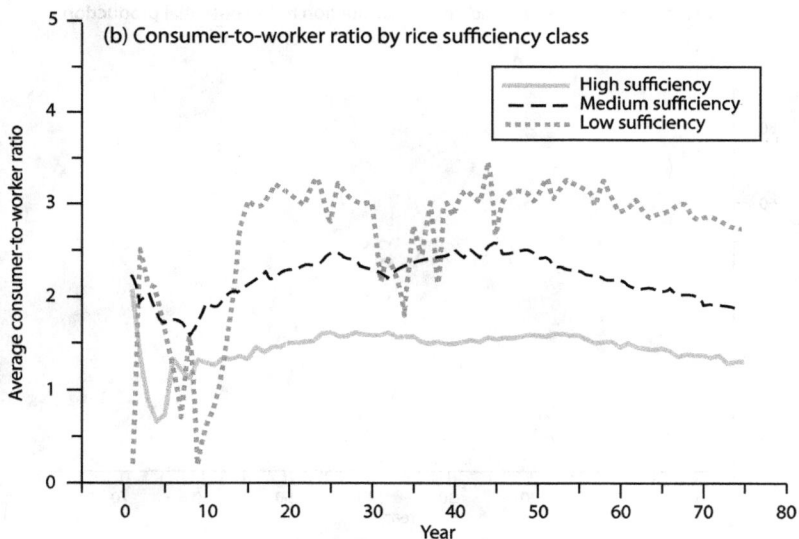

(b) Consumer-to-worker ratio by rice sufficiency class

Average consumer-to-worker ratio

High sufficiency
Medium sufficiency
Low sufficiency

FIGURES 6A AND 6B Rice sufficiency

the main reason for failing to achieve a high sufficiency level in the early years of the simulation is a high consumer-to-worker ratio. While households in such a state appear throughout the simulation, toward the end of the simulation they are explained by decreasing rice yields due to shortening fallows.

(a) Households that can meet the total cash demand

(b) Proportion of the 'realized' rubber production to the potential production

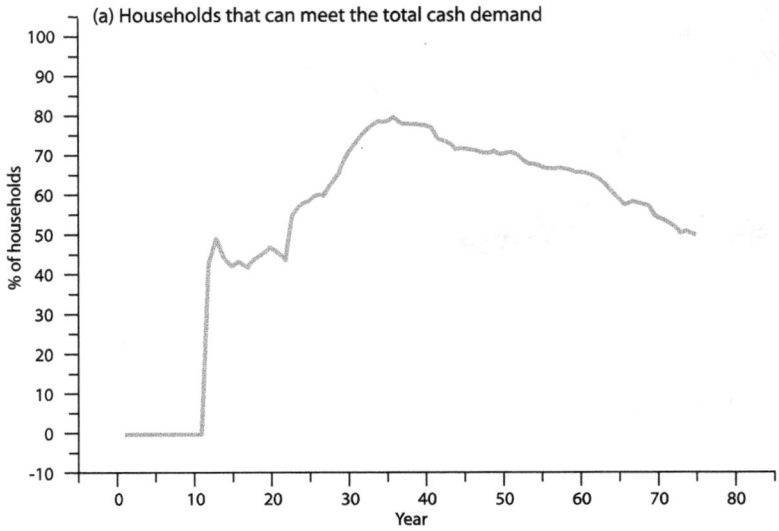

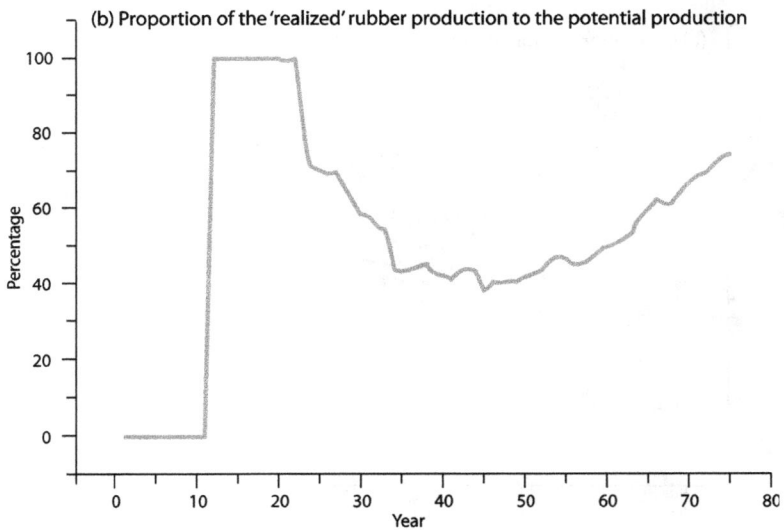

FIGURES 7A AND 7B Cash sufficiency

Cash Sufficiency

Rubber production in this model is geared to fulfilling the demand for cash to buy basic items and pay down cash debts. However, if a household cannot meet its total rice needs, then rubber production will be intensified to buy market rice to make up for the shortfall. On the production side, the factors that constrain rubber production include the labor available after cultivating swiddens and the availability of tappable rubber plots (mature and old). The model assumes that access to tappable rubber plots can be extended to others in share-tapping arrangements.

As in rice production, the analysis here focuses on the extent to which overall cash production can meet total cash demands. The results presented in figure 7a suggest that, during the period of years fifteen to seventy-five, an average of 37 percent of households fail to meet their cash demands. To explain this failure, I first examine whether cash production is constrained by the availability of tappable rubber, recollecting that rubber tapping is calculated on a worker-per-day-and-half-hectare basis, with each half-hectare block of rubber tappable for a maximum of 180 days per year. Since the extent of tappable rubber blocks at the village level is known, I can calculate the maximum potential rubber production (in kilograms). The extent of rubber exploitation can then be expressed in terms of the actual rubber production of all households divided by their potential production. Since rubber plots are essentially available for all households through share-tapping arrangements, the extent of rubber exploitation can be used as a proxy to indicate the availability of tappable rubber. The results presented in figure 7b show that in general the actual tapping is far below the potential. This suggests that the reason some households fail to meet their cash demands is a shortage of labor not a shortage of tappable rubber. The rice or cash shortfall for these households has to be fulfilled by borrowing from other households with surpluses.

Flexible Agricultural Strategy

The default model assumes that swidden cultivation is always given priority over tapping rubber, which reflects the agricultural strategy used by the Kantu' and by other communities in Borneo. This assumption implies a strong preference for growing one's own food. However, one may wonder whether this always constitutes an efficient strategy in a strict economic sense, as when a household has access only to less productive short-fallow plots. We may ask whether in such a situation labor could be better spent to tap rubber to earn cash later to be used to buy rice. The simulation explores this issue further by incorporating a measure of economic efficiency in land-use decision making in a modified version of the model. In the modified version, it is assumed that households attempt to pursue the most efficient way to obtain rice, whether growing rice in swiddens or tapping rubber and then using the proceeds to buy rice. In any one year, these two options are evaluated by considering the fluctuating resources of the household, particularly the size of its labor force, its fallow landholdings, and its rubber holdings. Over the years, households may switch from one strategy to another depending on their resources. The modified version of the model that applies this flexible agricultural strategy is called the "flexible model."

The flexible model differs from the default model only in the land-use decision-making procedures (see Sulistyawati 2001 for details). First, the model calculates the potential rice yield if the household decided to cultivate swiddens. This is done by searching for the best swidden plots potentially available for that particular year using the same procedures as in the default model. Since the vegetation types or fallow lengths of these plots are known, their potential yield can be calculated. To account for the value of vegetables and other nonrice cultigens commonly planted alongside the rice in a swidden, the model assumes that the value of these products equals 25 percent of the potential rice yield. Second, the model calculates the amount of cash potentially available from tapping rubber, given the extent of the household's mature rubber plots and the number of its workers. A household decides to tap rubber if the potential cash yield is equal to or greater than the monetary value of the potential swidden rice production and if the labor for this is also equal to or smaller than that for rice production. Otherwise, the household decides to cultivate rice. These rules imply that households select the land use that minimizes labor use.

This simulation contrasts the two agricultural strategies under two different rubber prices, one of which is the default price (Rp 160) and the other of which is double that price (Rp 320). Thus the overall comparison consists of four simulations: default-Rp160, default-Rp320, flexible-Rp160, and flexible-Rp320. The analysis focuses on the extent to which these contrasting strategies affect the extent of swidden cultivation, the rate of primary forest clearing, and the frequency of swiddens made in short fallow plots. The results are presented in figures 8 a–d.

Figure 8a indicates that the extent of rubber holdings in all simulations is similar, so this is unlikely to be the reason for any differences between the simulations. The flexible agriculture strategy makes only a small difference at the Rp 160-price level, as shown by the almost identical trends of the flexible_Rp160 and the default_Rp160 simulations. This suggests that at the Rp 160-price level, cultivating swiddens is generally a more efficient way to obtain rice. A noticeable difference between the flexible and the default strategies appears, however, when the rubber price is doubled. The comparison of the default_Rp320 and the flexible_Rp320 simulations indicates that, at this price level, a shift to tapping rubber often proves a more efficient strategy to obtain rice, which is reflected in a significant decrease in the

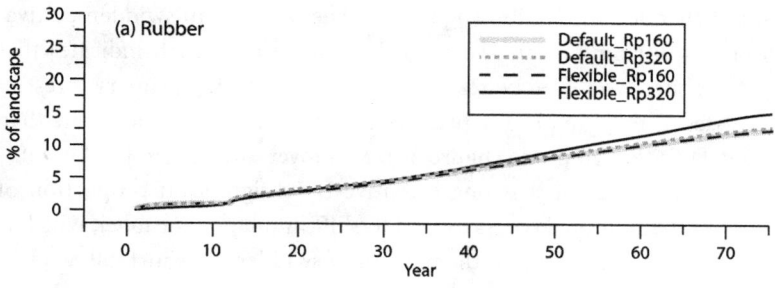

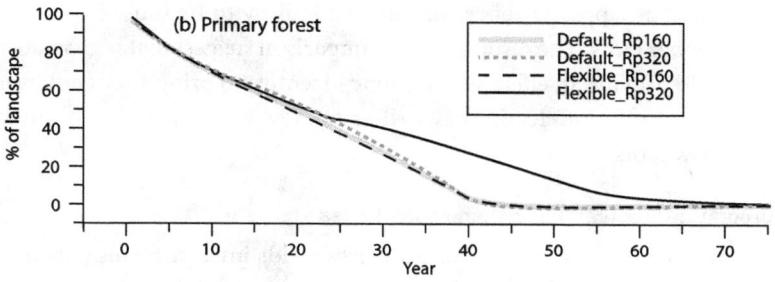

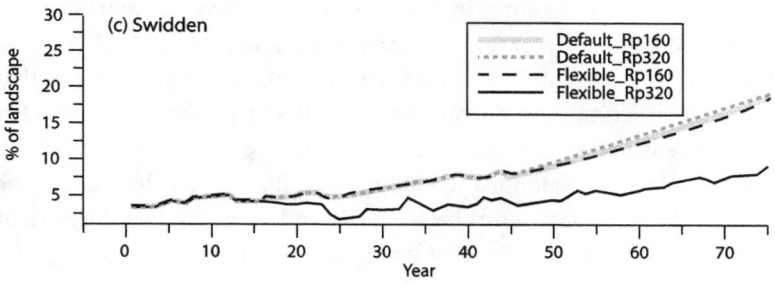

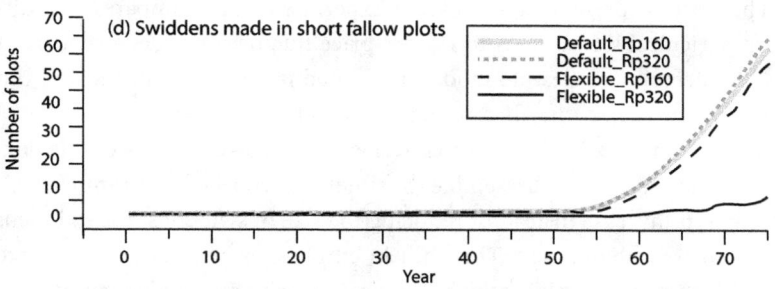

FIGURES 8A, 8B, 8C, AND 8D The implications of different agricultural strategies

extent of swidden cultivation (figure 8c). The decrease in swidden cultivation occurs during the early years of the simulation, which indicates that tapping rubber is an even better strategy than cultivating primary forest in many cases. Consequently, the primary forest declines at a slower rate than in the default_Rp320 mode (figure 8b). Moreover, the Rp 320 level is sufficiently high to provide a strong incentive for a significant proportion of households with limited access to land to shift into tapping rubber, which is reflected in the sharp drop in the making of swiddens in short fallow plots (8d). However, there are still a few households that cultivate short-fallow plots instead of tapping rubber, and these are likely to be those having no mature rubber plots. The results of this comparison suggest that high rubber prices could provide an effective economic incentive to bring about a slower rate of primary forest clearing, as well as a lesser frequency of cultivating short-fallow plots.

Incorporating Fluctuations in Commodity Prices

The experiment with different rubber prices yields interesting insights into the interconnections between commodity prices, access to land, and labor. The assumption of constant price levels, of course, is not realistic. In reality, the prices of an export commodity like rubber fluctuate (C. Barlow, Jayasurija, and Tan 1994: 10; Cramb 1988: 118), and the literature suggests that the level of agriculture production from smallholder rubber and swidden farmers is sensitive to this fluctuation. Cramb (1993) found that the Iban farmers in Sarawak responded to rubber-price fluctuations by varying the balance of the cultivation effort between rice and rubber. When the price of rubber is high, the farmers concentrate more on tapping rubber, decreasing the size of their swiddens.

This section presents a simulation experiment that compares two different agricultural responses to rubber-price fluctuations. The experiment uses the default and flexible models presented in the previous section, but they are now run under fluctuating rubber prices and are called default/fluctuating and flexible/fluctuating, respectively. The dynamics of this flexible model to some extent resemble the situation described by Cramb (1993), although in this case instead of reducing the size of swiddens, the swiddens are completely abandoned. This shift in emphasis when market prices are favorable to cash cropping, which is not as extensive as swidden cultivation, will leave more land in fallow, and this could temporarily ease the pressure on land. Comparing the dynamics of the default and flexible models thus also enables us to assess the extent to which temporary shifts to cash cropping ease the pressure on the land.

To simplify the price changes in this model, the rice price is held constant but the price of rubber varies according to a function that specifies the ratio between the rubber and the rice prices (figure 9a). The default price of rubber (Rp 160 per kilogram) is used as the base price. The fluctuating price is introduced only after year fifteen, which is when most households would have productive rubber plots. The simulated price fluctuates in approximately ten-year cycles with maximum and minimum rubber-rice ratios of 2.5 and 0.5, respectively. As a comparison, the highest and lowest ratios of rubber-to-rice prices (per kg) reported for the Jakarta market during 1968–95 were around 4.0 and 1.5, respectively (Sulistyawati 2001).

The results show that, as expected, the extent of swidden cultivation fluctuates inversely to rubber prices, though it never ceases completely (figure 9c). The periods of low rubber prices before about year forty are associated with the clearing of primary forest in the flexible model, and this eventually brings about the same rate of primary-forest decline as in the default mode (figure 9b). An occasional shift out of swidden cultivation into tapping rubber significantly alters the fallow regime and enables households to observe longer plot rotations (figure 9d). The trends in average fallow length suggest that the flexible strategy permits more environmentally sound practices in general. However, for households with limited access to land, these effects are only temporary (figure 9e). For these households, short fallowing can be avoided during high rubber-price years, but not after several successive low-price years. The timing and duration of low rubber-price years seem to be the critical factors in this respect. A prolonged low-price period will eventually trigger short fallowing practices.

The impact of different agricultural strategies on the level of rice sufficiency is shown in figure 9f (which shows the proportion of households with high rice sufficiency as defined in table 2). Note that the amount of rice in this measure refers to the rice produced in swiddens, excluding rice bought with the proceeds of rubber tapping. Thus the occasional drops in sufficiency observed in the flexible models result from foregoing swidden cultivation during years with high rubber prices. However, the most important point in figure 9f is that more households achieve high rice sufficiency even during years with low rubber prices in the flexible model than during the last twenty-five years of the default model, when land pressure is high. Temporary shifts into tapping rubber during the good-price years leave swidden lands in fallow for a longer period. Since land productivity increases under a longer fallow period, this gives households a better quality of land when they have to cultivate swiddens during the bad-price years.

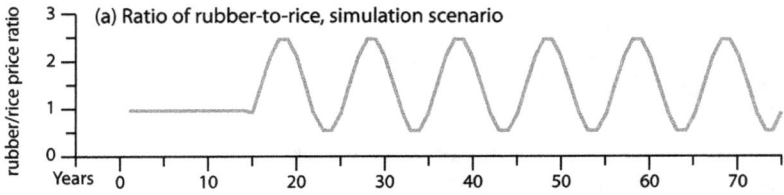

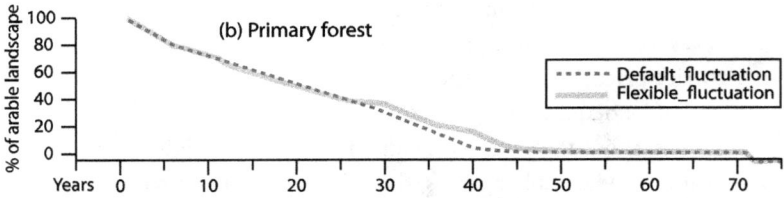

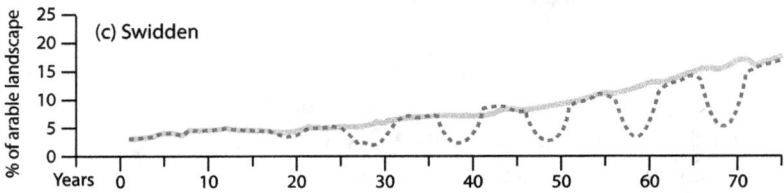

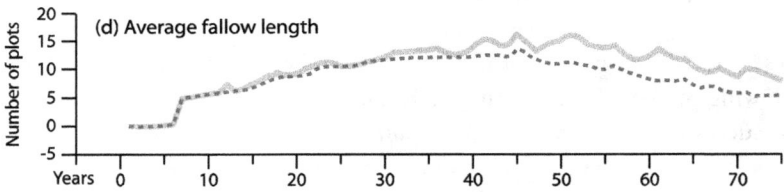

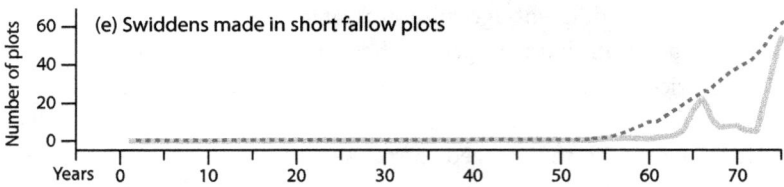

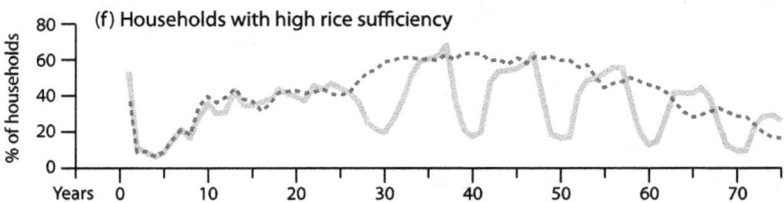

FIGURES 9A, 9B, 9C, 9D, 9E, AND 9F The implications of rubber price fluctuations

Lessons from the Simulations

This study has demonstrated how household-level dynamics interact with the social institutions regulating land tenure to affect the distribution of landholdings across households. This study shows that in situations in which land ownership can only be achieved by felling primary forest or inheriting the land from parents and in which household labor constitutes the main source of labor for cultivating swiddens, the main determinants of landholdings are demographic factors. Households with a greater chance to amass large landholdings are those with the following demographic advantages: they are formed early, they have many sons, and they experience fewer out-marriages. The first two factors are particularly important during the community's early development, when primary forest is still abundant. Once the primary forest runs out, the pattern of household proliferation determined by the pattern of postmarital residence becomes more important, since land inheritance is then the only mechanism to acquire land.

The emergence of differentiation in landholdings during the development of the community reinforces the point that communities do not constitute homogeneous units. I have demonstrated how differences in landholdings can lead to differences in fallowing practices, which eventually affect the ability to meet the household's consumption demands. In addition, the simulations with the flexible model under fluctuating rubber prices have shown that, while most households shift to tapping rubber when the rubber price is high, some of them continue to cultivate less productive short-fallow plots, apparently because of a lack of mature rubber plots. The results of these experiments emphasize the variation of responses across households because of differences in resource base, even when the community as a whole is subjected to homogeneous external driving forces such as price fluctuations.

This study provides a valuable lesson in exploring alternative development paths for swidden agricultural systems facing increasing population pressure. In an area with a limited capacity for territorial expansion, an increasing population can lead to a decreasing length of fallow, which eventually brings about decreasing swidden production. In this case, agricultural adaptations to improve cultivation technology (including intensification) and a greater involvement in cash cropping or other cash-earning activities seem unavoidable to ensure the survival of households. The simulations with the flexible model under fluctuating commodity prices have demonstrated that even without any technological improvement the potentially negative impacts associated with a decreasing length of fallow can be lessened or postponed simply by the increased tapping of rubber during periods with high rubber prices The greater involvement in cash cropping, however, always carries the risk of low returns when market prices drop. Therefore the simulation experiments demonstrate the benefit of maintaining swidden cultivation as an option in the farming system, rather than permanently replacing it with cash cropping. This allows farmers to switch back to swidden cultivation when the price of a cash crop is low, which moderates the impact of fluctuating commodity prices by ensuring some level of subsistence (Cramb 1993).

Limitations of the Model

As is evident from the ethnographic data, the swidden-cultivation system of the Kantu' is complex, as is the case in many other swidden-cultivation communities. To simulate this system, a degree of simplification is often unavoidable, and my model is no exception. The model partly or completely excludes a number of aspects of the Kantu' economy, including the cultivation of nonrice cultigens; other subsistence activities such as fishing, hunting, and the gathering of noncultigens in the forest; and swamp swiddens and pepper cultivation. Furthermore, the model assumes a closed community with no direct interactions with communities outside the longhouse. In reality, however, such interactions exist in many forms. Intervillage marriage is quite common (Dove 1985d; D. Freeman 1992). The model also excludes temporary labor migration, which is a pervasive feature in rural areas around the world, including most swidden agriculture communities (see Standing 1985; Wadley 1997). Labor migration affects the household economy primarily through reducing the availability of a household's labor for agricultural use and adding to the household's income from remittance.

Although some agricultural adaptation to external driving forces has been introduced in the model (as in the flexible simulations), many aspects of Kantu' life are assumed to be static throughout the seventy-five-year simulation period. But seventy-five years is a relatively long time for a land-use system to remain unchanged, since the sociocultural and economic circumstances are likely to alter and many assumptions may no longer hold after a certain period. One aspect of the model especially prone to change is the assumption of subsistence-oriented production, particularly when a swidden community is increasingly exposed to a market economy. In such a case, cash demand may extend beyond buying basic goods, as assumed by the model, to include the purchase of consumer goods and schooling. As the community becomes more involved in a market economy, profit considerations are also likely to become more important in land-use decision making. An indication that the Kantu' are responsive to new economic opportunities can be seen from the renewed interest in growing pepper in the study area following improvements in transportation and the surge of pepper prices in 1997.

Nevertheless, the model still offers a useful tool to study the interactions between the demographic, sociocultural, economic, and ecological factors associated with the swidden cultivation system of the Kantu'. This tool allows many aspects of Kantu' life to be explored in detail. The main challenges for its future development are, first, how to generalize from it, so that its lessons can be applied in other communities that have different social or land-tenure systems. The second challenge is to improve the model's capability of handling changing circumstances, to allow households to adapt to new circumstances.

Summary and Conclusions

The simulation study of the Kantu' (in the model's default mode) has shown that with less than 2.6 percent population growth per annum, the landscape changes dramatically over the seventy-five-year study period. Primary forest virtually disappears after forty-five years, and the landscape is progressively dominated by shrub land. From initially homogenous mature vegetation, the landscape is transformed into a mosaic of vegetation of many different successional states. As population density increases, the average length of fallow decreases, especially once the primary forest has disappeared. Such a finding is predictable without the aid of a simulation. This simulation model, however, demonstrates that the pressure for land is not uniformly felt across all households due to inequalities in landholdings; those with ample access to land can retain long fallows, but others are forced to shorten the fallow period. In this model, land inequality arises as a result of stochastic demographic events that are determined by social rules pertaining to marriage, land tenure, and the inheritance system. The emergence of land inequality is one of the most important findings of this study. This study demonstrates that one ramification of an increasing population is not simply the availability of less land but, more important, the availability of less land for some but not others.

The study has shown that most households are able to achieve high levels of rice sufficiency, although their number declines as population pressure intensifies. The lack of access to productive land is one reason why households fail to achieve high levels of rice sufficiency. The other reason is labor shortages induced by the demographic structure of the household, notably by a disproportionately large ratio of consumers to workers. Except for those with a severe labor shortage, most households can make up rice deficits through cash income from rubber tapping.

The simulations with a flexible agricultural strategy yield important insights into the alternative development paths of the swidden agriculture system under increasing land pressure. The simulation results show that shifting the balance of labor and land use between swidden cultivation and rubber tapping according to their economic efficiency at prevailing commodity prices can result in a more sustainable farming system.

This essay has demonstrated the use of formal models to study the impact of household demography on resource use in agricultural systems. The analysis presented here has shown that the position of a household in its life cycle determines the availability of labor resources, which in turn affect the extent of agricultural activity that the household can perform and thus the likelihood of fulfilling domestic demand. In their studies of the Brazilian Amazon, Walker and Homma (1996) and Perz (2002) also noted that household demography is a critical factor in determining the extent of agricultural activity.

Every decision made by households concerning the use of land resources is eventually reflected in the landscape. What happens in the landscape (e.g., to soil quality) will in turn affect the household's subsequent land-use decisions. The modeling approach used in this study makes explicit the connection between the dynamics of the household and the dynamics of the landscape, which indeed is a major strength of this model. By following the life history of each person and household, the model is able to capture the sociocultural aspects of the community that affect the agricultural system. The importance of cultural aspects in understanding swidden cultivation systems has been frequently emphasized (e.g., Dove 1981), but most studies examine too brief a period of time to properly examine these aspects. This model provides a framework to examine the relationship between culture and nature over a longer-term perspective. The simulation model allows us to investigate the long-term ecological consequences of a set of cultural practices. Modeling exercises like this also promote closer cooperation between ecologists and social scientists.

Notes

The Institutional Context of Biodiversity Conservation in Southeast Asia Project, supported by the John D. and Catherine T. MacArthur Foundation, provided the funding for this study. The author thanks the project's coprincipal investigators, Michael R. Dove and Percy E. Sajise, for providing excellent guidance during the writing of this essay. This study is part of the author's Ph.D. project carried out at the

Ecosystem Dynamics Group, Research School of Biological Science, Australian National University, which is funded by the Australian Aid (AusAID). The project was supervised by Ian R. Noble, Mike Roderick, and Habiba Gitay. Special thanks must be extended to them for their continuous guidance throughout the study. The author also thanks Musyarovah Zuhri, who helped to format the manuscript. An earlier version of this analysis has been published in Sulistyawati, Noble, and Roderick (2005).

1. The description of the Kantu' is entirely based on Dove's works unless otherwise stated; therefore it refers to the situation during the time of his study.

2. The longhouse Tikul Batu was abandoned in the early 1980s after epidemics of gastrointestinal disease killed many people. The residents of the longhouse were relocated to an area nearby, and they have been living in detached, single-family houses since then. Nevertheless, the term *longhouse* will continue to be used in this article as a gloss for "community."

3. When the term *attempt* is used to describe the rules of the model, it expresses the notion that the attempted action may or may not be carried out depending on the interaction with other related variables, which changes dynamically during the simulation.

4. The productivity for primary forest or long-fallow swiddens are approximate figures provided by Dove (1985d: 94); the rest are arbitrary numbers to reflect the decreasing productivity of swiddens cut from younger forest.

5. The model's rules restrict the role of land borrowing to balancing access to land among land-rich and land-poor households (see also Padoch 1982: 46–49). In reality, however, land borrowing can simply be a matter of obtaining a suitable location for the current season (see also Cramb 1989b: 284).

6. The per capita cash requirement (Rp. 11,665) is based on market prices in Putussibau, the capital of Kapuas Hulu, in 1978.

7. The assumption of low rubber productivity is based on Dove (1993c: 139) and Drake (1982: 143). Nevertheless, higher estimates of rubber production have also been reported (e.g., C. Barlow and Muharminto 1982: 102; Wadley 1997: 103). Under the current set of assumptions, potential (mature) rubber production is set at 720 kilograms per hectare per year, which is higher than the reported figure for Indonesian smallholders (525 kilograms per hectare per year) but lower than that of rubber estates (1,186 kilograms per hectare per year) (C. Barlow and Tomich 1991: 34).

THE ECOLOGICAL IMPLICATIONS OF
CENTRAL VERSUS LOCAL GOVERNANCE
The Contest over Integrated Pest Management in Indonesia

"Everything has been determined by the government: when to plant, what to plant, and how to plant"; this is how a farmer described the changes he has experienced since the introduction of high-yielding or "modern varieties" (MVs) of crops during the so-called green revolution. It constituted an abrupt change in his life to go from being an independent and free decision maker in farming to depending entirely on the state decisions. The introduction of that "seed-fertilizer-water revolution" to boost productivity (see H. Brookfield 2001) marked the beginning of the state's control over farmers' lives and farming practices. Farmers were forced to adopt a modern technology based on Western scientific knowledge that felt alien to them with its set of subsidies, facilities, and new bureaucratic institutions (see G. Hansen 1978; Birowo and Hansen 1981; Hardjono 1983; James Fox 1991, 1993; Sawit and Manwan 1991; Chang 1993). The sweeping green revolution programs have had wide-ranging repercussions for farmers' lives, repercussions extending well beyond the domain of cereals and even beyond that of agriculture itself (H. Brookfield 2001). The farmers' own acumen and initiative went largely ignored. Their dignity, capability, and knowledge became gradually erased through centralized resource management.

In the early 1990s, I studied farmers' mixed feelings and perspectives. On the one hand, they acknowledged the vast improvement of their incomes and living standards. The state's program in fact produced significantly higher rice productivity (see Conway and Barbier 1990; Conway and Pretty 1991; James Fox 1991, 1993). In terms of tons per hectare rice yields, farmers could obtain three times higher than in the period prior to the planting of the so-called miracle seeds. Yet there are costs. Not only has the farmers' freedom to decide been curtailed, but feelings of anxiety have replaced the "peace" they knew prior to the green revolution. Furthermore, rice sustainability is at risk, regional biotic diversity has been gradually diminished, and there have been unprecedented outbreaks of pests and diseases (Shiva 1988, 1991, 1993; Conway and Barbier 1990; Conway and Pretty 1991; James Fox 1991, 1993; Chang 1993). The intensified monoculture cultivation of crops and the injudicious use of pesticides have changed the farming landscape drastically by eliminating a diverse range of local varieties and natural pests. Such a modification of the farmers' environment has led to their alienation from the ecological conditions of their own fields.

Insects previously not considered pests have gradually become devastating threats to farmers' crops. Green Revolution extension officers taught the farmers the use of pesticides—introduced as "medicines"—to eradicate harmful insects, but they did not teach them about the impact of the pesticides on the insects' natural enemies. Therefore, the farmers did not understand why more "illnesses" were attacking their plants, even though they applied more pesticides (see Bentley 1992; Winarto 2004b). This unexpected result belied the government discourse of "using an umbrella before it rains," which was the rationale for the use of preventive "medicine." Reflecting on this puzzling condition, farmers described the green revolution as "the era of stupidity [masa pembodohan]," (Winarto, Maidi, and Darmowiyoto 1999: 68). The term *stupidity* refers to their mistake of only applying the government's recommended technology without a thorough understanding of their changed landscapes.

In contrast, farmers perceive the recent period following the introduction of the so-called Integrated Pest Management (IPM) program in Indonesia as "the period of cleverness [masa pencerdasan]" (Winarto, Maidi, and Darmowiyoto 1999: 72). This program was designed to correct the green revolution paradigm (see Moning 2006). Through a presidential decree (INPRES 3/1986), IPM was proclaimed the national pest-control strategy.[1] The sustainability of rice production through the control of the brown plant hopper (BPH, *Nilaparvata lugens*), which has threatened Indonesia's rice self-sufficiency since the mid-1970s, became its primary objective. The extinction of rice pest predators was now to be avoided by using pesticides more judiciously. To meet these objectives, the IPM program banned fifty-seven broad-spectrum pesticides and introduced a new, more effective juvenile hormone pesticide that prevented the BPH's development to reproductive maturity (James Fox 1991, 1993). Finally, the program aimed to make farmers masters and experts in their own fields by improving their knowledge of and skills in managing their new environment. Training both officials and

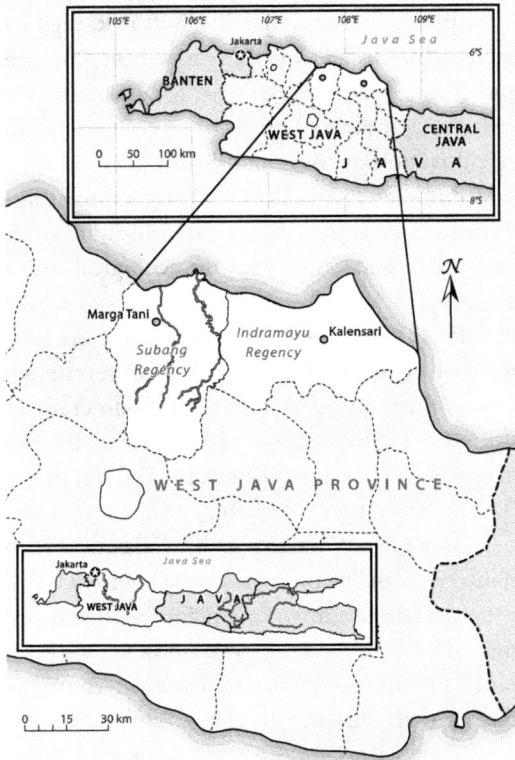

MAP 1 Location of IPM Villages in West Java, Indonesia

MAP 2 Location of IPM Villages in Lampung, Sumatra, Indonesia

farmers through knowledge transmission rather than technological transfer is the program's main strategy (Indonesian National IPM Program n.d.; van der Fliert 1993; Dilts and Hate 1996; Kenmore 1997; Röling and van der Fliert 1998; Pontius, Dilts, and Bartlett 2002; Gallagher 2003).

The IPM was launched to correct the green revolution paradigm by making farmers the masters of their own environment and thereby lessening the established hegemony of the state (see Dilts and Hate 1996; Kenmore 1997).[2] This was the desired "paradigm shift." Yet the program was still run by the state under the plant protection division of the Ministry of Agriculture. Furthermore, the government did not alter its commitment to higher rice production (James Fox 1993). Given these conflicts, to what extent have changes occurred in the agrarian relationship between the state, the people, and the environment? What sorts of changes have occurred? And how can they be measured? This essay discusses the extent of the changes experienced by farmers involved in the IPM programs by looking at the advantages they have gained and the constraints they have faced. I will examine examples of IPM in two villages on the north coast of West Java, Marga Tani in the regency of Subang and Kalensari in the regency of Indramayu (see map 1); and two villages in Central Lampung, Sumatra: Batanghari and Terbanggi Besar (see map 2).[3] The issues to be examined include the re-creation of the farmers' role in agriculture; the evolution of farming culture; and the implications for the macroecology, the extension service, and the state policy of crop production and protection.

Origins of IPM in Indonesia:
Introducing a Local Science of Conservation

The declaration of the IPM as the national strategy in food crop protection was a significant event in the history of the state's political ecology in Indonesia. From a modern agricultural management approach based on high-level inputs and technology, the Indonesian government moved toward use of a conservation strategy to protect crop production (Conway 1985; Shiva 1988, 1991, 1993; Conway and Barbier 1990; James Fox 1991; Chang 1993). The program was designed to counter unprecedented pest outbreaks, in particular the brown plant hopper, a pesticide-induced pest (FAO 1990; Kenmore 1992).[4] The BPH is in fact an extremely vulnerable rice pest with more than one hundred predators or parasites (James Fox 1993). Due to the practice of intense applications of pesticides, however, the BPH's natural enemies were eliminated in the first thirty days after planting the rice. Attempts at solving the problem through the laboratory, by continually releasing new varieties of rice purportedly resistant to BPH (Bahagiawati and Oka 1987; James Fox

1991), did not prove effective. The BPH, a "boom-and-bust-cycle" pest, can quickly react to modifications of the cultivars' genes by shifting their population structure to a different biotype. Hence it can overcome the defense mechanism of the newly bred cultivar (Chang 1993). At the same time, the Indonesian government faced a monetary crisis. At the end of the 1970s, the government subsidized all fertilizer requirements of the country's rice farmers. By the mid-1980s, however, the cost of this fertilizer subsidy had ballooned to about one-third of the entire development budget (James Fox 1993: 217), thus becoming unsustainable and forcing a reevaluation (also see Wardhani 1992).

In light of these problems, leading international and national scholars saw the transmission of scientific knowledge and the devolution of decision making to the farmers as the solution. The Harvard Institute of Development, in collaboration with the Center for Policy and Implementation Studies in Jakarta and supported by the Food and Agricultural Organization's Inter-country Program, worked with national experts to persuade the nation's leaders to transform the national policy of pest control and, at the same time, reduce the subsidies for both fertilizers and pesticides (FAO 1990; also see James Fox 1991; Wardhani 1992). Presidential Instruction No.3/1986 was the result, prioritizing biological over chemical controls by allowing only the judicious use of a limited range of pesticides and developing a diverse strategy of pest control. Whereas its manifest objective was to grow a healthy rice crop, its underlying objective was to shift the agricultural paradigm away from killing pests with pesticides toward managing the ecosystem by conserving the pests' natural enemies (FAO 1990, 1991; Indonesian National IPM Program n.d.; Kenmore 1992). Conserving the natural components of the rice ecosystem was the primary objective, to be achieved by placing the effort into the hands of local people, and relying upon their own empirical observation of their own habitat (see Dilts and Hate 1996; Pontius, Dilts, and Bartlett 2002; Gallagher 2003). The program was unusual in linking scientific and local knowledge and bringing "natural" conservation into the "cultural" domain.

Empowering Farmers: Devolving Centralized Decision Making

Farmers are good observers, as many scholars have noted (P. Richards 1985, 1986; Rhoades and Bebbington 1988, 1995); however, some difficult-to-observe phenomena lie beyond their understanding (Bentley 1992). Before the green revolution, small insects that preyed on other insects were not only not observed, they were culturally insignificant (see Bentley 1992). When farmers saw the health of their plants in jeopardy, cultural practices such as the use of herbs, chants, and traditional rituals were relied upon to make their plants "healthy" again. The changed landscape of the green revolution, with its attendant disease outbreaks, was not accompanied by knowledge of this altered habitat. Therefore, prey-predator relations, the role of insects as predators, and the unintended consequences of pesticides, were not part of farmers' knowledge (Winarto 1995, 2004b). This led IPM proponents to believe that the farmers' empowerment should be based on their mastering of scientific knowledge. Training both farmers and agricultural extension officers thus became the major pillar of the program, in the form of the so-called Farmer Field Schools, which have now spread globally (see Dilts and Hate 1996; Pontius, Dilts, and Bartlett 2002; CIP-UPWARD 2003; Winarto 2004a, 2005a; van den Berg and Jiggins 2007).

In line with the paradigm shift of devolving decision making to the hands of the farmers, training in the field schools was designed to enable farmers to make their own decisions based on an intimate knowledge of their modified landscapes. Therefore an experiential learning process became the basis of the training (Indonesian National IPM Program n.d.; Dilts and Hate 1996; Röling and van de Fliert 1998; Pontius, Dilts, and Bartlett. 2002). Farmers, much like scientists, now took weekly random samples of grain, carried out detailed observations, compared, engaged in experimentation, and collected insects. They recorded and wrote up their own observations and analyses, all novel skills in the world of Indonesian farming. By mastering these skills, the discovery of things previously unknown became possible. Farmers became "IPM experts in their own fields" or "experts in decentralized pest management through practical, field-based learning methods" (van den Berg and Jiggins 2007: 664; see also Indonesian National IPM Program n.d.; van de Fliert 1993; Winarto 1995, 2004b; Röling and van de Fliert 1998). This enrichment of farmers' local knowledge not only empowered them but also offered a promising path toward solving local problems.

Changing the farmers' perspectives and strategies and improving their confidence in only a single-season training session was challenging (Winarto 1996; Winarto and Choesin 1998; Winarto et al. 2000). From 1992 onward, therefore, IPM experts facilitated various other sorts of activities. Training for farmer-trainers, the encouragement of farmers' studies and experiments, and the organization of farmers' seminars and workshops all became ongoing activities within the Community-IPM program (see Pontius, Dilts, and Bartlett 2002; van den Berg and Jiggins 2007). The training of farmer-trainers provided an opportunity for talented farmers to become IPM farmer-facilitators, enabling them to train other farmers in the IPM schools, while also becoming the local extension and change agents. Toward the termination of the World

Bank's funding, an association of IPM alumni from the village level up to the national level was formed (e.g., Paiman n.d.; Kuswara n.d.; Hafisam n.d.; Susianto, Purwadi and Pontius n.d; Tim Bantuan Teknis FAO 1997; Kenmore 1997; Petani PHT Jawa Barat 1998; Musyawarah Petani PHT Indonesia 1999; Busyairi et al. 2000; Hidayat and Adinata 2001). This association provided a way to institutionalize the IPM alumni's activities, giving the farmers an official forum in which to speak up against ill-considered recommendations and policies from the government (see Media Jaringan Petani Indonesia 2000–2002; Indonesian Integrated Pest Management Farmers' Alliance and Department of Anthropology, University of Indonesia 2007). The benefits for the farmers included added self-confidence, dignity, and freedom of decision-making.

Regaining Lost Freedom, Returning Dignity

The farmers believed that IPM would return them to their "old condition of growing rice . . . without medicines" (Winarto 2004b: 358; H. Brookfield 2001). There was also a feeling that they were now free to decide whether or not to comply with the government's recommendations (Maidi and Darmowiyoto 1999). One farmer spoke of his new understanding that any newly introduced product or recommended technology would first require his own examination of its benefits.

THE FREEDOM TO DECIDE

Everyday struggles against environmental hazards and economic constraints oblige farmers to constantly search for the strategy that will produce the optimal harvest. The particular ecological conditions of their fields, coupled with scarce capital and labor, make personal adjustments in the recommended prescriptions of fertilizers and pesticides desirable. However, the possibility of such adjustments was greatly constrained within the green revolution's required technological packages and cultivation strategies. Prior to IPM, given the philosophy of gaining the highest possible yields with inorganic fertilizers and pesticides, cultivating rice without these two inputs seemed impossible. For example, a non-IPM farmer in Marga Tani did an experiment in his rice field in which he used organic fertilizers instead of inorganic ones, and the growth of his plants made him conclude that organic fertilizers were no longer appropriate in the changed soil conditions (Winarto 2004b). This story reveals the constraints farmers experienced in trying to return to low-level agricultural inputs in their modified landscapes.

The new paradigm of IPM, though, has opened up farmers' minds to the possibility that good yields are possible without relying on "medicines." Previously, the plants' visible condition, whether they looked healthy or sick, was the only factor in the farmers' decision-making process regarding pest and disease control. Spraying pesticides was their only option to cure their plants from "sicknesses," even if they had to borrow money or purchase products on credit. As far as possible, they avoided any risks that might result in the failure of the harvest (Winarto 2004b). The IPM training makes it possible to weigh more factors, such as the kinds of pests infesting their plants; the pest-related strengths and weaknesses of the planted crop varieties; the planting schedule; the growth of both the pest population and its predators'; the intensity and nature of pest damage; the pest's economic threshold;[5] and the pest's behavior and life cycle. A more careful calculation of expenses also became possible, including the amount of capital left, the cost of labor and pesticides in comparison to the benefits, and the price of preserving natural predators. Conserving the predators' habitat and saving costs became part of farmers' decision making, not simply the recovery of their plants' health or the elimination of pests. This is the most remarkable indicator of the extent to which IPM has been able to change the farmers' perspectives.

The responses of the IPM farmers to a white rice borer (WRB, *Scirpophaga innotata*) outbreak on the north coast of West Java (see Winarto 1995, 2004b; Busyairi et al. 2000) illustrates this growing complexity in farmers' decision making. At the time of the outbreak in 1989–90, farmers were still ignorant of the pest's life cycle and how it could cause such severe damage to their plants despite their pesticide applications. Their explanations thus referred to supernatural factors, the "bad" year in the traditional Javanese calendar, or the use of incorrect pesticides (Winarto 2004b). After learning of the WRB's life cycle and its nature of infestation, attention to life-cycle stages and appropriate pest control strategies became important. Farmers' understanding grew once they had a chance to respond to subsequent WRB outbreaks. They no longer discussed the kinds of "medicines" that would have the most powerful impact on pests. Instead, they discussed what strategies would prove the most effective and the least costly. "Now our way of thinking is different, isn't it?" reflected an IPM farmer as he tried to articulate the changes he had experienced after incorporating scientific learning into his own understanding.

Farmers are enmeshed in a network of social relationships in their community, and this affects who is deemed most worthy for IPM training, both by villagers and by the IPM officials. As the IPM training required farmers to write, IPM planners allowed only literate participants into the program. The training also necessitated participants' active decision making and the development of their own farming strategy, therefore farm owners and farm-cultivators were the school's first priority, not simple laborers (van de Fliert 1993; Winarto 2004b). Youth and, in some cases, women were only included on the basis of quotas. Women, young farmers, and laborers thus participated as "helpers," not as "decision makers." Yet in Marga Tani and some other sites, a few wage laborers, women, youth, and even interested non-farmers were also selected as participants. Would they be able to modify their landlords', parents', or husbands' decision making?

It was not easy for young IPM farmers in Marga Tani to argue against the conventional way of controlling pests practiced by their parents, parents-in-law, and other elders. Only after evidence of effectively controlling recurrent outbreaks of WRB did some older farmers acknowledge the youths' novel understanding and capabilities (Winarto 2004b). A poor farmer who used to be a wage laborer experienced the same thing. Other farmers would only listen to his diligent observations and discoveries after they had some empirical evidence that his findings were accurate. Women experienced similar constraints if their husbands were the decision makers in farming. In Batanghari, Central Lampung, a special rice IPM school for women was organized, followed by the recruitment of women into the soybean IPM school and a further program for rice. The women would become decision makers only if they had held this position all along due to absent husbands. If both husband and wife were IPM participants and the wife also spent considerable time in the fields, the two of them would usually jointly deliberate. But this was not the case for those women whose husbands held more dominant decision-making roles, particularly if the husbands had never been involved in IPM programs (Winarto et al. 2000). Despite these constraints, women could draw on their improved knowledge in whatever positions they had, either as laborers, kinfolk, or wives.

FARMER-SCIENTISTS: THE AGENCY OF CHANGE?

Within the communities in which IPM was introduced, new categories of farmers are emerging, namely as scientists, experts, researchers or "professors," and facilitators. Most recently, in Indramayu, West Java, the new category of farmer–plant breeders emerged in relation to their novel skills in plant breeding (Winarto and Ardhianto 2007). Such terms connote possession of modern scientific knowledge and reflect the fact that whereas carrying out detailed observations and examinations of crop pests was not previously part of the farmers' practices they are part of the IPM farmers' new activities. And the term *farmer-facilitator* reflects the fact that the role of agricultural extension workers is now in the farmers' own hands as well.

I first found the terms *farmer-experts* (*pakar petani*), *farmer-researchers* (*petani peneliti*), and *farmer-professors* (*profesor petani*) in Marga Tani in 1991–92, but the same terms also cropped up elsewhere, along with additional terms such as *field technical staff* (*petugas lapangan*) and *farmer–plant breeders*. These terms, reflecting farmers' involvement in agricultural studies and experiments, became indicators differentiating the IPM farmers from the non-IPM ones. The new language reflects the fact that farmers, as experts in their own right, began to produce the so-called *sains petani* (farmers' science) to solve their agricultural problems. From the farmers' perspective, they practice science by carrying out repetitive experiments and controlled comparisons, and they are committed to the pursuit of this science using local resources for the field studies, such as plastic bags, bamboo, mosquito nets, and so on (see Paiman n.d.; Kuswara n.d.; Hafisam n.d.; Susianto, Purwadi, and Pontius n.d.; Winarto 2004b; Tim Bantuan Teknis FAO 1997; Busyairi et al. 2000; Kenmore 1997; P. Ooi 1998; Media Jaringan Petani Indonesia 2000–2002; Moning 2006). Prior to IPM training, farmers did engage in everyday experiments, comparisons, and observations, though not as systematically. As also discovered by Frossard (1998a: 122), "Bringing together scientific method and indigenous knowledge in a novel way, increases their control of the local crop-development agenda" (also see Settle 1997; Winarto 1997b, 1998; Frossard 1998b).

The development of a farmers' science reflects a transformation in the relations between the producers of knowledge and its consumers. Knowledge is no longer just flowing from the experts to the farmers but is also produced by the latter and is exchanged through farmer-to-farmer communication. The use of the term *farmer-professor* reflects the farmers' perception of the similarity in how farmers and scientists learn. As argued by Frossard (1998a: 113) for the Philippines: "Peasant farmers become their own crop scientists, combining knowledge of local conditions (both ecological and social) with powerful scientific crop-production techniques in a way that increases the sustainability of the ecological system, and empowers farmers politically. This paradigm I refer to as 'peasant science.'" This new "peasant science" reveals not only farmers' capabilities in knowledge production but also the dynamics of so-called traditional or indigenous knowledge, as well as the blurring boundary between scientific and indigenous knowledge. Thus it is high time now to question the dichotomy of Western and indigenous domains of knowledge (see Agrawal 1998; M. Brookfield 1998; Choesin 2002; Olivier de Sardan 2005).

The presentation of the discoveries of *Sains Petani* has gradually become an integral part of farmers' meetings (Musyawarah Petani PHT Indonesia 1999), in which groups of farmers from various places share their findings.

The annual workshops held by the farmers' organization in Terbanggi Besar, Central Lampung, for example, focus on the promotion of knowledge production and exchange. In the first workshop, the participants presented their ideas for the kind of farmers' science they would carry out. Their discoveries were shared in the second workshop. Their studies concerned such things as the relation between daily temperatures and pest and disease outbreaks, the most effective light-trap control for the soybean seed borer, the use of hormones to trap male moths, and innovative tools for sowing and weeding in soybean fields. In Indramayu, the fruits of farmer science included rice and vegetable cross-breeding, botanic pesticides, organic fertilizers, bio-agency, and a farming strategy known as the "System of Rice Intensification." The farmers' local knowledge is always in the making through ongoing experiments and discoveries, and it is being enriched by diverse practices and products. The heterogeneity of farmers' activities provides an opportunity for changes in farming strategies to occur (Winarto 2004a, 2004b).

The Transformation of Farming Culture

For decades, Indonesian farmers used technical agrochemical inputs with little understanding of the nature of the substances involved. The situation was exacerbated by the official use of powerful but inaccurate metaphors such as "medicines" for pesticides—a cure for their plants' "sicknesses"— which stimulated their adoption. Based on this inferred analogy between human and plant health, the use of chemical pesticides was uniformly accepted as insurance of good harvests (see Winarto 1998, 2004b). But with IPM, farmers' knowledge evolved. Through daily observation, they now assess the benefits and efficacy of whatever chemical inputs they use by referring to their plants' performance with respect to select criteria: the visible health versus illness of the plants (referring to indications of pest or disease infestation), the number of stems, the color of the stems and leaves according to age, the uniform height of the rice, the number and form of panicles, and plant yields (Winarto 2004b). If they perceive the inputs as beneficial, they keep using them. If not, they modify them or find alternatives.[6] Trial and error has long been an integral part of farmers' efforts to test the efficacy of new inputs and the accuracy of new information. What differences did IPM make to this form of knowledge formation?

ENRICHING KNOWLEDGE AND GAINING CONFIDENCE

The fundamental difference IPM made was the introduction of novel ideas, such as the existence and role of natural enemies for insects. The metaphor of "farmers' friends and helpers" was introduced to restructure the farmers' view of these natural enemies. IPM also introduced the idea that the two categories of pest and predator are connected through prey-predator relations. Another significant change was the description of pesticides as "poisons" rather "medicines." Thus farmers learned that there are consequences to the unintended elimination of predators through excessive use of pesticides, and that it is possible to produce good harvests without pesticides, which can reduce farmers' expenses. This represented a completely new interpretive schema (see Winarto 2004a).

The actual nature and operation of the chemical pesticides and fertilizers remained a mystery, however. IPM training only stated that the "medicines" farmers used were in fact poisonous; and the experts maintained that this should suffice to lead the farmers away from their use (Peter Ooi, personal communication 1998). Still, the new interpretative schema provided a good starting point and a stimulus for the growth of further ideas and understandings. Simply by listening to what the trainers said in the IPM schools about the need to understand the stages of growth of a pest like the rice borer to control its predation, the IPM farmers in Marga Tani were able to develop a more complex understanding of their fields' ecology.

The farmers learned, for example, that in the WRB life cycle the rice borer moths are succeeded by a stage of egg clusters, which then hatch into larvae, which then turn into pupae before becoming moths again. In the next season, however, farmers discovered moths, egg clusters, and larvae in their fields all at the same time. This finding led them to criticize the theory of the rice borer's life cycle (*siklus hidup penggerek batang*), because they did not expect to simultaneously find different stages of the cycle. The farmers then reformulated the theory by concluding that multiple generations of the rice borer must be able to overlap in a single planting season. They concluded that the emergence of moths or the hatch of eggs did not necessarily happen at the same time of year. In the next season, however, they discovered that the moth flights of different generations of rice borers tended to occur at the same time, which led them to the amended conclusion that the metamorphosis of all pupae (from all generations) into moths is synchronized. From these observations they reached a further conclusion that two different patterns of infestation from the same pest are possible: that of overlapping generations and of discrete ones. They attributed the difference to the weather, in particular the rainfall pattern (Winarto 2004b). This is an example of how, starting with a basic theoretical concept like that of the life cycle, the farmers were able to expand their knowledge by incorporating data from everyday empirical observations. The IPM program did not alter the ways farmers acquired or developed their knowledge. However, its more systematic structuring of observation and experimentation increased the speed at which their knowledge developed. In only two years time, during the 1990–92 planting seasons, farmers in Marga Tani managed to gain a richer understanding of the WRB's behavior as a result of the successive outbreaks they experienced (see Winarto 2004b).

Nevertheless, not all new concepts and ideas are perceived as accurate and beneficial and, accordingly, some IPM participants question them. An example is the efficacy of prey-predator dynamics in solving the problems of unexpected pest outbreaks. Farmers questioned the lag time between the robust growth of pests and the slow growth of predators at the beginning of the planting season. They believed that relying on the role of predators in controlling pests at this stage would prove ineffective. Another example is the concept of pesticide-related pest resistance and resurgence, which is not immediately evident through empirical observation. This concept also contradicted their observation that pesticides can kill the BPH instantly or "rot" the rice borer's eggs (and prevent the development of the juveniles) and, hence, save their plants from further damage. It is a reality that many phenomena still cannot be directly, empirically observed by farmers given their existing tools and concepts. In addition, many government agencies have a vested interest in pursuing their own agendas, given government imperatives of food productivity and security. Belief in the use of pesticides as the sole effective strategy in securing a good harvest is still widely held. To what extent have significant changes in the farming culture thus occurred?

CHANGING THE CULTURE OR FORMING A SUBCULTURE?

There is a marked difference in the way in which the green revolution and the IPM programs have introduced novel ideas. During the green revolution, technical packages of inputs were disseminated to as many farmers and farming communities as possible. The government's goal was to persuade farmers, by force if necessary, to adopt the rice intensification programs in as short a time as possible. In contrast, the IPM program was designed to transmit knowledge on a small scale and improve the farmers' decision-making abilities without coercion. The expectation was that a small group of IPM-trained farmers would disseminate the new ideas to the rest of the community. The two programs thus radically differed in scale, the speed of knowledge transfer, and the extent to which changes were anticipated in farming culture. The success stories of IPM indicate that pesticide use has declined (Pincus 1991; Ruchijat and Sukmaraganda 1992; Suyanto, Budiyati, and Quizon 1994) and alternative pest-control strategies have been developed (Paiman n.d.; Kuswara n.d.; Hafisam n.d.; Susianto, Purwadi, and

Pontius n.d; Winarto 2004b; Tim Bantuan Teknis FAO 1997; Busyairi et al. 2000; Kenmore 1997; P. Ooi 1998; van den Berg and Jiggins 2007).[7] The question is, to what extent do these achievements reveal real changes in the farming paradigm?

Complete commitment to practicing the IPM strategy is apparent among some groups of IPM farmers (e.g., in Indramayu and Terbanggi Besar) who have tried hard to consciously replace the term *medicine* with *poison*. They see pesticides as not a panacea for their problems in securing good harvests, but often just the opposite. A new belief that spending capital on the purchase of pesticides is useless, depletes savings, and leaves farmers with high-interest loans and debts has become the basis of their decision making. Producing commodities "free of poisons" (*bebas racun*) or based on "organic farming, free of chemicals" has become their principal goal. For many other farmers, however, the term *medicine* remains intimately bound up with the idea of curing or protecting their plants. In switching to the term *poison* instead of *medicine*, farmers have to draw a symbol from an entirely different linguistic domain. In farmers' vocabularies, *poison* had previously been used only to refer to rodenticides. During my stay with different farmers, I found that they occasionally used both terms to refer to two different intentions: to make their plants healthy by using "medicine" and to kill plant pests by using "poison." In many cases, they still use the term *medicine* with the connotation of protecting their plants and curing their illnesses. Occasionally, they mix the two terms into one understanding: namely, "medicine" can be "poisonous" if they spray too much of it, just as if they themselves take too much medicine.

The adoption of the IPM program also varies in relation to the specific circumstances of pest and disease infestations. When there are no significant infestations, capital is short, or they face ill-considered suggestions from local bureaucrats—including the credit package complete with pesticides—farmers will employ the new schema of IPM and avoid the use of pesticides. However, when they were in a panic facing a high risk of harvest failure due to massive and unexpected pest or disease outbreaks, they then question the effectiveness of the new paradigm and fall back on their old way of thinking: "Please give us the most effective 'medicines' to kill the pests" and "What kinds of 'medicines' would effectively kill this pest?" Newly established beliefs and knowledge are easily jeopardized, therefore, especially without appropriate and timely extension assistance. This kind of inconsistency makes it difficult for non-IPM farmers to be fully convinced of their fellows' newly established "convictions." When I revisited Marga Tani and Indramayu in the late 1990s, and most recently in early 2006, the injudicious use of pesticides, including the mixed formulae (*oplosan obat*) of diverse brands of pesticides, was still practiced by non-IPM farmers and, in some cases, by some IPM farmers as well (see Winarto 2006).

There is even variation at the individual level in the adoption of the IPM program. A single farmer can have different interpretations and make different decisions in varied conditions. In Marga Tani, the same farmer might spray or not spray pesticides depending on the circumstances. In Batanghari, some IPM farmers sprayed their secondary crops, but refrained from doing so for the rice. Why? "Vegetables cannot be treated with IPM [sayuran tidak bisa di PHT]," or because of the perception of insufficient natural predators during the pest's outbreak. Thus these farmers were surprised to learn that farmers in Terbanggi Besar were successfully cultivating soybeans without pesticides. In the latter case, farmers have been able to develop various sorts of pest control for soybeans: mechanically trapping the moths of the seed borer (*Ethyella* spp.) in nets or using lights in the evening; or referring to the Javanese traditional calendar *pranata mangsa* for planting decisions (see Indrowuryanto 1999).[8]

In the past decade, repeated soybean harvest failures led the Central Lampung farmers to believe that a good harvest of soybeans depended on pesticide use. Only firsthand experience of a successful harvest without pesticides can alter this perception. The IPM Soybean Farmer Field School in 1998 improved their confidence that cultivating soybeans without pesticides was possible. "Seeing is believing"—in this way the farmers changed their minds (Winarto 2004b). Transmitting knowledge about farming in a narrative form is not effective without observable evidence. The IPM farmers experienced this firsthand when they tried to convince non-IPM farmers of the function of natural enemies or of the future threat reflected in rice borer egg clusters. Only persistent practice and talk ultimately made the non-IPM farmers pay attention.

Recognizing the constraints of the farmer-to-farmer transmission of knowledge, a systematic plan for transmitting the IPM principles through Farmer Field Schools was developed in Terbanggi Besar, and then the number of IPM farmers grew quickly. The more farmers trained through IPM schools, the easier it becomes for them to communicate with one another in the same IPM language. Assisting the IPM alumni after their formal training also became a major program, not only to improve their basic knowledge but also to strengthen their confidence and encourage them to consistently implement the IPM principles. Supporting one another in solving daily farming problems and coordinating responses to environmental perturbations and inappropriate governmental interventions were distinctive aspects of the IPM program in Terbanggi Besar. Yet I also discovered that part of the community—mostly those who were not full-time farmers and thus did not have time to carry out studies and experiments—refused to implement the IPM strategies.

In contrast, I did not find any significant changes in the knowledge of the wider part of the community in Batanghari, although IPM programs had been going on for five years in one hamlet. The circulation of IPM knowledge occurred mainly among the IPM participants, who came from two extended families. A plan to widen the circle of IPM participants to encompass the rest of the community, as in Terbanggi Besar, was missing (Winarto et al. 2000). And in Marga Tani, in the absence of any follow-up program after 1990 and following only one IPM training session, it proved difficult for the farmers to abandon their old paradigm of pest control. IPM farmers had to face the reality that hundreds of farmers in their communities were still ignorant of the novel ideas they had learned (Winarto 2004b, 2006). Moreover, they had to battle unprecedented pest and disease infestations, top-down recommendations and promotions, and inaccurate information (Douglas Dilts, personal communication 1991) about the use, functions, nature, or the proper mixture of pesticides received from elsewhere.

Nevertheless, changes in the farming culture seem underway, though it is still too early to say that a change in the national farmers' paradigm has occurred. Looking at farming communities across Java, the formation of a subculture based on the IPM paradigm seems to have emerged (Winarto 2004a). In the past few years, many more IPM farmers' meetings and networking programs have been carried out, either led by IPM facilitators or initiated by the farmers' groups themselves. IPM farmers' groups and networks have been formed at the village, district, regency, provincial and, recently, national levels. In these networks, the IPM alumni do not only share experiences but also strengthen their confidence and speak out against inappropriate governmental policies (Petani PHT Jawa Barat 1998; Musyawarah Petani PHT Indonesia1999). However, one key question remains: to what extent has the IPM paradigm affected the wider society, governmental policies, and the extension services? Has a significant change in decentralizing decision making occurred?

FARMERS AND THE NATION: STRIVING WITHIN
ENDURING POWER RELATIONS

The immediate results the farmers have received from their new practices of pest and disease management are reduced expenditures, because they no longer have to invest in costly pesticides, and the decrease of "illnesses" attacking their plants. The first result proves that reducing expenses is possible without sacrificing harvests. This proved to be a great incentive to find alternative ways to manage pests and diseases, and by understanding the negative impacts of "medicines" on natural predators, farmers are also able to answer the puzzle of why more pesticides lead to more illnesses attacking their plants. This has opened up their minds (see Maidi and Darmowiyoto 1999). But to what extent were the farmers able to alter their relationships with the state and the environment?

Macroecology: Changing or Persisting Landscapes?

When I left Marga Tani in 1992, farmers were still struggling to control a WRB outbreak. When I returned in 1996, the farmers confidently reported that the population of WRB had significantly declined in the previous four years, due to their pest-control strategies implemented in 1990–92. This explanation was reinforced by their observation that farmers in neighboring Indramayu were still struggling against high WRB populations. This positive feedback added to the farmers' confidence as masters of their agricultural environment. It does not mean, however, that they have been entirely free from other challenges and hazards, such as disease infestation and recurrent BPH outbreaks. The severe BPH outbreak on the north coast of West Java in the 1998 dry season raised a critical question regarding the extent to which IPM was actually repressing this pest.

The BPH outbreak in 1998 damaged a large number of rice fields on the north coast of West Java, which reactivated farmers' belief that only the intensive and frequent use of pesticides (and other substances such as diesel fuel, kerosene, and detergent) could save their harvests. Complicating the problem, the chemical companies—in collaboration with regional and local bureaucrats—supported the old beliefs in pesticides as a panacea by involving farmers in a demonstration promoting the efficacy of their products. The chemical companies—assisted by a previous IPM facilitator—came to Marga Tani, divided a number of rice fields into blocks, and asked the farmers to spray a particular brand of pesticide in each block. This is an example of how other institutions use a hazard for their own interests and by doing so preserve the green revolution's landscape. This undermines the IPM farmers' perception of their fields' performance as better than the rest. Ironically, those "good" IPM fields were selected as the demonstration plots. Why? The chemical companies' offer of free pesticides in a situation of economic crisis persuaded the IPM farmers to join, even though they were still critical of it.

The IPM farmers were able to explain why another outbreak of BPH had occurred. First, their fellow farmers continued to use pesticides injudiciously. Second, they noted the change in weather conditions from a long drought (the El Niño) in 1998 to heavy rainfall and high humidity in 1999 (the La Niña), which favored the growth of BPH. Third, the fields were largely planted with a local rice variety nonresistant to BPH. In the farmers' eyes, however, the problems with this variety were unusual. Despite the plant hopper outbreak, they did not favor planting the dominant high-yielding variety resistant to BPH. They had switched instead to planting a local variety (Muncul) in the dry season and, in some cases, in the rainy season as well. This variety was not an officially certified seed released by the government (see Winarto 1997b). When I revisited Marga Tani in 1996, some IPM farmers praised the good quality of the plants and claimed that the government now had to learn from them. They suggested that the state's laboratory should begin experimenting with their favorite native varieties. The return of diversity in rice varieties has indeed been a landmark development in this region. Not only have some traditional rice varieties returned (see Winarto 1997b) but also a range of the farmers' own sub-varieties, as discovered in Indramayu (Winarto and Ardhianto 2007; Indonesian Integrated Pest Management Farmers' Alliance 2007).[9]

I also found a return to diversity in soybean varieties in Central Lampung. The success the farmers had in Terbanggi Besar in protecting soybeans from severe damage by soybean seed borers (*Ethyella* spp.) after repeated harvest failures is only one of many stories. For a number of years, following the government's introduction of the soybean crash intensification program, farmers had not been able to successfully cultivate soybeans due to pest infestations. Hence the ability to cultivate soybeans without using pesticides (*tanam kedele bebas racun*)—with the initial assistance of a foreign entomologist—proved a breakthrough. Another success story was the farmers' soybean cross-breeding activity assisted by a local IPM facilitator. Working with the variety released by the local government laboratory, an IPM farmer in collaboration with a local IPM facilitator was able to develop a giant variety—measuring three meters tall—named "Amerikana." The IPM farmers in Terbanggi Besar also improved other varieties and developed appropriate cultivation strategies for them. This is a very promising step in the return of heterogeneity to the farmers' landscapes after the enforced change to a uniform monoculture of crops during the green revolution (see Shiva 1991; James Fox 1993; Chang 1993; Whitten and Settle 1998). Terbanggi Besar became known as the place to find promising seeds, and the IPM farmers thus became de facto consultants to farmers from other regions seeking advice (Winarto et al. 2000).

Another manifestation of farmers' creativity is the production of bio-pesticides from tree leaves, seeds, and roots. Bio-pesticides became an option as farmers learned to move away from chemical pesticides in controlling pests. Though the production of bio-pesticides was still in the early stages of development during the period of my observations, their use had begun to spread in Central Lampung through IPM farmers' meetings and field schools. For older farmers, this practice signaled a return to knowledge and practices predating the use of modern chemical "medicines." The Javanese transmigrants in Central Lampung likened this practice to their traditional Javanese way of controlling pests. However, some IPM farmers in Terbanggi Besar criticized the "poisonous" substances in bio-pesticides as contrary to the idea of cultivating crops without harmful substances. Still the practice represented a return to local inputs, rather than relying on the state's provision of manufactured ones (see Winarto et al. 2000).

An Evaluation of the Extension System

To what extent then have farmers' own management abilities affected the extension service, the state's means of providing top-down technical recommendations? "Since you left us in 1992, nobody has come to assist us. . . . We are like seeds, sown by the government, which were then left alone, the government did not even water us. . . . Hence, the seeds died off." These words greeted me from the Marga Tani farmers' leader when I returned to his hamlet in 1996 (see Winarto 2004b). His complaint notwithstanding, farmers' greater knowledge and sense of dignity had resulted in their decreased dependence on the extension workers. Previously the extension worker had been the only source of information on new technological inputs and the person from whom to seek advice. This was no longer the case after the IPM training. The present extension worker was perceived as incompetent since in the farmers' eyes he lacked the practical knowledge of daily work in the fields. "I know more than him about the white rice borer," said one IPM farmer who became knowledgeable about the pest through his own field observations. On the other hand, the farmers know that the extension agents earn a government salary to assist them, in particular during any crises in farming. The agents' frequent absences were thus seen as a violation of their responsibilities.

The participation of extension agents' in the IPM programs was intended to improve the field activities of the trainers. In the IPM schools I observed in Marga Tani and Batanghari, the extension agents did indeed attend classes and assist the IPM facilitators, or they became facilitators themselves. As soon as school ended, however, so did their weekly visits to the farmers, unless there were formal follow-up programs. In Terbanggi Besar, the farmers themselves managed all the IPM programs without any help from the extension agent. The farmers were the planners, the organizers, and the trainers.

In contrast to their minimal role in the IPM program, the extension agents' role in promoting orthodox technical inputs, like the government's credit packages, continued. The head of the extension agents in Batanghari told a group of farmers' representatives that they—the farmers—now had the right to determine the brands of pesticides they preferred; but they were not allowed to drop the pesticides from the credit scheme entirely. The IPM farmers' rejection of the policy was simply ignored. When farmers voiced unhappiness with the government's increase in the price of the credit package during the 1998–99 rainy season, the extension agent in Batanghari simply dropped the pesticide and herbicide packages in the village head's house without his consent. This agent also threatened to withhold the rest of the credit package if the farmers refused to accept the pesticides (Winarto et al. 2000). Earlier, during the 1990–91 planting season in Marga Tani, the village cooperative officials had refused to disburse the cash from the credit packages as a punishment for the farmers' refusal to accept the pesticides (Winarto 2004b).

Similar stories concerning the failures of extension agents were repeated in other places during the first decade of IPM. They have led me to question the efficacy of IPM in altering the traditional role of the extension agents in the top-down national policy of crop intensification. The traditional "training and visit system" of the extension services has not changed much. In practice, some extension agents assisted farmers in running the IPM meetings, but they did not help farmers in the fields. By attending the meetings and assisting the IPM facilitators, the extension agents' own knowledge would, of course, have been improved. And the government did agree to officially incorporate IPM principles into the curriculum at the Agricultural Extension Academy. Would this development bring the extension agents closer to the farmers' world and knowledge production or to the production of scientific knowledge? My own recent observations of the behavior of extension agents in the field (in Gunungkidul region in Yogyakarta in 2007) suggest not. The extension agents' learning process continues to diverge from the everyday learning activities of the IPM farmers. This is the principal critique that the farmers level against them.

In the early 1990s, IPM training had a more significant impact on the pest and disease monitors of the Directory of Food Crop Protection (DFCP) than on the agricultural extension agents. Thousands of these monitors were trained intensively, with the aim of changing their role from that of simply promoting the products of the chemical companies (Dilts, personal communication 1991) to that of the main transmitters of IPM knowledge to millions of farmers all over Indonesia. But the question remains as to what extent this objective has been accomplished.

Whatever the role of the government agents, it is clear that the farmer-facilitators have assumed a key role in transferring knowledge to other farmers and thus helping to open space for local innovative practices. For example, greater confidence in their own discoveries led farmers in Kalensari, Indramayu, to argue with the experts after they developed on their own an effective strategy to control the white rice stem borer, by eliminating the pest as early as possible in the nurseries without the use of pesticides (Dilts, personal communication 1996). Farmers have proved to be talented knowledge producers and disseminators.

Crop Production and Protection Strategies: Conflicting Perspectives?

The state's commitment to higher productivity has continued (James Fox 1993), and two contrasting paradigms have been at odds within the Ministry of Agriculture: the green revolution paradigm and the IPM paradigm. As a result, the persisting high priority on productivity has limited opportunities for significant changes in the relation between farmers and the state. Farmers continue to be both targets and objects. This is exemplified in the food intensification program, GEMA PALAGUNG 2001, an abbreviation for Gerakan Mandiri Padi, Kedelai dan Jagung Tahun 2001 (Self-Help Movement of Rice, Soybean, and Corn for the Year 2001).[10] The economic crisis in Indonesia in 1997–98 led the minister of agriculture to focus on improving productivity by increasing crop diversity, cropping areas, and yields for rice, soybean, and corn (Departemen Pertanian 1998).[11] The state heavily subsidized the production of miracle seeds for those three crops and the delivery of the required technological inputs. Except for the incorporation of IPM principles as one of the objectives of the program (Departemen Pertanian 1998), the underlying paradigm remained that of the green revolution. Some leading IPM experts saw this program as a setback, as a return to the first decade of the state's national program to intensify rice cultivation. The intensive and synchronized cropping pattern—of GEMA PALAGUNG 2001—not much different from the intensified cultivation of Green Revolution varieties in the previous three decades—provided a good habitat for pests to flourish.

At the local level, farmers still have to struggle against pressures to plant so-called miracle seeds and, most recently, "hybrid seeds." Indeed, the Indramayu plant breeders' talents and products were seen by the authorities as threatening national food security and thus they were deemed illegal. The farmers fought back with the screening of a film emphasizing their knowledge (Indonesian Integrated Pest Management Farmers' Alliance 2007), which helped them obtain support and recognition from the authorities. But farmers still experience hardships due to a variety of state policies (imported rice, the low price of grains, increased prices for agricultural inputs) and the unexpected problems of weather conditions and climate change. Do the IPM principles help them to survive these challenging circumstances better? On the one hand, their ability to find alternatives for expensive inputs does help them. Examples include the return to so-called traditional "medicines," experiments with botanical pesticides, the adoption of organic fertilizers, and the cross-breeding of local and modern varieties. On the other hand, they are still the objects of various profit-seeking initiatives by outside agents. The position of IPM within national food production policy reflects these conflicts. Though IPM was officially proclaimed the national food-crop protection strategy, in practice there is little government commitment to it. Farmers find themselves caught between two poles. A decade earlier, farmers felt obliged to accept the intensified credit-package scheme although they knew that it contradicted IPM teachings (see Winarto et al. 2000). More recent is the state's policy of importing rice and hybrid seeds although farmers know they have the capability to create local land races (see Winarto 2005b).

Farmers have spoken up against the government's intensification schemes and ill-conceived policies, but in the midst of economic crises and persisting unequal farmer-state power relations, they often fight a losing battle. In the fight against the national policy on agricultural credit, farmers' struggles initially failed to have significant impacts. Only in fora like the provincial- or national-level seminars of IPM farmers did the state's apparatus eventually listen and respond positively to the complaints regarding the mishandling of the credit scheme and the extension agents' failures (Petani PHT Jawa Barat 1998; Musyawarah Petani PHT Indonesia 1999). Also, on the positive side, there are promising signs that the farmers' discoveries in controlling pests can be scaled up. For example, the rice borer control strategy developed by the farmers in Indramayu has been integrated into the regional policy for controlling this pest. The head of the regency decided to adopt the farmers' findings and disseminate them throughout his regency (Pemerintah Daerah Tingkat II Indramayu 1996; Busyairi et al. 2000; Kenmore 1997; Winarto 1997a; P. Ooi 1998). Even his efforts still met with some resistance from regency bureaucrats. But the IPM facilitator—in collaboration with the other agricultural and local administrative officials—played a significant role in assisting the farmers in their efforts to disseminate their discoveries.

IPM and the Pendulum of National Policy: Conclusions

The IPM program in Indonesia has gained worldwide recognition as a break-through in agricultural development by incorporating the human aspects neglected in the green revolution paradigm. The replacement of human capability by the products of modern scientific technology led to the degradation of the agricultural ecology and to the destruction of farming culture. I argue that this disregard of individual farmers' potentials and erasure of people's knowledge is no longer sustainable. The farmers' description of the green revolution making them stupid was in fact a reality. They not only became ignorant of and alienated from their own environment but they also lost their sense of dignity and identity as capable decision makers. In contrast, IPM opened the door for the development of human resources in a field that for almost three decades had been overgoverned by the state. The creation of a policy atmosphere supportive of farmers' creativity, dignity, and confidence stimulated the growth of "seeds of knowledge." These are the true "miracle seeds" that are overlooked by the experts of the green revolution. IPM is thus much more than a modification of farm technology. It stimulates a different set of relations among people, the state, and the environment, which ultimately leads to a self-governing community that can help itself. This is the promise that IPM holds for the sustainability of farming land-scapes and the prosperity of the farmers. I argue strongly that this kind of resource management offers the most appropriate path toward a twenty-first-century ecology. Unfortunately the state, rejecting the broader paradigm shift promoted by IPM planners, is narrowly focused on simply incorporating IPM's policy of plant protection into its established agricultural bureaucracy, with the sole objective of achieving and sustaining security.

Indonesia's economic crises and threats to self-sustainability have been the leading causes for the simultaneous promotion of both the IPM strategy and the policy of high productivity. The struggles between the various parties involved reflect a pendulum swinging between two poles, high productivity and overgovernance, on one end, and ecological sustainability and self-governance on the other. In the absence of any imminent threats to national food production, the pendulum can swing toward the latter. Economic constraints such as increased pesticide prices and the absence of severe pest or disease outbreaks strengthen the farmers' motivation to find ways to avoid the intensive use of chemicals. On the other hand, climatic changes and a persisting injudicious use of pesticides induce pest and disease outbreaks and jeopardize the ecological sustainability of food-crop production. The economic crisis of 1998–99 also produced real threats to national food crop-production. In such circumstances, the pendulum swings to the green revolution end, with its emphasis on intensifying crop farming through increased subsidies of technical inputs and the provision of credit to farmers. The government pushes on them all the resources it has to boost crop productivity. Economic crises or the failure to reach the state's quota in national rice production thus cause a retreat from IPM principles, with predictable consequences. Pest outbreaks and farmer complaints about government policy have been reported in regions with increased crop-index diversity and cropping area. The harsh circumstances of economic and ecological crises, ill-considered policies, top-down coercive recommendations, and offstage rent-seeking behavior have all undercut efforts to empower farmers.

Believing that they should be the managers of their own environment, some farmers suggest that the ideal role of the government is to serve as the "herder of the flocks, the people." If the "herder" cannot lead the flocks, the latter will run everywhere. In the farmers' eyes, the "herder" should assist them in fulfilling their needs, alleviating their problems, and supporting their prosperity. The "herder" is not supposed to simply order them about. In some IPM farmers' eyes, the government should provide knowledge and explanation, not merely instructions and guidance. Such an ideal is reflected in the behavior of the IPM facilitators assisting the farmers in making decisions. The IPM facilitators not only do not undermine the farmers' autonomy but they also shoulder some of their responsibilities in developing a path toward prosperity. Though they are constrained in terms of access to knowledge and information, they can be brokers for whatever knowledge they do have access to. The underlying premises of IPM policy thus epitomize the ideal role of the "herders," the government agencies. IPM policy and practice might be said to exemplify the ideal character of peasant and state relations.

IPM can potentially play a significant role in altering the larger relations among the state, the farmer, and the environment: from a relationship characterized by subordination toward one characterized by a partnership; and from alienation from and ignorance of the environment to a closer and deeper understanding of it. If the state continues to dominate farmers' decision-making processes to meet its objective of high agricultural productivity, then IPM can help reverse some of the attendant damage by providing greater freedom for people to manage their own environment while still sustaining their economic productivity. Unfortunately, the pendulum often swings toward the end where the state overdetermines relations between people and their environment, not least because the IPM program has become the responsibility of the state. It is now time to give serious consideration to how a program like IPM can play a role in swinging the pendulum back toward the other end, where an active civil society governs its environment in collaboration with the state.

Notes

This essay is based on my studies from early 1990 to 1999 and through 2008. As part of my PhD program in anthropology at the Australian National University, I carried out research in 1990–92 on the IPM Farmer Field Schools and farmers' practices in Ciasem Baru (in Marga Tani) and Ciasem Tengah (in Marjim and several other hamlets), Ciasem District, Subang Regency on the north coast of West Java. I made return visits to these locales and the village of Kalensari and other villages in Indramayu between 1996 and 2008. Between June 1998 and March 1999 I carried out research in a number of villages in the regency of Central Lampung, Lampung Province, comparing the IPM program (in Balerejo, Selorejo, and Bumi Mas, Batanghari District) and an NGO's IPM program (in Karang Endah and Adijaya, Terbanggi Besar district). I am grateful for the support of the FAO Indonesian National IPM Program and the Ford Foundation for my studies in 1990–92. The Indonesian-FAO Inter-country Program, the Global FAO-IPM Facility, the "Institutional Context of Biodiversity Maintenance in Asia: Trans-national, Cross-sectoral, and Inter-disciplinary Approaches" project of the MacArthur Foundation, and SEAMEO-SEARCA supported my research in 1998–99. My return visits to the north coast of West Java were made possible by the support of the undergraduate program, Department of Anthropology, University of Indonesia; and KNAW (the Royal Netherlands Academy of Arts and Sciences), and the Indonesian Academy of Sciences. My sincere thanks to Michael R. Dove, Percy E. Sajise, Gil C. Saguiguit, Ben S. Malayang III, and Levita Duhaylungsod for their valuable suggestions and directions in preparing and writing up this essay. Thanks are also due to my colleagues, Iwan Tjitradjaja and Ezra M. Choesin, and my research assistants, Chamiyatus Sidqiyyah, Sri Widyastuti, A. Sri Handayani Ningsih, Fadli, Susanto Darmono, Iwan Meulia Pirous, Rhino Ariefiansyah, Hanantiwi Adtiyasari, and Imam Ardhianto.

1. IPM had already been incorporated into the Super Intensification Program (SUPRA INSUS), which was designed to boost rice productivity in particular regions, including the north coast of West Java.

2. See also Escobar (1995: 155) on integrated rural development as a strategy to correct the biases of the green revolution.

3. The north coast, irrigated by the Jatiluhur dam, is one of the rice baskets of West Java. The farmers in this region plant rice in both the rainy and the dry season, followed by either fallowing or secondary crop farming. In Marga Tani, farmers

do not plant secondary crops during the fallow period. In Central Lampung, the irrigation water available from small dams generally does not suffice for two rice planting seasons annually. The farmers in Batanghari and Terbanggi Besar can only cultivate rice once each year, followed by two seasons of secondary crops. In the 1998 planting seasons, however, farmers in some villages in Batanghari with land close to two irrigation canals were able to plant rice twice a year.

4. Much earlier, in the 1960s, experts in Kochi Prefecture, Japan, found that the secondary resurgence of plant hoppers and leafhoppers was due to regular insecticide applications. As early as 1950, Conway, Rao, and Wood recognized that the overuse of broad-spectrum insecticides stimulated secondary infestations and created catastrophic outbreaks (FAO 1990).

5. The pest's "economic threshold" is the pest density (or amount of plant damage) at which incremental costs of control just equal incremental crop returns (see Bottrell 1979).

6. See Fujisaka 1994 on the reasons for farmers' rejection of agricultural innovations.

7. See van den Berg and Jiggins (2007) for their compilation of studies of the impacts of IPM Farmer Field Schools around the world.

8. *Pranata mangsa* is based on the perceived movement of the sun and stars from the north to the south of the sky and vice versa. Movement to the north means that the dry season is coming, while movement to the south means that the rainy season is coming. In between these two seasons, there are two intermediary seasons when the sun is directly overhead. Farmers refer to these movements to decide when to start planting (see Indrowuryanto 1999).

9. See Frossard (1998a, 1998b) on Filipino farmers breeding seeds. On the Indramayu farmers' recent practices in plant-breeding, see Winarto and Ardhianto (2007) and Indonesian Integrated Pest Management Farmers' Alliance (2007).

10. This was the main program in the government's "Special Effort to Overcome the Food Production Crisis" (Upaya Khusus Penanggulangan Krisis Produksi Pangan) to increase food productivity (Departemen Pertanian 1998).

11. The national government officially set the production targets and specified all the methods and institutions for implementing this program, which included improving extension and facilitation; the provision of credits, fertilizers, and other production inputs; and the control of pest and disease during and after the harvest (Departemen Pertanian 1998: 2).

BIBLIOGRAPHY

Acciaioli, Greg. 2008. "Mobilizing Against the 'Cruel Oil': Dilemmas of Organizing Resistance Against Oil Palm Plantations in Central Kalimantan." In *Reflections on the Heart of Borneo*, ed. Gerard A. Persoon and Manon Osseweijer, 91–119. Tropenbos No. 24. Wageningen, The Netherlands: Tropenbos International.

Ackert, Andreas. 2003. "Comparing Coffee Production in Cameroon and Tanganyika, c. 1900 to 1960s." In *The Global Coffee Economy in Africa, Asia, and Latin America, 1500–1989*, ed. W. G. Clarence-Smith and S. Topik, 286–311. Cambridge: Cambridge University Press.

Adewoye, Omomiyi. 1986. "Legal Practice in Ibadan." *Journal of Legal Pluralism and Unofficial Law* 24:57–76.

Afiff, Suraya, and Celia Lowe. 2007. "Claiming Indigenous Community: Political Discourse and Natural Resource Rights in Indonesia." *Alternatives* 32:73–97.

Agbayani, Rene. 1990. "Laborers in Their Own Land." In *Struggles Against Development Aggression: Tribal Filipinos and Ancestral Domain*, ed. Ruffy Manaigod, 25–32. Quezon City: Tunay na Alyansa ng Bayan Alay sa Katutubo (TABAK).

Aglionby, Julia. 1995a. *Economic Issues in Danau Sentarum Wildlife Reserve*. Bogor: Wetlands International/PHPA/ODA.

——. 1995b. *Final Report of the Associate Professional Officer (Environmental Economist)*, vol. 1, *Economic Issues in Danau Sentarum Wildlife Reserve*. Bogor: Wetlands International/PHPA/ODA.

——. 1995c. "The Issue of Poisoning in DSWR: A Stakeholder Analysis." In *The Economics and Management of Natural Resources in Danau Sentarum Wildlife Reserve, West Kalimantan, Indonesia: A Collection of Papers*, ed. Julia Aglionby, 108–15. Bogor: Wetlands International/PHPA/ODA.

Agrawal, Arun. 1995. "Dismantling the Divide between Indigenous and Scientific Knowledge." *Development and Change* 26:413–39.

——. 1997. "Community in Conservation: Beyond Enchantment and Disenchantment." Working paper. Gainesville, Fla.: Conservation and Development Forum.

——. 1998. "Indigenous Knowledge: Some Critical Comments." *Antropologi Indonesia* 12 (55): 14–43.

——. 1999. "Community-in-Conservation: Tracing the Outlines of an Enchanting

Concept." In *A New Moral Economy for India's Forests: Discourses of Community Participation*, ed. Roger Jeffery and Nandini Sundar, 92–108. New Delhi: Sage.

——. 2000. "Adaptive Management in Transboundary Protected Areas: The Bialowieza National Park and Biosphere Reserve as a Case Study." *Environmental Conservation* 27 (4): 326–33.

——. 2005. *Environmentality: Technologies of Government and the Making of Subjects.* Durham, N.C.: Duke University Press.

Agrawal, Arun, and Clark C. Gibson. 1999. "Enchantment and Disenchantment: The Role of Community in Natural Resources Conservation." *World Development* 27: 629–649.

——, eds. 2001. *Communities and the Environment: Ethnicity, Gender, and the State in Community-Based Conservation*. Brunswick, N.J.: Rutgers University Press.

Agrawal, Arun, and Kent Redford. 2006. "Poverty, Development, and Biodiversity Conservation: Shooting in the Dark?" Working Paper No. 26. Bronx: Wildlife Conservation Society.

Aiken, S. Robert. 1973. "Images of Nature in Swettenham's Early Writings: A Prolegomenon to a Historical Perspective on Peninsular Malaysia's Ecological Problems." *Asian Studies* 11:135–49.

Aiken, S. Robert, and Colin H. Leigh. 1992. *Vanishing Rain Forests: The Ecological Transition in Malaysia*. Oxford: Clarendon.

Aiken, S. Robert, Colin H. Leigh, and C. H. Leinbach. 1982. *Development and Environment in Peninsula Malaysia*. Singapore: McGraw-Hill International.

An, L., J. Liu, Z. Ouyang, M. Linderman, S. Zhou, and H. Zhang. 2001. "Simulating Demographic and Socioeconomic Processes on Household Level and Implications for Giant Panda Habitats." *Ecological Modelling* 140:31–49.

Anderson, Benedict. 1983. *Imagined Communities: Reflections on the Origin and Spread of Nationalism*. New York: Verso.

Angelsen, A. 1995. "Shifting Cultivation and 'Deforestation': A Study from Indonesia." *World Development* 23:1713–29.

Appadurai, Arjun. 1996. *Modernity at Large: Cultural Dimensions of Globalization*. Minneapolis: University of Minnesota Press.

——. 2001. "Deep Democracy: Urban Governmentality and the Horizon of Politics." *Environment and Urbanization* 13 (2): 23–43.

Appell, George. 1985. "Land Tenure and Development among the Rungus of Sabah, Malaysia." In *Modernization and the Emergence of a Landless Peasantry: Essays on the Integration of Peripheries to Socioeconomic Centers*, ed. Appell, 111–55. Williamsburg, Va.: Department of Anthroplogy, College of William and Mary.

——. 1986. "Kayan Land Tenure and the Distribution of Devolvable Usufruct in Borneo." *Borneo Research Bulletin* 18:119–30.

——. 1988. "Emergent Structuralism: The Design of an Inquiry System to Delineate the Production and Reduction of Social Forms." In *Choice and Morality in Anthropological Perspective: Essays in Honor of Derek Freeman*, ed. Appell and Triloki N. Madan, 43–60. Albany: State University of New York Press.

———. 1991. "Resource Management Regimes among the Swidden Agriculturists of Borneo: Does the Concept of Common Property Adequately Map Indigenous Systems of Ownership?" Paper presented at the International Association for the Study of Common Property Conference, University of Manitoba, Winnipeg, September 28, 1991.

Archibold, Guillermo, and Sheila Davey. 1993. "Kuna Yala: Protecting the San Blas of Panama." In *The Law of the Mother: Protecting Indigenous Peoples in Protected Areas*, ed. Elizabeth Kemf, 52–57. San Francisco: Sierra Club Books.

Arcilla, J. S. S. 1979. "Jesuit Mission Policies in the Philippines (1859–1899)." *Philippine Studies* 2792:176–97.

Ashton, P. S., T. J. Givinish, and S. Appanah. 1988. "Staggered Flowering in the Dipterocarpaceae: New Insights into Floral Induction and the Evolution of Mast Fruiting in the Aseasonal Tropics." *American Naturalist* 132 (1): 44–66.

Bahagiawati, A., and Ida Nyoman Oka. 1987. "Perkembangan biotipe wereng coklat *Nilaparvata lugens*. stal. di Indonesia." In *Wereng coklat*, ed. J. Soejitno, Z. Harahap, and H. S. Suprapto, 31–42. Bogor: Badan Penelitian Tanaman Pangan.

Balée, William. 1993. "Indigenous Transformations of Amazonian Forests: An Example from Maranao, Brazil." *L'homme* 33(2–4): 231–54.

———. 1994. *Footprints of the Forest: Ka'apor Ethnobotany; The Historical Ecology of Plant Utilization by an Amazonian People*. New York: Columbia University Press.

Barlow, C. 1978. *The Natural Rubber Industry: Its Development, Technology, and Economy in Malaysia*. Kuala Lumpur: University of Malaya Press.

———. 1991. "Developments in Plantation Agriculture and Smallholder Cash-Crop Production." In *Indonesia: Resources, Ecology, and Environment*, ed. J. Hardjoro, 85–103. Singapore: Oxford University Press.

Barlow, C., and J. Drabble. 1990. "Government and the Emerging Rubber Industries in Indonesia and Malaya, 1900–40." In *Indonesian Economic History in the Dutch Colonial Era*, ed. Anne Booth, W. J. O'Malley, and Anna Weidemann, 187–209. New Haven, Conn.: Yale University Southeast Asia Studies.

Barlow, C., and S. K. Jayasurija. 1986. "Stages of Development in Smallholder Tree Crop Agriculture." *Development and Change* 17 (4): 635–58.

Barlow, C., S. K. Jayasuriya, and S. C. Tan. 1994. *The World Rubber Industry*. London: Routledge.

Barlow, C., and Muharminto. 1982. "The Rubber Smallholder Economy." *Bulletin of Indonesian Economic Studies* 18:86–119.

Barlow, C., and T. Tomich. 1991. "Indonesian Agricultural Development: The Awkward Case of Smallholder Tree Crops." *Bulletin of Indonesian Economic Studies* 27:9–53.

Barlow, H. S. 2000. "A Contribution towards a History of MNS." *Malaysian Naturalist* 54 (1): 16–23.

Barr, P. [1898] 1977. *Taming the Jungle: The Man Who Ruled Malaya*. London: Secker and Warburg.

Bartlett, Harley Harris. [1956] 2008. "Fire, Primitive Agriculture, and Grazing in the

Tropics." In *Southeast Asian Grasslands: Understanding a Folk Landscape*, ed. Michael R. Dove, 77–117. New York: New York Botanical Gardens Press.

Bates, Diane, and Thomas K. Rudel. 2000. "The Political Ecology of Conserving Tropical Rain Forests: A Cross-National Analysis." *Society and Natural Resources* 13:619–34.

Bauer, P. T. 1948. *The Rubber Industry: A Study in Competition and Monopoly*. London: Longmans, Green.

———. [1946] 1961a. "The Economics of Planting Density in Rubber Growing." In *Readings in Malayan Economics*, ed. T. H. Silcock, 236–41. Singapore: Eastern Universities Press.

———. [1944] 1961b. "Some Aspects of the Malayan Rubber Slump, 1929–33." In *Readings in Malayan Economics*, ed. T. H. Silcock, 185–200. Singapore: Eastern Universities Press.

———. [1944] 1961c. "The Working of Rubber Regulations.". In *Readings in Malayan Economics*, ed. T. H. Silcock, 242–67. Singapore: Eastern Universities Press.

Beek, Walter E. A. van, and Pieteke M. Banga. 1992. "The Dogon and Their Trees." In *Bush Base, Forest Farm: Culture, Environment, and Development*, ed. Elizabeth Croll and David Parkin, 57–75. London: Routledge.

Belgrave, W. N. C. 1930. "The Effects of Cover Crops on Soil Moisture." *Agricultural Journal* 18:492–94.

Benjamin, G. 2000. "On Being Tribal in the Malay World." In *Tribal Communities in the Malay World–Historical, Cultural, and Social Perspectives*, ed. Benjamin and C. Chou, 7–76. Leiden: IIAS/ISAS.

Bennagen, Ponciano. 1985. "The Continuing Struggle for Survival and Self-Determination among Philippine Ethnic Minorities." Paper presented at the Philippine Social Science Center Forum on Social Sciences and Government, Quezon City, March 16, 1985.

Bentley, Jeffery W. 1992. "Alternatives to Pesticides in Central America: Applied Studies of Local Knowledge." *Culture and Agriculture* 44:10–13.

Berg, Henk van den, and Janice Jiggins. 2007. Investing in Farmers: The Impacts of Farmer Field Schools in Relation to Integrated Pest Management." *World Development* 35 (4): 663–86.

Berkes, Fikret. 1999. *Sacred Ecology: Traditional Ecological Knowledge and Resource Management*. Philadelphia: Taylor and Francis.

Berkes, Fikret, and Carl Folke, eds. 1998. *Linking Social and Ecological Systems: Management Practices and Social Mechanisms for Building Resilience*. Cambridge: Cambridge University Press.

Berry, Sara. 1988. "Concentration without Privatization? Some Consequences of Changing Patterns of Rural Land Control in Africa." In: *Land and Society in Contemporary Africa*, ed. R. E. Downs and S. P. Reyna, 53–75. Hanover, N.H.: University Press of New England.

———. 1993. *No Condition Is Permanent: The Social Dynamics of Agrarian Change in Sub-Saharan Africa*. Madison: University of Wisconsin Press.

———. 2004. "Reinventing the Local? Privatization, Decentralization, and the Politics of Resource Management: Examples from Africa. *African Study Monographs* 25 (2): 79–101.

Beusekom, Monica M. van. 2002. *Negotiating Development: African Farmers and Colonial Experts at the Office du Niger, 1920–1960*. Oxford: James Currey.

Biernatzki, William. 1973. "Bukidnon Datuship in the Upper Pulangi River Valley." In *Bukidnon Politics and Religion*, ed. A. de Guzman and E. Pacheco, 21–33. Quezon City: Ateneo de Manila University.

Bird-David, Nurit. 1990. "The Giving Environment: Another Perspective on the Economic System of Gatherer-Hunters. *Current Anthropology* 31 (2): 189–96.

Birowo, A. T., and G. E. Hansen. 1981. "Agricultural and Rural Development." In *Agricultural and Rural Development in Indonesia*, ed. Hansen, 1–27. Boulder, Colo.: Westview Press.

Blaikie, Piers. 1985. *The Political Economy of Soil Erosion in Developing Countries*. New York: Longman.

Blaikie, Piers, and Harold Brookfield. 1987. *Land Degradation and Society*. London: Methuen.

Bloch, Maurice. 1989. "The Symbolism of Money in Imerina." In *Money and the Morality of Exchange*, ed. Jonathan Parry and Bloch, 165–90. Cambridge: Cambridge University Press.

———. 1995. "People into Places: Zafimaniry Concepts of Clarity." In *The Anthropology of Landscape: Perspectives on Place and Space*, ed. Eric Hirsch and Michael O'Hanlon, 63–77. Oxford: Clarendon.

Bloch, Maurice, and Jonathan Parry. 1989. "Introduction: Money and the Morality of Exchange." In *Money and the Morality of Exchange*, ed. Parry and Bloch, 1–32. Cambridge: Cambridge University Press.

Boeck, Filip de. 1994. "Of Trees and Kings: Politics and Metaphor among the Aluund of Southwestern Zaire." *American Ethnologist* 21 (3): 451–73.

Boomgaard, Peter. 1998a. *Paper Landscapes: Explorations in the Environmental History of Indonesia*. Leiden: KITLV Press.

———. 1998b. "The VOC Trade in Forest Products in the Seventeenth Century." In *Nature and the Orient: The Environmental History of South and Southeast Asia*, ed. Richard H. Grove, V. Damodaran, and S. Sangwan, 375–95. New Delhi: Oxford University Press.

———. 2001. *Frontiers of Fear: Tigers and People in the Malay World, 1600–1950*. New Haven, Conn.: Yale University Press.

Boon, G. G, ed. 1971. *Agricultural Directory of Malaysia*. Kuala Lumpur: University of Malaya Agricultural Graduates Alumni.

Botkin, D. 1990. *Discordant Harmonies: A New Ecology for the Twenty-First Century*. Oxford: Oxford University Press.

Bottrell, D. R. 1979. *Integrated Pest Management*. Washington, D.C.: Council on Environmental Quality.

Bouman, M. A. 1952 (1922). "Gegevens uit Smitau en Boven-Kapoeas." *Adatrechtbundels* 44: 47–86.

Bouquet, Mary. 1995. "Exhibiting Knowledge: The Trees of Dubois, Haeckel, Jesse, and Rivers at the Pithecanthropus Centennial Exhibition." In *Shifting Contexts: Transformations in Anthropological Knowledge*, ed. Marilyn Strathern, 31–55. London: Routledge.

Bourdieu, Pierre. 1977. *Outline of a Theory of Practice*. Trans. Richard Nice. Cambridge: Cambridge University Press.

Bourque, L. Nicole. 1995. "Developing People and Plants: Life-Cycle and Agricultural Festivals in the Andes." *Ethnology* 34 (1): 75–87.

Brandon, K., K. Redford, and S. Sanderson. 1998. *Parks in Peril: People, Politics, and Protected Areas*. Washington, D.C.: Nature Conservancy.

Braun, B. 2002. *The Intemperate Rainforest: Nature, Culture, and Power on Canada's West Coast*. Minneapolis: University of Minnesota Press.

Bray, David Barton, Jose Luís Plaza Sánchez, and Ellen Contreras Murphy. 2002. "Social Dimensions of Organic Coffee Production in Mexico: Lessons for Eco-Labeling Initiatives." *Society and Natural Resources* 15 (5): 429–46.

Brechin, Steve, P. Wilschusen, C. Fortwangler, and P. West. 2003. *Contested Nature: Promoting International Biodiversity with Social Justice in the Twenty-First Century*. Albany: State University of New York Press.

Brookfield, Harold. 1994. "Change and the Environment." In *Transformation with Industralization in Peninsular Malaysia*, ed. Brookfield with Lorene Doube and Barabara Banks, 168–86. Kuala Lumpur: Oxford University Press.

———. 2001. *Exploring Agrodiversity*. New York: Columbia University Press.

Brookfield, Harold, L. Potter, and Y. Byron. 1995. *In Place of the Forest: Environment and Socio-Economic Transformation in Borneo and the Eastern Malay Peninsula*. Tokyo: United Nations University Press.

Brookfield, Harold, and Y. Byron. 1990. "Deforestation and Timber Extraction in Borneo and the Malay Peninsula: The Record since 1965." *Global Environmental Change* 1 (1): 42–56.

Brookfield, Muriel. 1998. "Indigenous Knowledge: A Long History and an Uncertain Future." *Antropologi Indonesia* 12 (55): 5–13.

Brosius, J. Peter. 1997a. "Endangered Forest, Endangered People: Environmentalist Representations of Indigenous Knowledge." *Human Ecology* 25 (1): 47–69.

———. 1997b. "Transcripts, Divergent Paths: Resistance and Acquiescence to Logging in Sarawak, East Malaysia." *Comparative Studies in Society and History* 39 (3):468–510.

———. 1999a. "Anthropological Engagements with Environmentalism." *Current Anthropology* 40 (3): 277–310.

———. 1999b. "Green Dots, Pink Hearts: Displacing Politics from the Malaysian Rain Forest." *American Anthropologist* 10 (1): 36–57.

———. 2003. "Defining Success in Conservation: Toward an Interdisciplinary Dialogue." Opening comments delivered to "Defining Success in Conservation: Toward an Interdisciplinary Dialogue," conference held at the University of Georgia, Athens, October 11, 2003.

Brosius, J. Peter, Anna L. Tsing, and Charles Zerner. 1998. "Representing Communities: Histories and Polities of Community-Based Natural Resource Management." *Society and Natural Resources* 11:157–68.

——, eds. 2005. *Communities and Conservation: History and Politics of Community-Based Natural Resource Management*. Walnut Creek, Calif.: AltaMira.

Brown, S., A. J. R. Gillespie, and A. E. Lugo. 1991. "Biomass of Tropical Forests of South and Southeast Asia." *Canadian Journal for Forest Research* 21 (1): 111–17.

Bruenig, J. F. 1985. "Deforestation and Its Ecological Implications for the Rain Forests in South East Asia." In *The Future of Tropical Rain Forests in South-East Asia*, ed. Commission on Ecology, 17–26. Gland, Switzerland: IUCN.

Bryant, Raymond L. 1994. "Shifting the Cultivator: The Politics of Teak Regeneration in Colonial Burma." *Modern Asian Studies* 28 (2): 225–50.

——. 1996. "Romancing Colonial Forestry: The Discourse of 'Forestry as Progress' in British Burma." *Geographical Journal* 162:169–78.

——. 1998. "Resource Politics in Colonial South-East Asia: A Conceptual Analysis." In *Environmental Challenges in South-East Asia*, ed. Victor T. King, 27–51. Richmond, Surrey, U.K.: Curzon.

Bryant, Raymond L., and Sinead Bailey. 1997. *Third World Political Ecology: An Introduction*. London: Routledge.

Bryant, Raymond L., Jonathan Rigg, and Peter Stott, eds. 1993. "The Political Ecology of Southeast Asia's Forests: Trans-disciplinary Discourses." Special issue, *Global Ecology and Biogeography Letters* 3 (4–6).

Buckley, C. B. [1902] 1965. *An Anecdotal History of Old Times in Singapore*. Kuala Lumpur: University of Malaya Press.

Bukri, Hasan Sayuti, et al., eds. 1981. *Sejarah Daerah Lampung*. Jakarta: Proyek Penelitian dan Pencatatan Kebudayaan Daerah, Pusat Penelitian Sejarah dan Budaya, Departemen Pendidikan dan Kebudayaan.

Burger, Julian. 1987. *Report from the Frontier: The State of the World's Indigenous Peoples*. London: Zed.

Burkill, H. M. 1971. "A Plea for the Inviolacy of Taman Negara." *Malayan Nature Journal* 24:206–9.

Burkill, I. H. [1936] 1966. *A Dictionary of the Economic Products of the Malay Peninsula*. 2 vols. Kuala Lumpur: Government of Malaysia and Singapore.

Burns, Peter. 1989. "The Myth of Adat." *Journal of Legal Pluralism and Unofficial Law* 28:1–127.

——. 2004. *The Leiden Legacy: Concepts of Law in Indonesia*. Leiden: KITLV.

Busyairi, Mufid A., et al. 2000. *Membangun pengetahuan emansipatoris: Kasus riset aksi petani di Indramayu, studi kehidupan dan gerakan pengendalian hama penggerek batang padi putih*. Jakarta: FAO and Lapesdam NU.

Buttel, Frederick H. 1992. "Environmentalization: Origins, Processes, and Implications for Rural Social Change." *Rural Sociology* 57 (1): 1–27.

Caldwell, Malcolm. 1977. "War, Boom, and Depression." In *Malaya: The Making of a Neo Colony*, eds. Mohamed Amin and Caldwell, 38–63. Nottingham: Spokesman Books.

Cant, R. G. 1973. *An Historical Geography of Pahang*. Kuala Lumpur: Malaysian Branch of the Royal Asiatic Society.

Carey, Iskandar. 1976. *Orang Asli: The Aboriginal Tribes of Peninsular Malaysia*. Kuala Lumpur: Oxford University Press.

Carpenter, Carol. 1989. "Food for Thought about Village-Court Relations." Paper presented at the annual meeting of the American Anthropological Association, Washington, D.C., November 15–19, 1989.

Castree, Noel. 2005. *Nature*. London: Routledge.

Castro, Peter. 1990. "Sacred Groves and Social Change in Kirinyaga, Kenya." In *Social Change and Applied Anthropology*, ed. M. Chaiken and A. Fleuret, 277–98. Boulder, Colo.: Westview.

Cernea, Michael. 2000. "Risks, Safeguards, and Reconstruction: A Model for Population Displacement and Resettlement. In *Risks and Reconstruction: The Experiences of Resettlers and Refugees*, ed. Cernea and C. McDowell, 11–55. Washington, D.C.: World Bank Publications.

Chang, Te-Tzu. 1993. "Sustaining and Expanding the 'Green Revolution' in Rice." In *South-East Asia's Environmental Future: The Search for Sustainability*, ed. H. Brookfield and Y. Byron, 201–9. Tokyo: United Nations University Press.

Chapin, M. 2004. "A Challenge to Conservationists." *World Watch*, November/December, 17–31.

Chapman, F. Spencer. [1949] 1997. *The Jungle Is Neutral*. Singapore: Times Book International.

Chayanov, A. V. 1966. *The Theory of Peasant Economy*. Ed. Daniel Thorner, Basile Kerblay, and R. E. F. Smith. Homewood, Ill.: R. D. Irwin.

Choesin, Ezra M. 2002. "Connectionism: Alternatif dalam memahami dinamika pengetahuan lokal dalam globalisasi." *Antropologi Indonesia* 69:1–9.

CIP-UPWARD (International Potato Center, User's Perspectives with Agriculture and Development). 2003. *Farmer Field Schools: From IPM to Platforms for Learning and Empowerment*. Los Baños, Laguna, Philippines: International Potato Center.

"Circular Notice to Officers." 1928. North Borneo Company Archive file #815, Kota Kinabalu, Sabah.

Clay, Jason. 1991. "Cultural Survival and Conservation: Lessons from the Past Twenty Years." In *Biodiversity: Culture, Conservation, and Ecodevelopment*, ed. Margery L. Oldfield and Janis B. Alcorn, 248–73. Boulder, Colo.: Westview.

Clements, F. E. 1916. *Plant Succession: An Analysis of the Development of Vegetation*. Washington, D.C.: Carnegie Institution Press.

Clifford, Hugh. 1897. "A Journey through the Malay States of Terengganu and Pahang." *Geographical Journal* 9:1–37.

Cohen, A. P. 1985. *The Symbolic Construction of Community*. London: Tavistock Publications.

Colchester, Marcus. 1994. "Salvaging Nature: Indigenous Peoples, Protected Areas, and Biodiversity." Discussion paper. Geneva: United Nations Research Institute for Social Development (UNRISD).

——. 1995. "Indigenous Peoples' Rights and Sustainable Resource Use in South and Southeast Asia." In *Indigenous Peoples of Asia*, ed. R. H. Barnes et al., 59–76. Ann Arbor, Mich.: Association for Asian Studies.

Colchester, Marcus, and Christian Erni, eds. 1999. *From Principles to Practice: Indigenous Peoples and Protected Areas in South and Southeast Asia*. Copenhagen: International Work Group for Indigenous Affairs (IWGIA) and the Forest Peoples Programme.

Colding, J., and C. Folke. 2001. "Social Taboos: 'Invisible' Systems of Local Resource Management and Biological Conservation." *Ecological Applications* 11 (2): 584–600.

Cole, Franklin S. 1956. *The Bukidnon of Mindanao*. Chicago: Natural History Museum.

Colfer, Carol J. Pierce, with Nancy Peluso and See Chung Chin. 1997. *Beyond Slash and Burn: Building an Indigenous Management of Borneo's Tropical Rain Forests*. Bronx: New York Botanical Garden.

Colfer, Carol J. Pierce, Reed L. Wadley, and P. Venkateswarlu. 1999. "Understanding Local People's Use of Time: A Pre-condition for Good Co-management." *Environmental Conservation* 26:41–52.

Comaroff, John. 1989. "Images of Empire, Contests of Conscience: Models of Colonial Domination in South Africa." *American Ethnologist* 16 (4): 661–85.

Conklin, Beth A., and Laura R. Graham. 1995. "The Shifting Middle Ground: Amazonian Indians and Eco-politics." *American Anthropologist* 97 (4): 695–710.

Connerton, Paul. 1989. *How Societies Remember*. Cambridge: Cambridge University Press.

Conway, Gordon R. 1985. "Agroecosystem Analysis." *Agricultural Administration* 20:31–55.

Conway, Gordon R., and E. B. Barbier. 1990. *After the Green Revolution: Sustainable Agriculture for Development*. London: Earthscan Publications.

Conway, Gordon R., and Jules N. Pretty. 1991. *Unwelcome Harvest: Agricultural Pollution*. London: Earthscan Publications.

Cooke, Fadzillah Majid. 1999. *The Challenge of Sustainable Forests: Forest Resource Policy in Malaysia, 1970–1995*. Honolulu: University of Hawai'i Press.

——. 2002. "Vulnerability, Control and Oil Palm in Sarawak: Globalization and a New Era?" *Development and Change* 33 (2): 189–211.

Coomes, O. T., F. Grimard, and G. J. Burt. 2000. "Tropical Forest and Shifting Cultivation: Secondary Forest Fallow Dynamics among Traditional Farmers of the Peruvian Amazon." *Ecological Economics* 32:109–24.

Corrigan, P. 1994. "State Formation." In *Everyday Forms of State Formation: Revolution and the Negotiation of Rule in Modern Mexico*, eds. Gilbert M. Joseph and Daniel Nugent, xvii–xix. Durham, N.C.: Duke University Press.

Couillard, Marie-Andrée. 1984. "The Malays and the Sakai: Some Comments on Their Social Relations in the Malay Peninsula." *Kajian Malaysia: Journal of Malaysian Studies* 2:81–108.

Courtenay, P. P. 1979. "Some Trends in the Malaysian Plantation Sector, 1963–73." In

Issues in Malaysian Development, ed. J. C. Jackson and M. Rudner, 148–53. Kuala Lumpur: Heineman.

Cramb, R. A. 1988. "The Commercialization of Iban Agriculture." In *Development in Sarawak*, ed. Cramb and R. H. W. Reece, 105–34. Oakleigh South, Vic.: Centre of Southeast Asian Studies.

——. 1989a. "Explaining Variations in Bornean Land Tenure: The Iban Case." *Ethnology* 28:277–300.

——. 1989b. "The Use and Productivity of Labour in Shifting Cultivation: An East Malaysian Case Study." *Agricultural Systems* 29:97–115.

——. 1993. "Shifting Cultivation and Sustainable Agriculture in East Malaysia." *Agricultural Systems* 42:209–26.

Creed, Gerald W., ed. 2006. *The Seductions of Community: Emancipations, Oppressions, Quandaries.* Santa Fe, N.M.: School of American Research Press.

Cronon, William. 1983. *Changes in the Land: Indians, Colonists, and the Ecology of New England.* New York: Hill and Wang.

——. 1991. *Nature's Metropolis: Chicago and the Great West.* New York: W. W. Norton.

——. 1996. "The Trouble with Wilderness; or, Getting Back to the Wrong Nature." In *Uncommon Ground: Rethinking the Human Place in Nature*, ed. Cronon, 69–90. New York: W. W. Norton.

Crosby, Alfred. 1972. *The Columbian Exchange: Biological and Cultural Consequences of 1492.* Westport, Conn.: Greenwood.

Crumley, Carole L., ed. 1994. *Historical Ecology: Cultural Knowledge and Changing Landscapes.* Santa Fe, N.M.: School of American Research Press.

Cullen, Vincent S. S. 1973. "Bukidnon Animism and Christianity." In *Bukidnon Politics and Religion*, ed. A. de Guzman and E. Pacheco, 34–46. Quezon City: Ateneo de Manila University.

Curran, L. M., I. Caniago, G. D. Paoli, D. Astianti, M. Kusneti, M. Leighton, C. E. Nirarita, H. Haeruman. 1999. "Impact of El Niño and Logging on Canopy Tree Recruitment in Borneo." *Science* 286 (5447): 2184–88.

Curry, M. R. 1995. "Rethinking Rights and Responsibilities in Geographic Information Systems: Beyond the Power of Imagery." *Cartographic and Geographic Information Systems* 22 (1): 58–60.

Dale, V. H., S. Brown, R. A. Haeuber, N. T. Hobbs, N. J. Huntly, R. J. Naiman, H. E. Reibsame, M. G. Turner, and T. J. Valone. 2000. "Ecological Principles and Guidelines for Managing the Use of Land." *Ecological Applications* 10 (3): 639–70.

Darlington, Susan M. 2003. "Practical Spirituality and Community Forests: Monks, Rituals, and Radical Conservation in Thailand." In *Nature in the Global South: Environmental Projects in South and Southeast Asia*, ed. Paul R. Greenough and Anna L. Tsing, 346–66. Durham, N.C.: Duke University Press.

Davidson, Jamie S., and David Henley, eds. 2007. *The Revival of Tradition in Indonesian Politics: The Deployment of Adat from Colonialism to Indigenism.* London: Routledge.

Davis, S. H., and K. Ebbe, eds. 1995. *Traditional Knowledge and Sustainable Develop-*

ment: Proceedings of a Conference. Proceedings of the conference for the International Year of the World's Indigenous People, Washington, D.C., September 27–28, 1993. Washington, D.C.: World Bank.

Davison, Julian, and Vinson H. Sutlive Jr. 1991. "The Children of *Nising*: Images of Headhunting, Male Sexuality in Iban Ritual, and Oral Literature." In *Female and Male in Borneo: Contributions and Challenges to Gender Studies*, vol. 1, ed. Sutlive, 153–230. Williamsburg, Va.: Borneo Research Council.

Dean, Warren. 1987. *Brazil and the Struggle for Rubber: A Study in Environmental History.* Cambridge: Cambridge University Press.

Deleuze, Gilles, and Félix Guattari. 1987. *A Thousand Plateaus: Capitalism and Schizophrenia.* Trans. Brian Massumi. Minneapolis: University of Minnesota Press.

Demsetz, Harold. 1967. "Toward a Theory of Property Rights." *American Economic Review* 62:347–59.

Denevan, W. M. 1992. "The Pristine Myth: The Landscape of the Americas in 1492." *Annals of the Association of American Geographers* 82 (3): 369–85.

Dennis, Rona. 1997. *Where to Draw the Line: Community-Level Mapping in and around the Danau Sentarum Wildlife Reserve.* Bogor: Wetlands International/PHPA/ODA.

Dentan, Robert K. 1997. "The Persistence of Received Truth: How Ruling Class Malays Construct Orang Asli Identity." In *Indigenous Peoples and the State: Politics, Land, and Ethnicity in the Malayan Peninsula and Borneo*, ed. R. Winzeler, 98–134. New Haven, Conn.: Yale University Press.

Departemen Pertanian. 1998. "Panduan operasional gerakan mandiri padi, kedelai dan jagung tahun 2001 (gema palagung 2001)." Unpublished manuscript, Jakarta.

Dilts, Douglas Russell, and Simon Hate. 1996. "IPM Farmer Field Schools: Changing Paradigms and Scaling-Up." *Agricultural Research and Extension Network* 59b:1–4.

Director of Agriculture. 1958. "Letter from the Director of Agriculture to the West Coast, January 6, 1958." Ranau Native Court Office.

Director of Lands and Surveys. 1957. "Letter from the Director of Lands and Surveys to All Residents, October 17, 1957." Ranau District Office Records.

Djalins, Upik. 2007. "Colonial Knowledge and the Native Scholar: Supomo, *Adat* Land Rights, and Agrarian Reorganization in Surakarta, 1900–1920s." Master's thesis, Yale University.

Dodge, Nicholas. 1981. "The Malay-Aborigine Nexus under Malay Rule." *Bijdragen Tot de Taal, Land en Volkenkunde* 137:1–16.

Doolittle, Amity. 1999. "Controlling the Land: Property Rights and Power Struggles in Sabah, Malaysia, 1881–1996." PhD diss., Yale University.

——. 2001. "From Village Land to 'Native Reserve': Changes in Property Rights in Sabah, 1950–1996." *Human Ecology* 29 (1): 69–98.

——. 2003. "Colliding Discourses: Western Land Laws and Native Customary Rights in North Borneo, 1881–1928." *Journal of Southeast Asian Studies* 34 (1): 97–126.

——. 2005. *Property and Politics in Sabah, Malaysia: Native Struggles over Land Rights.* Seattle: University of Washington Press.

BIBLIOGRAPHY 313

———. 2006. "Resources, Ideologies, and Nationalism: The Politics of Development in Sabah, Malaysia." In *Development Brokers and Translators*, ed. David Mosse and Davis Lewis, 51–74. Bloomfield, Conn.: Kumarian.

———. 2007. "Native Land Tenure, Conservation, and Development in a Pseudo-democracy: Natural Resource Conflicts in Sabah, Malaysia." *Journal of Peasant Studies* 34 (3): 474–97.

Douglas, Mary. 1966. *Purity and Danger: An Analysis of the Concepts of Pollution and Taboo*. London: Routledge and Kegan Paul.

Dove, Michael R. 1981. "Swidden Systems and Their Potential Role in Agricultural Development: A Case-Study from Kalimantan." *Prisma* 21:81–100.

———. 1983a. "Forest Preference in Swidden Agriculture." *Tropical Ecology* 24 (1): 122–42.

———. 1983b. "Theories of Swidden Agriculture and the Political Economy of Ignorance." *Agroforestry Systems* 1 (3): 85–89.

———. 1984. "The Chayanov Slope in a Swidden Society: Household Demography and Extensive Agriculture in Western Kalimantan." In *Chayanov, Peasants, and Economic Anthropology*, ed. E. Paul Durrenberger, 97–132. New York: Academic.

———. 1985a. "The Agroecological Mythology of the Javanese and the Political Economy of Indonesia." *Indonesia* 39:1–36.

———. 1985b. "The Kantu' System of Land Tenure: The Evolution of Tribal Land Rights in Borneo." In *Modernization and the Emergence of a Landless Peasantry: Essays on the Integration of Peripheries to Socioeconomic Centers*, ed. G. Appell, Studies in Third World Societies No. 33, 159–82. Williamsburg, Va.: Department of Anthropology, College of William and Mary.

———. 1985c. "The Perception of Peasant Land Rights in Indonesian Development: Causes and Implications." In *Land, Trees, and Tenure*, ed. John Raintree, 265–71. Nairobi: ICRAF.

———. 1985d. *Swidden Agriculture in Indonesia: Subsistence Strategies of the Kalimantan Kantu'*. Berlin: Mouton.

———. 1988. "The Ecology of Intoxication among the Kantu' of West Kalimantan." In *The Real and Imagined Role of Culture in Development: Case Studies from Indonesia*, ed. Dove, 139–82. Honolulu: University of Hawai'i Press.

———. 1993a. "The Responses of Dayak and Bearded Pig to Mast-Fruiting in Kalimantan: An Analysis of Nature-Culture Analogies." In *Tropical Forests, People, and Food*, ed. C. M. Hladik et al., 113–23. Paris: UNESCO.

———. 1993b. "A Revisionist View of Tropical Deforestation and Development." *Environmental Conservation* 20 (1): 17–25.

———. 1993c. "Smallholder Rubber and Swidden Agriculture in Borneo: A Sustainable Adaptation to the Ecology and Economy of the Tropical Forest." *Economic Botany* 47 (2): 136–47.

———. 1994. "Transition from Native Forest Rubbers to *Hevea Brasiliensis* (Euphorblaceae) among Tribal Smallholders in Borneo." *Economic Botany* 48 (4): 382–96.

———. 1996. "Rice-Eating Rubber and People-Eating Governments: Peasant versus

State Critiques of Rubber Development in Colonial Indonesia." *Ethnohistory* 43 (1): 33–63.

——. 1998. "Living Rubber, Dead Land, and Persisting Systems in Borneo: Indigenous Representations of Sustainability." *Bijdragen* 154 (1): 20–54.

——. 2000. "The Life-Cycle of Indigenous Knowledge, and the Case of Natural Rubber Production." In *Indigenous Environmental Knowledge and Its Transformations*, ed. Roy Ellen, Peter Parkes, and Alan Bicker, 213–51. Amsterdam: Harwood.

——. 2001. "Inter-disciplinary Borrowing in Environmental Anthropology and the Critique of Modern Science." In *New Directions in Anthropology and Environment: Intersections*, ed. Carole L. Crumley, with A. Elizabeth van Deventer and Joseph J. Fletcher, 90–110. Walnut Creek, Calif.: AltaMira.

——. 2003. "Forest Discourse in South and Southeast Asia: A Comparison with Global Discourses." In *Nature in the Global South: Environmental Projects in South and Southeast Asia*, ed. Paul Greenough and Anna L. Tsing, 103–23. Durham, N.C.: Duke University Press.

——. 2006. "Indigenous People and Environmental Politics." *Annual Review of Anthropology* 35 (1):191–208.

——. 2008. *Southeast Asian Grasslands: Understanding a Folk Landscape*. New York: New York Botanical Gardens Press.

——. 2011. *The Banana Tree at the Gate: The History of Marginal Peoples and Global Markets in Borneo*. New Haven: Yale University Press.

Dove, Michael R., and Carol Carpenter, eds. 2007. *Environmental Anthropology: A Historical Reader*. Malden, Mass.: Blackwell.

Dove, Michael R., Amity A. Doolittle, and Percy E. Sajise, eds. 2005. *Conserving Nature in Culture: Case Studies from Southeast Asia*. New Haven, Conn.: Yale University Press.

Drabble, J. 1973. *Rubber in Malaya, 1867–1922: The Genesis of the Industry*. Kuala Lumpur: Oxford University Press.

——. 1979. "Peasant Smallholders in the Malayan Economy: An Historical Study with Special Reference to the Rubber Industry." In *Issues in Malaysian Development*, ed. J. C. Jackson and M. Rudner, 69–99. Kuala Lumpur: Heineman.

Drake, R. A. 1982. "The Material Provision of Mualang Society in Hinterland Kalimantan Barat, Indonesia." PhD diss., Michigan State University.

Dressler, W. 2005. "Disentangling Tagbanua Lifeways, Swidden, and Conservation on Palawan Island." *Human Ecology Review* 12 (1): 21–24.

Duckett. J. E. 1976. "Plantations as a Habitat for Wild Life in Peninsular Malaysia with Particular Reference to Oil Palm (*Elaeis Guineensis*)." *Malayan Nature Journal* 29 (3): 176–82.

Dudley, Richard. 2000. "The Fisheries of Dana Sentarum." *Borneo Research Bulletin* 31:261–306.

Duhaylungsod, Levita. 1993. "Ancestral Domain and Cultural Identity: Missing Notes in Development." *PSSC Social Science Information* 20 (4): 19–23, 41.

——. 1998. "Rethinking Subsistence through Cultural Economies of Indigenous Peoples." *Diliman Review* 4 (2): 14–22.

Duhaylungsod, Levita, and David Hyndman. 1992a. "Creeping Resource Exploitation in the T'boli Homeland: Political Ecology of the Tasaday Hoax." In *The Tasaday Controversy: An Assessment of the Evidence*, ed. T. Headland, 59–75. Washington, D.C.: American Anthropological Association.

———. 1992b. "Where All That Glitters Is Not Gold: Crossroads of Mining Exploitation in the T'boli Homeland." *Bulletin of Concerned Asian Scholars* 24 (3): 3–15.

1993. *Where T'boli Bells Toll: Political Ecology Voices behind the Tasaday Hoax.* Copenhagen: International Work Group for Indigenous Affairs.

———. 1995. "Enduring Systems of Subsistence and Simple Reproduction in the Philippines." *Pilipinas* 25:21–48.

Dunn, Frederick L. [1975] 1982. *Rainforest Collectors and Traders: A Study of Resource Utilization in Modern and Ancient Malaya.* Kuala Lumpur: Malayan Branch of the Royal Asiatic Society.

Durrenberger, Paul E., ed. 1984. *Chayanov, Peasants, and Economic Anthropology.* New York: Academic Press.

DWNP (Department of Wildlife and National Parks), Malaysia. 1987. *Taman Negara Master Plan.* Kuala Lumpur: Ministry of Science, Technology and the Environment.

DWNP (Department of Wildlife and National Parks), MNS (Malayan Nature Society), MMA, and WWF (World Wildlife Fund). 1986. *Tourist Development Proposals for the Kuala Tahan Area of Taman Negara.* Selangor: World Wildlife Fund Malaysia.

Dwyer, Peter D. 1996. "The Invention of Nature." In *Redefining Nature: Ecology, Culture, and Domestication*, ed. Roy F. Ellen and Katsuyoshi Fukui, 157–86. Oxford: Berg.

Eastman, J. R. 1997. *IDRISI for Windows: User's Guide.* Worcester, Mass.: Clark University Press.

Echols, John M., and Hassan Shadily. 1992. *Kamus Indonesia-Inggris.* 3rd ed. Jakarta: PT Gramedia.

Edgerton, Robert. 1982. "Frontier Society in the Bukidnon Plateau, 1870–1941." In *Philippine Social History*, ed. A. McCoy and E. de Jesus, 361–90. Quezon City: Ateneo de Manila University.

"Editorial." 1975. *Malayan Naturalist* 1 (3): 1.

Edmunds, David, and Eva Wollenberg. 2001. "A Strategic Approach to Multistakeholder Negotiations." *Development and Change* 32 (2): 231–53.

Ellen, Roy F. 1999. "Forest Knowledge: Forest Transformation." In *Transforming the Indonesian Uplands:Marginality, Power, and Production*, ed. Tania Murray Li, 131–57. Amsterdam: Harwood.

———, ed. 2007. *Modern Crises and Traditional Strategies: Local Ecological Knowledge in Island Southeast Asia.* New York, NY: Berghahn Books.

Endicott, Kirk. 1970. *An Analysis of Malay Magic.* Singapore: Oxford University Press.

———. 1997. "Batek History, Interethnic Relations, and Subgroup Dynamics." In *Indigenous Peoples and the State: Politics, Land, and Ethnicity in the Malay Peninsula and Borneo*, ed. Robert Winzeler, 30–50. New Haven, Conn.: Yale University Press.

———. 2000. "The Batek of Malaysia." In *Endangered Peoples of Southeast and East Asia: Struggles to Survive and Thrive*, ed. Leslie Sponsel, 101–22. Westport, Conn.: Greenwood.

Enthoven, J. J. K. 1903. *Bijdragen tot de geographie van Borneo's Westerafdeeling*. Leiden: Brill.

Escobar, Arturo. 1995. *Encountering Development: The Making and Unmaking of the Third World*. Princeton, N.J.: Princeton University Press.

———. 1996. "Constructing Nature: Elements for a Postructural Political Ecology." In *Liberation Ecologies: Environment, Development, Social Movements*, ed. Richard Peet and Michael Watts, 46–68. London: Routledge.

———. 1998. "Whose Knowledge, Whose Nature? Biodiveristy, Conservation, and the Political Ecology of Social Movements." *Journal of Political Ecology* 5:53–82.

———. 1999. "After Nature: Steps to an Anti-essentialist Political Ecology." *Current Anthropology* 40 (1): 1–30.

Etkins, N. 2002. "Local Kowledge of the Biotic Diversity and Its Conservation in Rural Hausaland, Northern Nigeria." *Economic Botany* 56 (1): 73–88.

Evans, Ivor H. N. 1937. *The Negritos of Malaya*. Cambridge: Cambridge University Press.

Fairhead, James, and Melissa Leach. 1995. "Reading Forest History Backwards: The Interaction of Policy and Local Land Use in Guinea's Forest-Savanna Mosaic, 1893–1993." *Environment and History* 1:55–91.

———. 1996. *Misreading the African Landscape: Society and Ecology in a Forest-Savanna Mosaic*. Cambridge: Cambridge University Press.

FAO (Food and Agricultural Organization). 1990. *Mid-term Review of FAO Intercountry Program for the Development and Application of Integrated Pest Control in Rice in South and South East Asia*. Jakarta: Food and Agricultural Organization.

———. 1991. *Mid-term Review Mission: Training and Development of Integrated Pest Management in Rice-Based Cropping Systems*. Jakarta: Food and Agricultural Organization.

Farber, Paul Lawrence. 2000. *Finding Order in Nature: The Naturalist Tradition from Linnaeus to E. O. Wilson*. Baltimore, Md.: Johns Hopkins University Press.

FD (Forest Department) (Jabatan Hutan). 1924. *Report of the FMS Forest Administration 1923*. Kuala Lumpur: Government Press.

———. 1933. *Report of the FMS Forest Administration 1932*. Kuala Lumpur: Government Press.

———. 1959. *Report of the FMS Forest Administration 1958*. Kuala Lumpur: Government Press.

———. 1972. *Report of the FMS Forest Administration 1968*. Kuala Lumpur: Government Press.

Fedick, Scott L., ed. 1996. *The Managed Mosaic: Ancient Maya Agriculture and Resource Use*. Salt Lake City: University of Utah Press.

———. 2003. "In Search of the Mayan Forest." In *In Search of the Rain Forest*, ed. Candace Slater, 133–66. Durham, N.C.: Duke University Press.

Ferguson, James. 1990. *The Anti-Politics Machine: "Development," Depoliticization, and Bureaucratic Power in Lesotho.* Cambridge: Cambridge University Press.

Fernandez, Perfecto. 1980. "Towards a Definition of National Policy on the Recognition of Ethnic Law within the Philippine Legal Order." *Philippine Law Journal* 55 (4): 383–93.

Ferrera, E. L. 2007. "Descent Rules and Strategic Transfers: Evidence from Matrilineal Groups in Ghana." *Journal of Development Economics* 83:280–301.

Fiedler, Peggy L., Peter S. White, and Robert A. Lediy. 1997. "The Paradigm Shift in Ecology and Its Implications for Conservation." In *The Ecological Basis of Conservation: Heterogeneity, Ecosystems, and Biodiversity*, ed. Steward T. A. Pickett et al., 83–92. New York: Chapman and Hall.

Fliert, Elske van de. 1993. *Integrated Pest Management: Farmer Field Schools Generate Sustainable Practices: A Case Study in Central Java Evaluating ipm Training.* Wageningen, The Netherlands: Agricultural University Wageningen.

Folke, Carl, Johan Colding, and Fikret Berkes. 2003. "Synthesis: Building Resilience and Adaptive Capacity in Socio-ecological Systems. In *Navigating Social-Ecological Systems: Building Resilience for Complexity and Change*, ed. Berkes, Colding, and Folke, 352–87. Cambridge: Cambridge University Press.

Folke, Carl, Thomas Hahn, Per Olsson, and Jon Norberg. 2005. "Adaptive Governance of Social-ecological Systems." *Annual Review of Environment and Resources* 30:441–73.

Ford Foundation. 1998. *Forestry for Sustainable Rural Development: A Review of Ford Foundation-Supported Community Forestry Programs in Asia.* New York: Ford Foundation.

Foresta, Hubert de. 1996. "Transcript of the Damar Agroforest Conference in Bandarlampung." Unpublished manuscript, World Agroforestry Centre, Bogor, Indonesia.

Foresta, Hubert de, and S. Petit. 1997. "Precious Wood from the Agroforests of Sumatra: Where Timber Provides a Solid Source of Income." *Agroforestry Today* 9 (4): 18–20.

Forman, Richard T. T., and Michel Godron. 1986. *Landscape Ecology.* New York: John Wiley.

Fortes, M. 1971. Introduction to *The Developmental Cycle in Domestic Groups*, ed. J. Goody, 1–14. Cambridge: Cambridge University Press.

Forsberg, Vivian. 1988. "T'boli Medicine and the Supernatural." Master's thesis, Fuller Theological Seminary, Menlo Park.

Forsyth, Tim. 2003. *Critical Political Ecology: The Politics of Environmental Science.* London: Routledge.

Foucault, Michel. 1993. "Power as Knowledge." In *Social Theory: The Multicultural and Classic Readings*, ed. Charles Lemert, 518–23. Boulder, Colo.: Westview.

Fowler, Cynthia T. 2003. "The Ecological Implications of Ancestral Religion and Reciprocal Exchange in a Sacred Forest in Karendi (Sumba, Indonesia)." *Worldviews: Environment, Culture, Religion* 7 (3): 303–29.

Fox, James. 1971. "Sister's Child as Plant: Metaphors in an Idiom of Consanguinity." In *Rethinking Kinship and Marriage*, ed. Rodney Needham, 219–52. London: Tavistock.

———. 1988. "Origin, Descent, and Precedence in the Study of Austronesian Societies." Paper presented at "De Wisselleerstoel Indonesische Studien," Leiden University, March 17, 1988.

———. 1991. "Managing the Ecology of Rice Production in Indonesia." In *Indonesia: Resources, Ecology, and Environment*, ed. J. Hardjono, 61–84. Singapore: Oxford University Press.

———. 1993. "Ecological Policies for Sustaining High Production in Rice: Observations on Rice Intensification in Indonesia." In *South-East Asia's Environmental Future: The Search for Sustainability*, ed. H. Brookfield and Y. Byron, 210–24. Tokyo: United Nations University Press.

Fox, Jefferson. 1990. "Diagnostic Tools for Social Forestry." In *Keepers of the Forest: Land Management Alternatives in Southeast Asia*, ed. Mark Poffenberger, 119–33. West Hartford, Conn.: Kumarian.

Fox, Jefferson, Krisnawati Suryanata, Peter Hershock, Albertus Hadi Pramono. 2008. "Mapping Boundaries, Shifting Power: The Social-Ethical Dimensions of Participatory Mapping." In *Contentious Geographies: Environmental Knowledge, Meaning, Scale*, ed. M. Goodman, M. Boykoff, and K. Evered, 203–18. Hampshire, U.K.: Ashgate.

Frake, Charles. 1955. "Social Organization and Shifting Cultivation among the Sindangan Subanon." PhD diss., Yale University.

———. 1996. "Pleasant Places, Past Times, and Sheltered Identity in Rural East Anglia." In *Senses of Place*, ed. Steven Feld and Keith H. Basso, 229–57. Santa Fe, N.M.: School of American Research Press.

Frank, Andre Gunder. 1991. *The Underdevelopment of Development*. Stockholm: Bethany.

Frazer, Sir James George. [1890] 1951. 2 vols. *The Golden Bough: A Study in Magic and Religion*. Abridged ed. New York: Avenel.

Freeman, Derek. 1955. *Iban Agriculture: A Report on the Shifting Cultivation of Hill Rice by the Iban of Sarawak*. London: Her Majesty's Stationery Office.

———. 1967. "Shaman and Incubus." *Psychoanalytic Study of Society* 4:315–43.

———. 1970. *Report on the Iban*. London: Athlone.

———. 1992. *The Iban of Borneo*. Kuala Lumpur: S. Abdul Majeed.

Freeman, J. R. 1999. "Gods, Groves, and the Culture of Nature in Kerala." *Modern Asian Studies* 33 (2): 257–302.

Freudenburg, William R., Robert Gramling, Shirley Laska, and Kai T. Erikson. 2009. "Organizing Hazards, Engineering Disasters? Improving the Recognition of Political-Economic Factors in the Creation of Disasters." *Social Forces* 87 (2): 1015–38.

Fried, Stephanie Gorson. 2000. "Tropical Forests Forever? A Contextual Ecology of Bentian Rattan Agroforestry Systems." In *People, Plants, and Justice: The Politics of*

Nature Conservation, ed. Charles Zerner, 204–33. New York: Columbia University Press.

———. 2003. "Writing for Their Lives: Bentian Dayak Authors and Indonesian Development Discourse." In *Culture and the Question of Rights: Forests, Coasts, and Seas in Southeast Asia*, ed. Charles Zerner, 142–83. Durham, N.C.: Duke University Press.

Friedman, Jonathan. 1998. "The Heart of the Matter." *Identities* 4 (3–4): 561–69.

Frossard, David. 1998a. " 'Peasant Science': A New Paradigm for Sustainable Development?" *Research in Philosophy and Technology* 17:111–26.

———. 1998b. "Asia's Green Revolution and Peasant Distinctions: Between Science and Authority." In *Representing Natural Resource Development in Asia: "Modern" Versus "Postmodern" Scholarly Authority*, ed. Michael Fischer, 31–42. Centre for Social Anthropology and Computing (CSAC) Monograph No. 11. Canterbury, U.K.: CSAC. Available at http://lucy.ukc.ac.uk/PostModern/David_Frossard_TOC.html.

Fryer, D. W. 1964. "The Plantation Industries: The Estates." In *Malaysia: A Survey*, ed. Gungwu Wang, 227–45. New York: Frederick A. Praeger.

Fujisaka, Sam. 1994. "Learning from Six Reasons Why Farmers Do Not Adopt Innovations Intended to Improve Sustainability of Upland Agriculture." *Agricultural Systems* 46:409–25.

Gadgil, Madhav, and Ramachandra Guha. 1993. *This Fissured Land: An Ecological History of India*. Berkeley: University of California Press.

Gagnon, Alexander S., Andrew B. G. Bush, and Karen E. Smoyer-Tomic. 2001. "Dengue Epidemics and the El Nino Southern Oscillation." *Climate Research* 19:35–43.

Gallagher, Kevin. 2003. "Fundamental Elements of a Farmer Field School." LEISA 19 (1): 5–6.

Ganjanapan, Anan. 1998. "The Politics of Conservation and the Complexity of Local Control over Forests in the Northern Thai Highlands." *Mountain Research and Development* 18 (1): 71–82.

Garvan, John. 1931. *The Manobos of Mindanao*. Vol. 23, Memoirs of the National Academy of Sciences. Washington, D.C.: US Government Printing Office.

Gaspar, Karl. 1997. "The Mindanao Lumad Social Movement." *Mindanao Focus* 15 (1): 3–47.

Gatmaytan, Augusto B. 2005. "Advocacy as Translation: Notes on the Philippine Experience." In *Communities and Conservation: History and Politics of Community-Based Natural Resource Management*, ed. J. Peter Brosius, Anna Ling Tsing, and Charles Zerner, 459–76. Walnut Creek, Calif.: AltaMira.

Gautam, Madhur, Uma Lele, Hariadi Kartodihardjo, Azis Khan, Ir. Erwinsyah, and Saeed Rana. 2000. *Indonesia: The Challenges of World Bank Involvement in Forests*. Evaluation Country Case Study Series. Washington, D.C.: World Bank.

Geertz, Clifford. 1963. *Agricultural Involution: The Processes of Ecological Change in Indonesia*. Berkeley: University of California Press.

———. 1973. *The Interpretation of Cultures: Selected Essays*. New York: Basic.

———. 1983. *Local Knowledge*. New York: Basic.

Ghimire, Krishna B., and Michel P. Pimbert. 1997. "Social Change and Conservation:

An Overview of Issues and Concepts." In *Social Change and Conservation: Environmental Politics and Impacts of National Parks and Protected Areas*, ed. Ghimire and Pimbert, 1–45. London: Earthscan.

Giddens, Anthony. 1984. *The Constitution of Society: Outline of the Theory of Structuation*. Berkeley: University of California Press.

Giesen, Wim. 1987. *Danau Sentarum Wildlife Reserve: Inventory, Ecology, and Management Guidelines; World Wildlife Fund Report for the Directorate of Forest Protection and Nature Conservation (PHPA)*. Bogor: World Wildlife Fund.

Gillson, Lindsey, Michael Sheridan, and Dan Brockington. 2003. "Representing Environments in Flux: Case Studies from East Africa." *Area* 34 (4): 371–89.

Gilruth, P. T., S. E. Marsh, and R. Itami. 1995. "A Dynamic Spatial Model of Shifting Cultivation in the Highlands of Guinea, West Africa." *Ecological Modeling* 79:179–97.

GM (Government of Malaysia). 1966. *First Malaysia Plan, 1966–1970*. Kuala Lumpur: Government Press.

———. 1968. *Malaysia 1967: Buku Rasmi Tahunan*. Kuala Lumpur: Government Press.

———. 1976. *The Third Malaysia Plan, 1976–1980*. Kuala Lumpur: Government Press.

———. 1991. *The Sixth Malaysia Plan, 1991–1995*. Kuala Lumpur: Government Press.

———. 1995. *National Forestry Policy, 1978, Revised 1992*. Kuala Lumpur: Percetakan Asli.

———. 1996. *The Seventh Malaysia Plan, 1996–2000*. Kuala Lumpur: National Press.

Gobbi, José A. 2000. "Is Biodiversity-Friendly Coffee Financially Viable? An Analysis of Five Different Coffee Production Systems in Western El Salvador." *Ecological Economics* 33 (2): 267–81.

Gómez-Pompa, Arturo, Jose Salvador Flores, and Mario Aliphat Fernandez. 1990: "The Sacred Cacao Groves of the Maya." *Latin American Antiquity* 1 (3): 247–57.

Gómez-Pompa, Arturo, and Andrea Kaus. 1992. "Taming the Wilderness Myth." *BioScience* 42 (4): 271–79.

Goodland, Robert. 1982. *Tribal Peoples and Economic Development: Human Ecologic Considerations*. Washington, D.C.: World Bank.

Gordon, Alec. 1993. "Smallholder Commercial Cultivation and the Environment: Rubber in Southeast Asia." In *Asia's Environmental Crisis*, ed. Michael C. Howard, 135–53. Boulder, Colo.: Westview.

Gouyon, A., H. de Foresta, and P. Levang. 1993. "Does 'Jungle Rubber' Deserve Its Name? An Analysis of Rubber Agroforestry Systems in Southeast Sumatra." *Agroforestry Systems* 22:181–206.

Government of Indonesia. 2004. *Statistik karet Indonesia*. Jakarta: Badan Pusat Statistik.

Gray, Andrew. 1995. "The Indigenous Movement in Asia." In *Indigenous Peoples of Asia*, ed. R. H. Barnes, Gray, and Benedict Kingsbury, 35–58. Ann Arbor, Mich.: Association for Asian Studies.

Grove, Richard. 1995. *Green Imperialism: Colonial Expansion, Tropical Island Edens, and the Origins of Environmentalism, 1600–1860*. Cambridge: Cambridge University Press.

Grueso, Libia, Carlos Rosero, and Arturo Escobar. 1996. "The Organizational Process of Black Communities in the Colombian Pacific." In *Culture of Politics, Politics of Culture: Revisioning Latin American Social Movements*, ed. Sonia E. Alvarez, Evelina Dagnino, and Escobar, 196–217. Boulder, Colo.: Westview.

Gruner, Sheila. 2007. "Contested Territories: Development, Displacement, and Social Movements in Colombia." In *Development's Displacements: Ecologies, Economies, and Cultures at Risk*, ed. Peter Vandeergest, Pablo Idahaso, and Pablo S. Bose, 155–86. Vancouver: University of British Columbia Press.

Gudeman, Stephen. 1986. *Economics as Culture: Models and Metaphors of Livelihood*. London: Routledge.

Guha, Ramachandra. 1989. *The Unquiet Woods: Ecological Change and Peasant Resistance in Himalayas*. Berkeley: University of California Press.

Guha, Ranajit. 1963. *A Rule of Property for Bengal*. Paris: Mouton.

Gunderson, Lance H., and C. S. Holling, eds. 2001. *Panarchy: Understanding Transformations in Human and Natural Systems*. Washington, D.C.: Island Press.

Gunderson, Lance H., C. S. Holling, and Stephen S. Light, eds. 1995 *Barriers and Bridges to the Renewal of Ecosystems and Institutions*. New York: Columbia University Press.

Gupta, Akhil. 1995. "Blurred Boundaries: The Discourse of Corruption, the Culture of Politics, and the Imagined State." *American Ethnologist* 22 (2): 375–402.

———. 1998. *Postcolonial Developments: Agriculture in the Making of Modern India*. Durham, N.C.: Duke University Press.

Guyot, D. 1971. "The Politics of Land: Comparative Development in Two States of Malaysia." *Pacific Affairs* 40 (3): 368–400.

Hadikusuma, H. 1988. *Masyarakat dan Adat Budaya Lampung*. Bandung: Mandar Maju.

Hafisam, M. n.d. "IPM in Padaherang Sub-district, Ciamis District, West Java." Unpublished manuscript, Jakarta.

Haila, Yrjö, and Chuck Dyke. 2006. *How Nature Speaks: The Dynamics of the Human Ecological Condition*. Durham, N.C.: Duke University Press.

Haines, W. B. [1934] 1940. *The Use and Control of Natural Undergrowth*. Kuala Lumpur: Rubber Research Institute.

Hakkenberg, Christopher. 2008. "Biodiversity and Sacred Sites: Vernacular Conservation Practices in Northwest Yunnan, China." *Worldviews* 12: 74–90.

Hansen, G. E. 1978. "Bureaucratic Linkages and Policy-Making in Indonesia: BIMAS Revisited." In *Political Power and Communications in Indonesia*, ed. K. W. Jackson and L. W. Pye, 322–42. Berkeley: University of California Press.

Hansen, J. D., M. K. Luckert, S. Minae, and F. Place. 2005. "Tree Planting under Customary Tenure Systems in Malawi: Impacts of Marriage and Inheritance Pattern." *Agricultural Systems* 84:99–118.

Hardjono, Joan. 1983. "Rural Development in Indonesia: The 'Top-Down' Approach." In *Rural Development and the State*, ed. D. A. M. Leam and D. P. Chaudhri, 38–65. London: Methuen.

Harley, J. B. 1989. "Deconstructing the Map." *Cartographica* 26 (2): 1–20.

Harper, T. 1997. "Forest Politics in Colonial Malaya." *Modern Asian Studies* 31 (1): 1–29.

Harris, Marvin. 1966. "The Cultural Ecology of India's Sacred Cattle." *Current Anthropology* 7:51–59.

Harrison, C. W. [1923] 1985. *An Illustrated Guide to the Federated Malay States.* Singapore: Oxford University Press

Harwell, Emily. 1997. *Law and Culture in Resource Management: An Analysis of Local Systems for Resource Management in the Danau Sentarum Wildlife Reserve, West Kalimantan, Indonesia.* Bogor: Wetlands International, Indonesia Programme.

———. 2000. "The Un-natural History of Culture: Ethnicity, Tradition, and Territorial Conflicts in West Kalimantan, Indonesia, 1800–1997." PhD diss., Yale University.

Heersink, Christiaan G. 1994. "Selayar and the Green Gold: The Development of the Coconut Trade on an Indonesian Island (1820–1950)." *Journal of Southeast Asian Studies* 25 (1): 47–69.

Hefner, Robert W. 1985. *Hindu Javanese: Tengger Tradition and Islam.* Princeton, N.J.: Princeton University Press.

Heri, V. 1996. *Laporan Data Hukum Adat: Berupa Peraturan Nelayan di Suaka Margasatwa Danau Sentarum, Kalimantan Barat, Indonesia.* Bogor: Wetlands International-ODA.

Heroepoetri, Arimbi, and Arif Aliadi, eds. 1994. *Peran Serta Masyarakat dalam Pelestarian Hutan.* Jakarta: Walhi.

Hewitt, Kenneth, 1983. "The Idea of Calamity in a Technocratic Age." In *Interpretations of Calamity: From the Viewpoint of Human Ecology*, ed. Hewitt, 3–32. Boston: Allen and Unwin.

Hidayat, Bambang, and Kustiwa S. Adinata. 2001. "Farmers in Indonesia: Escaping the Trap of Injustice." Paper presented at the Programme Advisory Committee (PAC) Meeting, Ayutthaya, Thailand, November 26–28.

Hill, Stephanie, and Brad Coombes. 2004. "The Limits to Participation in Disequilibrium Ecology: Maori Involvement in Habitat Restoration with Te Urewera National Park." *Science as Culture* 13 (1): 37–74.

Hirtz, Frank. 2003. "It Takes Modern Means to Be Traditional: On Recognizing Indigenous Cultural Communities in the Philippines." *Development and Change* 34 (5): 887–914.

Hislop, J. A. 1961. "Protection of Wild Life in the Federation of Malaya." In *Nature Conservation in Western Malaysia*, ed. J. Wyatt-Smith and P. R. Wycherley, 136–42. Kuala Lumpur: Malayan Nature Society.

Hitchcock, Robert K., and John D. Holm. 1993. "Bureaucratic Domination of Hunter-Gatherer Societies: A Study of the San in Botswana." *Development and Change* 24:305–38.

Ho, R. 1970. "Land Ownership and the Economic Prospects of Malayan Peasants." *Modern Asian Studies* 4 (1): 83–92.

Hohfeld, Wesley. 1913. "Fundamental Legal Conceptions as Applied in Judicial Reasoning I." *Yale Law Journal* 23 (1): 16–59.

Holling, Crawford S., ed. 1978. *Adaptive Environmental Assessment and Management*. London: John Wiley.

———. 1995. "What Barriers? What Bridges?" In *Barriers and Bridges to the Renewal of Ecosystems and Institutions*, ed. Lance H. Gunderson, Holling, and Stephen S. Light, 3–34. New York: Columbia University Press.

Holling, Crawford S., Paul Taylor, and Michael Thompson. 1991. "From Newton's Sleep to Blake's Fourfold Vision: Why the Climax Community and the Rational Bureaucracy Are Not the Ends of the Ecological and Social-Cultural Roads." *Annals of Earth* 9 (3): 19–21.

Hooker, M. B. 1996. "The Orang Asli and the Laws of Malaysia, with Special Reference to Land." *Akademika* 48:21–50.

Horne, Elinor Clark. 1974. *Javanese-English Dictionary*. New Haven, Conn.: Yale University Press.

Hornborg, Alf. 1996. "Ecology as Semiotics: Outlines of a Contextualist Paradigm for Human Ecology." In *Nature and Society: Anthropological Perspectives*, ed. Philippe Descola and Gísli Pálsson, 45–62. London: Routledge.

Hoskins, Janet, ed. 1996. *Headhunting and the Social Imagination in Southeast Asia*. Stanford, Calif.: Stanford University Press.

Hubback, T. R. 1932. *Report of the Wildlife Commission of Malaya*. 3 vols. Singapore: Government Printing Office.

Hughes, David McDermott. 2006. *From Enslavement to Environmentalism: Politics on a Southern African Frontier*. Seattle: University of Washington Press.

Hurst, P. 1990. *Rainforest Politics: Ecological Destruction in South-East Asia*. London: Zed.

Hyndman, David, and Levita Duhaylungsod. 1990. "The Development Saga of the Tasaday: Gentle Yesterday, Hoax Today, Exploited Forever?" *Bulletin of Concerned Asian Scholars* 22 (4): 38–54.

———. 1992. "Behind and Beyond the Tasaday: The Untold Struggle over T'boli Resources." In *Mindanao: Land of Unfulfilled Promise*, ed. Michael Turner, Ron May, and Lulu Turner, 249–68. Manila: New Day Publishers.

———. 1993. "Political Movements and Indigenous Struggles in the Philippines." *Fourth World Bulletin* 3 (1): 4–7.

———. 1994. "Ethnonationalism in the Southern Philippines." *Society and Nature* 2 (2): 171–94.

Independent Commission on International Humanitarian Issues. 1987. *Indigenous Peoples: A Global Quest for Justice*. London: Zed.

Indonesian Integrated Pest Management Farmers' Alliance, Indramayu Regency, and the Undergraduate Program of the Department of Anthropology, University of Indonesia. *Bisa dèwèk* (We can do it ourselves). 2007. Documentary film, http://www.youtube.com/rhino ariefiansyah.

Indonesian National IPM Program. n.d. *Farmers as Experts*. Jakarta: Indonesian National IPM Program, and Food and Agricultural Organization.

Indrowuryanto. 1999. "Pranata mangsa dalam aktivitas pertanian di Jawa." In *Petani merajut tradisi era globalisasi: Pendayagunaan sistem pengetahuan lokal dalam pembangunan*, ed. Kusnaka Adimihardja, 17–40. Bandung: Humaniora Utama Press Bandung.

IPBN (Indigenous People's Biodiversity Network). 2000. "Indigenous Peoples, Indigenous Knowledge, and Innovations and the Convention on Biological Diversity." http://www.ecouncil.ac.cr/rio/focus/report/english/ipbn.htm.

Jackson, J. C. 1965. "Chinese Agricultural Pioneering in Singapore and Johor, 1800–1917." *Journal of the Malaysian Branch of the Royal Asiatic Society* 38 (1): 77–105.

Jamison, A. 1996. "The Shaping of the Global Environmental Agenda: The Role of Non-governmental Organizations." In *Risk, Environment, and Modernity: Towards a New Ecology*, ed. Scott Lash, Branislaw Szerszynski, and Brian Wynnen, 224–45. London: Sage.

Jarvis, D. I., C. Padoch, and H. D. Cooper, eds. 2007. *Managing Biodiversity in Agricultural Ecosystems*. New York: Columbia University Press.

JHEOA (Jabatan Hal-Ehwal Orang Asli), Malaysia. 1961. *Statement of Policy Regarding the Administration of the Orang Asli of Peninsular Malaysia*. Kuala Lumpur: Department of Orang Asli Affairs.

Jorgenson, Anders Baltzer. 1989. "A Natural View: Pwo Karen Notions of Plants, Landscapes, and People." *Folk* 31:21–51.

Joseph. K. T. 1964. "Problems of Agriculture." In *Malaysia: A Survey*, ed. Gungwu Wang, 274–92. London: Pall Mall.

Jung, Courtney. 2003. "The Politics of Indigenous Identity: Neoliberalism, Cultural Rights, and the Mexican Zapatistas." *Social Research* 70 (2): 433–62.

———. 2008. *The Moral Force of Indigenous Politics: Critical Liberalism and the Zapatistas*. Cambridge: Cambridge University Press.

Kanwil Kehutanan Propinsi Tingkat Lampung. 1995. "Pembangunan Hutan Melalui Peningkatan Kesejahteraan Masyarakat Sekitar Hutan 'Pendekatan Pola krui-Lampung Barat.'" Paper presented at the "Kebun Damar di Krui Lmapung Sebagai Model Hutan Rakyat" seminar, Bandar Lampung, Indonesia, June 6.

Kartikasari, Sri Nurani. 2008. *The Conservation of BioDiversity: The Perception of Key Local Stakeholders in Gorontalo, Indonesia*. PhD diss., Lincoln University, Canterbury, New Zealand.

Kater, C. 1883. "Iets over de Batang Loepar Dajakhs in de 'Westerafdeeling van Borneo.'" *Indische Gids* 5:1–14.

Kathirithamby-Wells, Jeyamalar. 1995. "Socio-political Structures and the Southeast Asian Eco-system: An Historical Perspective up to the Mid-Nineteenth Century." In *Asian Perceptions of Nature: A Critical Approach*, ed. Ole Bruunand and Arne Kalland, 25–47. London: Curzon Press.

———. 2003. *Nature and Nation: Forests and Development in Peninsular Malaysia*. Copenhagen: NIAS.

Kato, Tsuyoshi. 1989. "Different Fields, Similar Locusts: Adat Communities and the Village Law of 1979 in Indonesia." *Indonesia* 47:89–114.

———. 1991. "When Rubber Came: The Negeri Sembilan Experience." *Southeast Asian Studies* 29 (2): 109–57.

———. 1994. "The Emergence of Abandoned Paddy Fields in Negeri Sembilan, Malaysia." *Southeast Asian Studies* 32 (2): 145–72.

Kedit, Peter. 1993. *Iban Bejalai*. Kuala Lumpur: Sarawak Literary Society

Keller, Eva. 2008. "The Banana Plant and the Moon: Conservation and the Malagasy Ethos of Life in Masoala." *American Ethnologist* 35(4): 650–64.

Kenmore, Peter E. 1992. *Indonesia's Integrated Pest Management: A Model for Asia;* FAO *Inter-Country Programme for Integrated Pest Control in Rice in South and Southeast Asia*. Rome: FAO.

———. 1997. "A Perspective on IPM." *LEISA* 13 (4): 8–9.

Khera, H. S. 1976. *The Oil Palm Industry of Malaya: An Economic Study*. Kuala Lumpur: Penerbit Universiti Malaya.

King, A. W. 1939. "Plantation Agriculture in Malaya, with Notes on the Trade of Singapore." *Geographical Journal* 93 (2): 136–48.

King, Victor. 1985. *The Maloh of West Kalimantan*. Dortrecht: Foris.

———. 1998. *Environmental Challenges in South East Asia*. Copenhagen: Curzon Press.

Kingsbury, Benedict. 1995. " 'Indigenous Peoples' as an International Legal Concept." In *Indigenous Peoples of Asia*, ed. R. H. Barnes, Andrew Gray, and Kingsbury, 13–34. Ann Arbor, Mich.: Association for Asian Studies.

———. 1998. " 'Indigenous Peoples' in International Law: A Constructivist Approach to the Asian Controversy." *American Journal of International Law* 92 (3): 414–57.

Kingston, Jeffrey B. 1991. "Manipulating Tradition: The State, *Adat*, Popular Protest, and Class Conflict in Colonial Lampung." *Indonesia* 52:21–45.

Kirsch, Stuart. 2006. *Reverse Anthropology: Indigenous Analysis of Social and Environmental Relations in New Guinea*. Stanford, Calif.: Stanford University Press.

———. 2008. "Social Relations and the Green Critique of Capitalism in Melanesia." *American Anthropologist* 110 (3): 288–98.

Kitayama, K. 1987. "Vegetation of Mount Kinabalu." *Review of Forest Culture* 8:103–13.

Kitchener, H. J. 1961. "A Bleak Prospect for *Seladang* or Malayan Gaur." In *Nature Conservation in Western Malaysia, 1961*, ed. J. Wyatt-Smith and P. R. Wycherley, 197–201. Kuala Lumpur: Malayan Nature Society.

Kleiman, P. J. A., R. B. Bryant, and D. Pimentel. 1996. "Assessing Ecological Sustainability of Slash-and-Burn Agriculture through Soil Fertility Indicators." *Agronomy Journal* 88:122–27.

Kratoska, P. 1975. "The Chettier and the Yeoman: British Cultural Categories and Rural Indebtedness in Malaya." Occasional Paper No. 32. Singapore: Institute of Southeast Asian Studies.

———. 1983. "Ends That We Cannot See: Malay Reservations in British Malaya." *Journal of South East Asian Studies* 14:149–68.

Krech, S., III. 1999. *The Ecological Indian: Myth and History*. New York: W. W. Norton.

Kuper, A. 2003. "The Return of the Native." *Current Anthropology* 44:389–402.

Kurus, Bilson. 1994. "Review on the Role and Functions of the Local Chieftain System in Sabah." *Borneo Review* 5 (1): 1–21.

Kuswara, Engkus. n.d. "The Impact of IPM in Marga Sub-district, Tabanan District, Bali." Unpublished manuscript, Jakarta.

Kusworo, Ahmad. 1997. "Kajian Beberapa Kebijakan Pemerintah Menyangkut Repong Damar di Pesisir Krui, Lampung Barat." Unpublished manuscript, WATALA (Friends for Nature and the Environment), Bandarlampung, Indonesia.

Langdon, Steve J. 2003. "Relational Sustainability: Indigenous Northern North American Logic of Engagement." Paper presented at the Annual Meeting of American Anthropological Association, Chicago, November 19–23.

Lansing, J. Stephen. 1991. *Priests and Programmers: Technologies of Power in the Engineered Landscapes of Bali*. Princeton, N.J.: Princeton University Press.

——. 2006. *Perfect Order: Recognizing Complexity in Bali*. Princeton, N.J.: Princeton University Press.

Latin. n.d. "Pembahasan SK KDTI." Internal discussion transcript. Bogor: Lembaga Alam Tropika Indonesia.

——. 1999. "Bahan Bacaan Peserta Loka *Adat*: Bogor, 14–16 Mei 1999." Unpublished manuscript, Lembaga Alam Tropika Indonesia, Bogor.

Latour, Bruno. 1988. *Science in Action: How to Follow Scientists and Engineers through Society*. Cambridge, Mass.: Harvard University Press.

——. 2004. "Why Has Critique Run out of Steam? From Matters of Fact to Matters of Concern." *Critical Inquiry* 30:225–48.

Laungaramsri, Pinkaew. 2001. *Redefining Nature: Karen Ecological Knowledge and the Challenge to the Modern Conservation Paradigm*. Chennai: Earthworm Books.

Lawrence, D., D. R. Peart, and M. Leighton. 1998. "The Impact of Shifting Cultivation on a Rainforest Landscape in West Kalimantan: Spatial and Temporal Dynamics." *Landscape Ecology* 13:135–48.

Leach, Melissa, and James Fairhead. 2001. "Fashioned Forest Pasts: Occluded Histories? International Environmental Analysis in West African Locales." *Development and Change* 31:35–59.

Leach Melissa, Robin Mearns, and Ian Scoones. 1999. "Environmental Entitlements: Dynamics and Institutions in Community-Based Natural Resource Management." *World Development* 27 (2): 225–47.

LeBar, F. 1972. *Ethnic Groups of Insular South East Asia*. New Haven, Conn.: Human Relations Area Files.

Lebbie, A., and R. Guries. 1995. "Ethnobotanical Value and Conservation on Sacred Groves of the Kpaa Mende in Sierra Leone." *Economic Botany* 49(3): 297–308.

Lee, G. Y. 1988. "Household and Marriage in a Thai Highland Society." *Journal of the Siam Society* 76:162–73.

Lefebvre, Henri. 1991. *The Production of Space*. Trans. Donald Nicholson-Smith. Oxford: Blackwell.

Lemert, Charles, ed. 1993. *Social Theory: The Multicultural and Classic Readings*. Boulder, Colo.: Westview.

Lenoir, Timothy 1997. *Instituting Science: The Cultural Production of Scientific Disciplines*. Stanford, Calif.: Stanford University Press.

Leonen, Marvic. 1993. "A Review of Philippine Environment Policy for Community-Based Resource Management." Paper presented at the Fourth Annual Common Property Conference, Manila, June 16–19.

Li, Tania Murray. 1996. "Images of Community: Discourse and Strategy in Property Relations." *Development and Change* 27 (3): 501–27.

———. 1999. "Marginality, Power, and Production: Analyzing Upland Transformation." In *Agrarian Transformations in Upland Indonesia*, ed. Li, 1–44. London: Harwood.

———. 2000. "Articulating Indigenous Identity in Indonesia: Resource Politics and the Tribal Slot." *Comparative Studies of History and Society* 42 (1): 149–79.

———. 2001. "Boundary Work: Community, Market, and State Reconsidered." In *Communities and the Environment: Ethnicity, Gender, and the State in Community-Based Conservation*, ed. Arun Agarwal and Clark Gibson, 157–80. New Brunswick, N.J.: Rutgers University Press.

———. 2005. "Beyond the 'State' and Failed Schemes." *American Anthropologist* 107 (3): 383–94.

———. 2007. *The Will to Improve: Governmentality, Development, and the Practice of Politics*. Durham, N.C.: Duke University Press.

———. 2010. "Indigeneity, Capitalism, and the Management of Dispossession." *Current Anthropology* 51(3): 385–414.

Lim Heng Seng. 1998. "Towards Vision 2020: Law, Justice, and the Orang Asli." In *Warga pribumi menghadapi cabaran pembangunan* [*Indigenous Peoples Confronting the Challenges of Development*], ed. Hasan Mat Nor, 118–142. Bangi, Selangor: Universiti Kebangsaan, Malaysia.

Lim Teck Ghee. 1974. "Malayan Peasant Small Holders and the Stevenson Restriction Scheme, 1922–1928." *Journal of the Malaysian Branch of the Royal Asiatic Society* 47 (2): 105–22.

———. 1976. *Origins of a Colonial Economy: Land and Agriculture in Perak, 1874–1897*. Penang: Penerbit Universiti Sains Malaysia.

———. 1977. *Peasants and Their Agricultural Economy in Colonial Malaya, 1874–1941*. Kuala Lumpur: Oxford University Press.

Lowe, Celia. 1999. "Cultures of Nature: Mobility, Identity, and Biodiveristy Conservation in the Togian Islands, Indonesia." PhD diss., Yale University.

———. 2006. *Wild Profusion: Biodiversity Conservation in an Indonesian Archipelago*. Princeton, N.J.: Princeton University Press.

Lubis, Zulkifli. 1997. "Repong Damar: Kajian tentang Pengambilan Keputusan dalam Pengelolana Laha Hutan di Pesisir Krui, Lampung Barat." CIFOR Working Paper No. 20. Bogor: Center for International Forestry Research.

Lye Tuck-Po. 1997. "Knowledge, Forest, and Hunter-Gatherer Movement: The Batek of Pahang, Malaysia." PhD diss., University of Hawai'i, Manoa.

———. 2000. "Forest, Bateks, and Degradation: Environmental Representations in a Changing World." *Tonan Ajia Kenkyu* 38:165–84.

———. 2002. "Forest Peoples, Conservation Boundaries, and the Problem of 'Modernity' in Malaysia." In *Tribal Communities in the Malay World: Historical, Cultural, and Social Perspectives*, ed. Geoffrey Benjamin and Cynthia Chou, 160–84. Leiden: IIAS.

———. 2004. *Changing Pathways: Forest Degradation and the Batek of Pahang, Malaysia.* Lanham, Md.: Lexington.

———. 2005. "Uneasy Bedfellows: Contrasting Models of Conservation in Malaysia." In *Conserving Nature in Culture: Case Studies from Southeast Asia*, ed. Michael R. Dove, Amity A. Doolittle, and Percy E. Sajise, 83–118. New Haven, Conn.: Yale University Press.

Lynch, Owen, Jr. 1982. "Native Title, Private Rights, and Tribal Land Laws: An Introductory Survey." *Philippine Law Journal* 57 (2):268–306.

———. 1984. "The Invisible Filipinos: Indigenous and Migrant Citizens within the 'Public Domains.'" *Philippine Law Register* 5:18–22.

———. 1986. "Philippine Law and Upland Tenure." In *Man, Agriculture and the Tropical Forests*, ed. S. Fujisaka, P. Sajise, and R. del Castillo, 269–92. Bangkok: Winrock International Institute for Agricultural Development.

Lynch, Owen, Jr., and Emily Harwell. 2001. *Whose Resources? Whose Common Good?* Jakarta: ELSAM.

MacAndrews, Colin. 1986. *Land Policy in Modern Indonesia.* Boston: Oelgeschalger, Gunn, and Hain.

Mackie, C. 1986. "Disturbance and Succession Resulting from Shifting Cultivation in an Upland Rainforest in Indonesian Borneo." PhD diss., Rutgers University.

Maidi and Darmowiyoto. 1999. "Masa pembodohan dan masa pencerdasan: Dinamika budaya petani, suatu pengalaman." Paper presented at the "Memasuki abad ke-21: Antropologi Indonesia menghadapi krisis budaya bangsa" conference, Japanese Research Center and Faculty of Social and Political Sciences, Kampus Universitas Indonesia, Depok, May 6–8.

Manuel, Arsenio E. 1973. *Manuvuꞌ Social Organization.* Quezon City: Community Development Research Council, University of the Philippines.

Manuta, Jessie B. 2001. "Negotiating the Political Economy of Dispossession and Commodification: Reclaiming and Regenerating the Ancestral Domains of the Lumad of Mindanao, Southern Philippines." PhD diss., University of Delaware.

Marcus, George E., and Michael M. J. Fischer. 1986. *Anthropology as Cultural Critique: An Experimental Moment in the Human Sciences.* Chicago: University of Chicago Press.

Mary, F., and G. Michon. 1987. "When Agroforests Drive Back Natural Forests: A Socio-Economic Analysis of a Rice-Agroforest System in Sumatra." *Agroforestry Systems* 5:27–55.

Mathews, Andrew. 2005. "Power/Knowledge, Power/Ignorance: Forest Fires and the State in Mexico." *Human Ecology* 33 (6): 795–820.

Maxwell, Allen R. 1992. "Balui Reconnaissances: The Sihan of the Menamang River." *Sarawak Museum Journal* 43 (64): 1–45.

Maxwell, George. 1982. *In Malay Forests*. 3rd. ed. Singapore: Eastern Universities Press.

Mayer, Enrique. 2002. *The Articulated Peasant: Household Economies in the Andes*. Boulder, Colo.: Westview.

McCabe, Terrence. 2004. *"Cattle Bring Us to Our Enemies": Turkana Ecology, Politics, and Raiding in a Disequilibrium System*. Ann Arbor: Univeristy of of Michigan Press.

McCarthy, John F. 2006. *The Fourth Circle: A Political Ecology of Sumatra's Rainforest Frontier*. Stanford, Calif.: Stanford University Press.

McKeown, Francis A. 1983. "The Merakai Iban: An Ethnographic Account, with Special Reference to Dispute Settlement." PhD diss., Monash University.

McKinley, Robert. 1979. "Zaman dan Masa, Eras and Periods: Religious Evolution and the Permanence of Epistemological Ages in Malay Culture." In *The Imagination of Reality: Essays in Southeast Asian Coherence Systems*, ed. A. L. Becker and A. Yengoyan, 303–24. Norwood, N.J.: Ablex.

McNaughton, E. J. 1936. "Plantation Rubber in Malaya." *British Malaya* 9 (4): 89–93.

Media Jaringan Petani Indonesia (Media for the Network of Indonesian Farmers). 2000–2002. *Petani Indonesia (Indonesian Farmers)*. Jakarta Selatan: Ikatan Petani Pengendalian Hama Terpadu Indonesia (Indonesian Integrated Pest Management Farmers' Alliance).

Meek, Charles. K. 1946. *Land Law and Custom in the Colonies*. London: Oxford University Press.

Memorandum by A. B. C. Francis, n.d. [1920?]. Colonial Office 874/985, Colonial Office Records Public Records Office, London.

Merry, Sally. E. 1988. "Legal Pluralism." *Law and Society Review* 22 (5): 867–96.

——. 1991. "Law and Colonialism." *Law and Society Review* 25 (4): 890–922.

Mertz, Olie, et al. 2008. "A Fresh Look at Shifting Cultivation: Fallow Length an Uncertain Indicator of Productivity." *Agricultural Systems* 96: 75–84.

Mguni, Siyakha. 2009. "Natural and Supernatural Convergences: Trees in Southern African Rock Art." *Current Anthropology* 50(1): 139–48.

Michon, Genevieve, H. De Foresta, Kusworo, and P. Levang. 2000. "The Damar Agroforests of Krui: Justice for Forest Farmers." In *People, Plants, and Justice: The Politics of Nature Conservation*, ed. Charles Zerner, 159–203. New York: Columbia University Press.

Miles, Douglas. 1976. *The Cutlass and the Crescent Moon: A Case Study in Social and Political Change in Outer Indonesia*. Sydney: Center for Asia Studies, University of Sydney.

Ministry of Forestry, Republic of Indonesia. 1998. *Keputusan Menteri Kehutanan*. Decree of the Ministry of Forestry, Republic of Indonesia 47/KPTS-II/1998.

Ministry of Science, Environment, and Technology, Malaysia. See MOSTE.

Mohd, Mohammad Khan. 1968. "Specialist Wildlife Training from the Standpoint of an Asian Trainee." *Proceedings of the Conference on Conservation of Nature and Natural Resources in Tropical South-East Asia* (Bangkok, November–December 1965). Morges, Switzerland: International Union for the Conservation of Nature.

———. 1977. "On the Population and Distribution of the Malayan Elephant." *Malayan Nature Journal* 30:1–14.

———. 1988. "Animal Conservation Strategies." In *Key Environments: Malaysia*, ed. Earl of Cranbrook, 251–72. Oxford: Pergamon.

Moniaga, Sandra. 1993. "Toward Community-Based Forestry and Recognition of Adat Property Rights in the Outer Islands of Indonesia." In *Legal Frameworks for Forest Management in Asia: Case Studies of Community/State Relations*, ed. J. Fox, 131–50. Honolulu: East-West Center.

Moning, Elias Tana. 2006. "Reinventing Indigenous Knowledge: A Crucial Factor for an IPM-Based Sustainable Agricultural Development." PhD diss., University of Massachussetts, Amherst.

Moore, Donald S. 1993. "Contesting Terrain in Zimbabwe's Highlands: Political Ecology, Ethnography, and Peasant Resource Struggles." *Economic Geography* 69:390–401.

———. 1997. "Remapping Resistance: 'Grounds for Struggle' and the Politics of Place." In *Geographies of Resistance*, ed. Steve Pile and Michael Keith, 87–106. New York: Routledge.

———. 1998. "Subaltern Struggles and the Politics of Place: Remapping Resistance in Zimbabwe's Eastern Highlands." *Cultural Anthropology* 13 (3): 344–81.

Moore, Sally Falk. 1986a. "Legal Systems of the World: An Introductory Guide to Classifications, Typological Interpretations, and Bibliographical Resources." In *Law and the Social Sciences*, ed. Leon Lipson and Stanton Wheeler, 11–61. New York: Russell Sage Foundation.

———. 1986b. *Social Facts and Fabrication: "Customary" Law on Kilimanjaro, 1880–1980*. Cambridge: Cambridge University Press.

Morgan, R. P. C. 1974. "Estimating Regional Variations in Soil Erosion Hazard in Peninsular Malaysia." *Malayan Nature Journal* 28 (2): 94–106.

Mosko, Mark S. 2009. "The Fractal Yam: Botanical Imagery and Human Agency in the Trobriands." *Journal of the Royal Anthropological Institute* (n.s. 15): 679–700.

Mosse, David. 1997. "The Symbolic Making of a Common Property Resource: History, Ecology, and Locality in a Tank-Irrigated Landscape in South India." *Development and Change* 28 (3): 467–504.

———. 2004. "Is Good Policy Unimplementable? Reflections on the Ethnography of Aid Policy and Practice." *Development and Change* 35 (4): 639–71.

MOSTE (Ministry of Science, Environment, and Technology, Malaysia). 1997. "Assessment of Biological Diversity in Malaysia." Unpublished report. Ministry of Science, Environment, and Technology, Kuala Lumpur.

Murray, Martin J. 1992. " 'White Gold' or 'White Blood'? The Rubber Plantations of Colonial Indochina, 1910–40." *Journal of Peasant Studies* 19 (3–4): 41–67.

Musyawarah Petani PHT Indonesia. 1999. "Laporan musyawarah petani PHT Indonesia [Conference of the Indonesian IPM Farmers], 16–21 Juli, 1999." Unpublished manuscript. Jakarta: Indonesian FAO-Inter Country Program.

Myers. N. 1979. *The Sinking Ark: A New Look at the Problem of Disappearing Species.* Oxford: Pergamon.

Nadasky, Paul. 2003. *Hunters and Bureaucrats: Power, Knowledge, and Aboriginal-State Relations in the Southwest Yukon*. Vancouver: University of British Columbia Press.

"The National Forest Policy." 1980. *Malayan Forester* 43 (1): 1–7.

Nazarea, Virginia D. 1998. *Cultural Memory and Biodiversity*. Tucson: University of Arizona Press.

Nazaruddin, M. Jali. 1990. "The Javanese of Kuala Langat: Participation in Development Programmes." In *Margins and Minorities: The Peripheral Areas and Peoples of Malaysia*, ed. V. T. King and M. J. G. Parnwell, 60–76. Hull, U.K.: Hull University Press.

Neidel, David. 2006. "The Garden of Forking Paths: History, Its Erasure, and Remembrance in Sumatra's Kerinci Seblat National Park." PhD diss., Yale University.

Nelson, Fred, and Arun Agrawal. 2008. "Patronage or Participation? Community-Based Natural Resource Management in Sub-Saharan Africa." *Development and Change* 39(4): 557–85.

Netting, Robert McC. 1990. "Links and Boundaries: Reconsidering the Alpine Village as Ecosystem." In *The Ecosystem Approach in Anthropology*, ed. Emilio F. Moran, 229–45. Ann Arbor: University of Michigan Press.

Neumann, Roderick P. 1992. "Political Ecology of Wildlife Conservation in the Mount Meru Area of North-East Tanzania." *Land Degradation and Rehabilitation* 3 (3): 85–98.

——. 1997. "Primitive Ideas: Protected Area Buffer Zones and the Politics of Land in Africa." *Development and Change* 28 (3): 559–82.

——. 1998. *Imposing Wilderness: Struggles over Livelihood and Nature Preservation in Africa*. Berkeley: University of California Press.

Ng, F. 1988. "Forest Tree Biology." In *Key Environments: Malaysia*, ed. Earl of Cranbrook, 102–25. Oxford: Pergamon.

Nicholas, Colin. 2000. *The Orang Asli and the Contest for Resources: Indigenous Politics, Development, and Identity in Peninsular Malaysia*. Kuala Lumpur: Center for Orang Asli Concerns.

Niezen, R. 2003. *The Origins of Indigenism: Human Rights and the Politics of Identity*. Berkeley: University of California Press.

Noble, I. R., and R. Dirzo. 1997. "Forests as Human-Dominated Ecosystems." *Science* 277:522–25.

Nonini, Donald M. 1992. *British Colonial Rule and the Resistance of the Malay Peasantry, 1900–1957*. New Haven, Conn.: Yale University Southeast Asia Studies.

Nor, Salleh Mohd, and Francis Ng. 1983. "Conservation in Malaysia." *Planter* 59:493–90.

Nugent, D. 1994. "Building the State, Making the Nation: The Bases and Limits of State Centralization in 'Modern' Peru." *American Anthropologist* 96 (2): 333–69.

Obeyesekere, Gananath. 1992. *The Apotheosis of Captain Cook: European Mythmaking in the Pacific*. Princeton, N.J.: Princeton University Press.

Odum, Eugene P. 1953. *Fundamentals of Ecology*. Philadelphia: Saunders College.

Okoth-Ogenda, H. W. 1979. "The Imposition of Property Law in Kenya." In *The*

Imposition of Law, ed. Susan Burman and Barbara Harrell-Bond, 147–65. New York: Academic.

Olivier de Sardan, Jean-Pierre. 2005. *Anthropology and Development: Understanding Contemporary Social Change.* Trans. Antoinette Tidjani Alou. London: Zed.

Ongkili, James. 1972. *Modernization in East Malaysia, 1960–1970.* Kuala Lumpur: Oxford University Press.

Ooi Jim Bee. 1959. "Rural Development in Tropical Areas, with Special Reference to Malaya." *Journal of Tropical Geography* 12:46–65.

———. 1976. *Peninsular Malaysia.* London: Longmans.

Ooi, Peter A. C. 1998. "Farmer Participation in IPM Action Research." Paper presented at the International Conference of IPM, Guangzhou, China, June 15–20.

Ortner, Sherry. 1995. "Resistence and the Problem of Ethnographic Refusal." *Comparative Studies in Society and History* 37 (1): 173–93.

Padoch, Christine. 1982. *Migration and Its Alternatives among the Iban of Sarawak.* The Hague: Martinus Nijhoff.

———. 1995. "Creating the Forest: Dayak Resource Management in Kalimantan." In *Society and Non-timber Forest Products in Tropical Asia*, ed. Jefferson Fox, 3–11. Honolulu: East-West Center.

Page, H. J. 1938. "The Role of Cover Crops in Rubber Growing." *Planter* 19:1–3.

Pahang Natural Resource Board. n.d. "River Cleaning in Pahang." Information Paper 16, PHN Phg 198/55. Kuala Lumpur: Arkib Negara (National Archives).

Paiman. n.d. "Gerung Sub-district: An IPM Case Study from West Lombok District, West Nusa Tenggara Province." Unpublished manuscript, Jakarta.

Parker, Eugene. 1992. "Forest Islands and Kayapo Resource Management in Amazonia: A Reappraisal of the Apete." *American Anthropologist* 94 (2): 406–28.

Pearsall, Judy, ed. 1999. *The Concise Oxford Dictionary.* 10th ed. Oxford: Oxford University Press.

Peck, Gunter. 2006. "The Nature of Labor: Fault Lines and Common Ground in Environmental and Labor History." *Environmental History* 11 (2): 212–38.

Peet, Richard, and Michael Watts, eds. 1996. *Liberation Ecologies: Environment, Development, and Social Movements.* London: Routledge.

Peluso, Nancy L. 1992. *Rich Forests, Poor People: Resource Control and Resistance in Java.* Berkeley: University of California Press.

———. 1993. "Coercing Conservation? The Politics of State Resource Control." *Global Environmental Change* 3 (2): 199–218.

———. 1995. "Whose Woods Are These? Counter-mapping Forest Territories in Kalimantan, Indonesia." *Antipode* 27 (4): 383–406.

———. 1996. "Fruit Trees and Family Trees in an Anthropogenic Forest: Ethics of Access, Property Zones, and Environmental Change in Indonesia." *Comparative Studies in Society and History* 38:510–48.

———. 1998. "Legal Pluralism and Legacies of 'Customary Rights' in Indonesian and Malaysian Borneo." *Common Property Resource Digest* 47:10–13.

———. 2009. "Rubber Erasures, Rubber Producing Rights: Making Racialized Territories in West Kalimantan, Indonesia." *Development and Change* 40 (1): 47–80.

Peluso, Nancy L., and Emily Harwell. 2001. "Territory, Custom, and Cultural Politics of Ethnic War in West Kalimantan, Indonesia." In *Violent Environments*, ed. Peluso and Michael Watts, 83–116. Ithaca, N.Y.: Cornell University Press.

Peluso, Nancy L., and Peter Vandergeest. 2001. "Genealogies of the Political Forest and Customary Rights in Indonesia, Malaysia, and Thailand." *Journal of Asian Studies* 60 (3): 761–812.

Peluso, Nancy L., and Michael Watts, eds. 2001. *Violent Environments*. Ithaca, N.Y.: Cornell University Press.

Pelzer, Karl. 1945. *Pioneer Settlement in the Asiatic Tropics: Studies in Land Utilization and Agricultural Colonization in Southeastern Asia*. New York: American Geographical Society.

———. 1978. *Planter and Peasant: Colonial Policy and the Agrarian Struggle in East Sumatra, 1863–1947*. Gravenhage: Martinus Nijhoff.

Pemerintah Daerah Tingkat II Indramayu. 1996. "Getap-Sitampan: Gerakan pemantapan produksi tanaman pangan; Program lima tahun (1996–2001) Kabupaten DT.II Indramayu." Unpublished manuscript, Indramayu.

Penot, Eric. 2007. "From Shifting Cultivation to Sustainable Jungle Rubber: A History of Innovations In Indonesia." In *Voices from the Forest: Integrating Indigenous Knowledge into Sustainable Upland Farming*, ed. Malcolm Cairns, 577–99. Washington, D.C.: Resources for the Future.

Pepper, David. 1993. *Eco-socialism: From Deep Ecology to Social Justice*. London: Routledge.

Peres, C., and John Terbough. 1995. "Amazonian Nature Reserves: An Analysis of the Defensible Status of Existing Conservation Units and Design Criteria for the Future." *Conservation Biology* 9:34–46.

Persoon, Gerard A., and Padmapani L. Perez. 2008. "The Relevant Context: Environmental Consequences of Images of the Future." In *Against the Grain: The Vayda Tradition in Human Ecology and Ecological Anthropology*, ed. B. Walters et al., 287–305. Lanham, Md.: Lexington.

Perz, Stephen G. 2002. "Household Demography and Land Use Allocation among Small Farms in the Brazilian Amazon." *Human Ecology Review* 9 (2): 1–16.

Petani PHT Jawa Barat. 1998. "Musyawarah daerah III petani PHT Jawa Barat." Unpublished manuscript, Jakarta.

Peters, Charles. 1994. *Forest Resources of the Danau Sentarum Wildlife Reserve: Strategies for the Sustainable Exploitation of Timber and Non-timber Products*. Bogor: Wetlands International/PHPA/ODA.

Peters, Pauline. 1994. *Dividing the Commons: Politics, Policy, and Culture in Botswana*. Charlottesville: University Press of Virginia.

Phelan, Peter. 1988. "Native Land in Sabah: Its Administrators: Headmen and Native Chiefs." *Sabah Society Journal* 8 (4): 475–501.

Pickering, A. 1995. *The Mangle of Practice: Time, Agency, and Science*. Chicago: University of Chicago Press.

Pickett, Stuart T. A., and R. S. Ostfeld. 1995. "The Shifting Paradigm in Ecology." In *A*

New Century for Natural Resources Management, ed. Richard L. Knight and Sarah F. Bates, 261–78. Washington, D.C.: Island.

Pickett, Stuart, V. Parker, and P. Fiedler. 1992. "The New Paradigm in Ecology: Implications for Conservation Biology above the Species Level." In *Conservation Biology: The Theory and Practice of Nature Conservation, Preservation, and Management*, ed. Fiedler and K. Jain, 65–88. New York: Chapman and Hall.

Pickett, Stuart, and P. White, eds. 1985. *The Ecology of Natural Disturbance and Patch Dynamics*. London: Academic.

Pickles, John, ed. 1995. *Ground Truth: The Social Implications of Geographic Information Systems*. New York: Guilford.

Pincus, Jonathan. 1991. "Farmer Field School Survey: Impact of IPM Training on Farmers' Pest Control Behavior." Unpublished manuscript, Jakarta.

Pontius, John, Russel Dilts, and Andrew Bartlett. 2002. *From Farmer Field School to Community IPM: Ten Years of IPM Training in Asia*. Bangkok: FAO Community IPM Programme, and Food and Agriculture Organization of the United Nations Regional Office for Asia and the Pacific.

Posey, Darrell A. 1985. "Indigenous Management of Tropical Forest Ecosystems: The Case of the Kayapó Indians of the Brazilian Amazon." *Agroforestry Systems* 3 (2): 139–58.

Potter, Lesley. 2008. "The Oil Palm Question in Borneo." In *Reflections on the Heart of Borneo*, ed. Gerard A. Persoon and Manon Osseweijer, 69–90. Tropenbos 24. Wageningen, Netherlands: Tropenbos International.

Pringle, Robert. 1968. "Rajahs and Rebels: The Ibans of Sarawak under Brooke Rule, 1841–1941." PhD diss., Cornell University.

Pulido, Laura. 1998. "Ecological Legitimacy and Cultural Essentialism." In *The Struggle for Ecological Legitimacy: Environmental Justice Movements in the United States*, ed. Daniel Faber, 293–311. New York: Guilford.

Puri, R. 2005. "Postabandonment Ecology of Penan Forest Camps: Anthropological and Ethnobiological Approaches to the History of a Rain Forested Valley in East Kalimantan." In *Conserving Nature in Culture: Case Studies from Southeast Asia*, ed. Michael R. Dove, Percy E. Sajise, and Amity A. Doolittle, 1–82. New Haven, Conn.: Yale University Southeast Asia Studies.

Raffles, Hugh. 2002. *In Amazonia: A Natural History*. Princeton, N.J.: Princeton University Press.

Ramakrishnan, P. S. 1992. *Shifting Agriculture and Sustainable Development*. Paris: UNESCO.

Rambo, A. Terry. 1982. "Orang Asli Adaptive Strategies: Implications for Malaysian Natural Resource Development Planning." In *Too Rapid Rural Development: Perceptions and Perspectives from Southeast Asia*, ed. Colin MacAndrews and Chia Lin Sen, 251–99. Athens: Ohio University Press.

Rappaport, Roy A. 1968. *Pigs for the Ancestors: Ritual in the Ecology of a New Guinea People*. New Haven, Conn.: Yale University Press.

———. 1979. *Ecology, Meaning, and Religion*. Richmond, Calif.: North Atlantic.

Ras, J. J. 1968. *Hikajat Bandjar: A Study in Malay Historiography*. The Hague: Martinus Nijhoff.

Redford, Kent H. 1991. "The Ecologically Noble Savage." *Cultural Survival Quarterly* 15 (1): 46–48.

Resurreccion, Bernadette. 1999. "Transforming Nature, Redefining Selves." PhD diss., The Hague Institute of Social Studies.

Restrepo, Eduardo, and Arturo Escobar. 2005. " 'Other Anthropologies and Anthropology Otherwise': Steps to World Anthropologies Framework." *Critique of Anthropology* 25 (2): 99–129.

Rhoades, R. E., and Anthony Bebbington. 1988. "Farmers as Experimenters." *ILEIA* 4 (3): 328.

——. 1995. "Farmers Who Experiment: An Untapped Resource for Agricultural Research and Development." In *The Cultural Dimension of Development: Indigenous Knowledge Systems*, ed. D. Michael Warren, L.Jan Slikkerveer, and David Brokensha, 296–307. London: Intermediate Technology Publications.

Richards, Anthony. 1981. *An Iban-English Dictionary*. Oxford: Clarendon.

Richards, Paul. 1985. *Indigenous Agricultural Revolution: Ecology and Food Production in West Africa*. London: Hutchinson.

——. 1986. *Coping with Hunger: Hazard and Experiment in African Rice-Farming Systems*. London: Allen and Unwin.

Ridley, H. N. 1893. "On the Dispersal of Seeds by Mammals." *Journal of the Royal Asiatic Society Singapore Branch* 25:11–32.

——. 1922. *The Flora of the Malay Peninsula*. London: Reeves.

Rival, Laura. 1993. "The Growth of Family Trees: Understanding Huaorani Perceptions of the Forest." *Man* 28 (4): 635–52.

Robbins, Paul. 2004. *Political Ecology: A Critical Introduction*. Malden, Mass.: Blackwell.

Robbins, Paul, and Alistair Fraser. 2003. "A Forest of Contradictions: Producing the Landscapes of the Scottish Highlands." *Antipode* 35 (1): 95–118.

Robbins, Paul, Kendra McSweeney, Thomas Waite, and Jennifer Rice. 2006. "Even Conservation Rules Are Made to Be Broken: Implications for Biodiversity." *Environmental Management* 37 (2): 162–69.

Robin, Libby. 2000. "Yellowstone, Edens, and Local Place." Paper presented at the "Land, Place, Culture, Identity" workshop, Institute of Advanced Studies, University of Western Australia, Nedlands, July 2–7.

Rocheleau, Dianne, and David Edmunds. 1997. "Women, Men, and Trees: Gender, Power, and Property in Forest and Agrarian Landscapes." *World Development* 25 (8): 1351–71.

Rocheleau, Dianne, and Laurie Ross. 1995. "Trees as Tools, Trees as Text: Struggles over Resources in Zambrana-Chacuey, Dominican Republic." *Antipode* 27:407–28.

Rocheleau, Dianne E., Philip E. Steinberg, and Patricia A. Benjamin. 1995. "Environment, Development, Crisis, and Crusade: Ukambani, Kenya, 1890–1990." *World Development* 23 (6): 1037–51.

Rocheleau, Dianne, Barbara Thomas-Slayter, and Esther Wangari, eds. 1996. *Feminist Political Ecology: Global Issues and Local Experiences*. London: Routledge.

Rodil, Rodis. 1990. "The Lumad Side on the Issue of Autonomy in Mindanao." PSSC *Social Science Information* 18 (3): 4–14.

Röling, Nils, and Elske van de Fliert. 1998. "Introducing Integrated Pest Management in Rice in Indonesia: A Pioneering Attempt to Facilitate Large-Scale Change." In *Facilitating Sustainable Agriculture: Participatory Learning and Adaptive Management in Times of Environmental Uncertainty*, ed. Röling and M. A. E. Wagemakers, 153–280. Cambridge: Cambridge University Press.

Rose, Carol. 1991. "Rethinking Environmental Controls: Management Strategies for Common Resources." *Duke Law Journal* 1991 (1): 1–38.

——. 1994. *Property and Persuasion: Essays on the History, Theory, and Rhetoric of Ownership*. Boulder, Colo.: Westview.

Ross-Larson, Bruce. 1976. *The Politics of Federalism: Syed Kechik in East Malaysia*. Singapore: Bruce Ross-Larson.

Rouquette, J. 1995. *Honey Harvesting: Developing Alternative Sources of Income in the of Danau Sentarum Wildlife Reserve, Kalimantan Barat, Indonesia*. Bogor: Wetlands International/PHPA/ODA.

Roy Chowdhury, R. 2007. "Household Land Management and Biodiversity: Secondary Succession in a Forest-Agriculture Mosaic in Southern Mexico." *Ecology and Society* 12 (2): 31 (online), http://www.ecologyandsociety.org/vol12/iss2/art31/.

Ruchijat, E., and T. Sukmaraganda. 1992. "National Integrated Pest Management in Indonesia: Its Successes and Challenges." In *Integrated Pest Management in the Asia-Pacific Region*, eds. Peter A. C. Ooi et al., 329–47. Wallingford, U.K.: CAB International.

Rudner, M. 1976. "Malayan Rubber Policy: Development and Anti-development during the 1950s." *Journal of South-East Asian Studies* 7 (2): 235–59.

Ruwiastuti, Maria. 1999. "Masyarakat *Adat* dan hak Atas Sumber-sumber Alam." In Bahan Bacaan Peserta Loka *Adat*: Bogor, 14–16 Mei 1999. Lembaga Alam Tropika Indonesia. Bogor, Indonesia.

Sack, Robert David. 1986. *Human Territoriality: Its Theory and History*. Cambridge: Cambridge University Press.

Safitri, Myrna. 1999. "Hak dan Akses Masyarakat Lokal Pada Sumberdaya Hutan Kajian Peraturan perundang-undangan Indonesia." In Bahan Bacaan Peserta Loka *Adat*: Bogor, 14–16 Mei 1999. Lembaga Alam Tropika Indonesia: Bogor Indonesia.

Sahlins, Marshall. 1971. "The Intensity of Production in Primitive Societies: Social Inflections of the Chayanov Slope." In *Studies in Economic Anthropology*, ed. George Dalton, 30–51. Washington, D.C.: American Anthropological Association.

——. 1972. *Stone Age Economics*. Chicago: Aldine-Atherton.

Said, Edward W. 1978. *Orientalism*. New York: Vintage.

Salafsky, N., Barbara L. Dugelby, and John W. Terborgh. 1993. "Can Extractive Reserves Save the Rain Forest? An Ecological and Socioeconomic Comparison of Nontimber Forest Product Extraction Systems in Petén, Guatemala, and West Kalimantan, Indonesia." *Conservation Biology* 7 (1): 39–52.

Sandin, Benedict. 1967. *The Sea Dayaks of Borneo before White Rajah Rule*. Ann Arbor: University of Michigan Press.

Sandker, Marieke, Aritta Suwarno, and Bruce M. Campbell. 2007. "Will Forests Remain in the Face of Oil Palm Expansion? Simulating Change in Malinau, Indonesia." *Ecology and Society* 12 (2): 37 (online), http://www.ecologyandsociety.org/vol12/iss2/art37/.

Sather, Clifford. 1990. "Trees and Tree Tenure in Paku Iban Society: The Management of Secondary Forest Resources in a Long-Established Iban Community." *Borneo Review* 1 (1): 16–40.

Sawit, M. H., and I. Manwan. 1991. "The New Supra Insus Rice Intensification Program: The Case of the North Coast of West Java and South Sulawesi." *Bulletin of Indonesian Economic Studies* 27 (1): 81–103.

Schaik, Carel P. van, John W. Terborgh, and S. Joseph Wright. 1993. "The Phenology of Tropical Forests: Adaptive Significance and Consequences for Primary Consumers." *Annual Review of Ecology and Systematics* 24:353–77.

Schärer, Hans. 1963. *Ngaju Religion: The Conception of God among a South Borneo People*. Trans. Rodney Needham. The Hague: Martinus Nijhoff.

Schebesta, Paul. [1928] 1973. *Among the Forest Dwarfs of Malaya*. London: Oxford University Press.

Schlegel, Stuart. 1979. *Tiruray Subsistence: The Transformation from Shifting Cultivation to Plow Agriculture*. Manila: Ateneo de Manila University.

Schmidgall-Tellings, A. Ed, and Alan M. Stevens. 1981. *A Contemporary Indonesian-English Dictionary*. Athens: Ohio University Press.

Schneider, Jane. 1990. "Spirits and the Spirit of Capitalism." In *Religious Orthodoxy and Popular Faith in European Society*, ed. Ellen Badone, 24–53. Princeton, N.J.: Princeton University Press.

Schneider, Jürg. 1995. *From Upland to Irrigated Rice: The Development of Wet-Rice Agriculture in Rejang Musi Southwest Sumatra*. Berlin: Reimer.

Schot, Wim E. M. van der. 1986. "The Batek of Taman Negara: A Hunters/Gatherers Community in a Malay National Park." PhD diss., Universiteit van Amsterdam.

Schroeder, Richard. 1993. "Shady Practices: Gender and the Political Ecology of Resource Stabilization in Gambia Garden/Orchards." *Economic Geography* 69 (4): 349–65.

——. 1999. *Shady Practices: Agroforestry and Gender Politics*. Berkeley: University of California Press.

——. 2005. "Community, Forestry, and Conditionality in the Gambia." In *Communities and Conservation: Histories and Politics of Community-Based Natural Resource Management*, ed. J. Peter Brosius, Anna Lowenhaupt Tsing, and Charles Zerner, 207–29. Walnut Creek, Calif.: AltaMira.

Schroeder, Richard, and Randy Neumann. 1995. "Manifest Ecological Destinies: Local Rights and Global Environmental Agenda." *Antipode* 27 (4): 325–42.

Schroeder, Richard, and Krisnawati Suryanata. 1996. "Gender and Class Power in Agroforestry Systems: Case Studies from Indonesia and West Africa." In *Liberation*

Ecologies: Environment, Development, Social Movements, ed. Richard Peet and Michael Watts, 188–204. London: Routledge.

Schultes, R. E. 1987. "Members of Euphorbiaceae in Primitive and Advanced Societies." *Botanical Journal of the Linnaean Society* 94:79–95.

Schwartzman, S., A. Moreira, and D. Nepstad. 2000. "Rethinking Tropical Forest Conservation: Perils in Parks." *Conservation Biology* 14 (5): 1351–57.

Schwartzman, S., D. Nepstad, and A. Moreira. 2000. "Arguing Tropical Forest Conservation: People versus Parks." *Conservation Biology* 14 (5): 1370–74.

Scoones, I. 1999. "New Ecology and the Social Sciences: What Prospects for a Fruitful Engagement?" *Annual Review of Anthropology* 28:479–507.

Scott, James C. 1976. *The Moral Economy of the Peasant: Rebellion and Subsistence in Southeast Asia*. New Haven, Conn.: Yale University Press.

———. 1998. *Seeing Like a State: How Certain Schemes to Improve the Human Condition Have Failed*. New Haven, Conn.: Yale University Press.

Sen, Amartya. 1981. *Poverty and Famines: An Essay on Entitlement and Deprivation*. Oxford: Clarendon.

Senftleben, Wolfgang. 1978. *Background to Agricultural Land Policy in Malaysia*. Wiesbaden: Otto Harrassowitz.

Settle, William H. 1997. "Science by Farmers and Science by Researchers: Challenges and Opportunities." Paper presented at the International Symposium on Integrated Pest Management in Rice-Based Ecosystems, October 20–24, Guangzhou, China.

Sham Sani. 1993. *Environment and Development in Malaysia: Changing Concerns and Approaches*. Kuala Lumpur: Institute for Strategic and International Studies, Malaysia.

Shamsul, A. B. 1986. *From British to Bumiputera Rule: Local Politics and Rural Development in Peninsula Malaysia*. Singapore: Institute for Southeast Asian Studies.

Sheil, Douglas, Anne Casson, Erik Meijaard, Meine van Noordwijk, Joanne Gaskell, Jacqui Sunderland-Groves, Karah Wertz, and Markku Kanninen. 2009. "The Impacts and Opportunities of Oil Palm in Southeast Asia: What Do We Know and What Do We Need to Know?" Occasional Paper No. 51. Bogor: Center for International Forestry Research.

Sheridan, Michael. 2008. "The Dynamics of African Sacred Groves: Ecological, Social, and Symbolic Processes." In *African Sacred Groves: Ecological Dynamics and Social Change*, ed. Sheridan and C. Nyamweru, 9–41. Oxford: James Currey.

Sheridan, Michael, and C. Nyamweru, eds. 2008. *African Sacred Groves: Ecological Dynamics and Social Change*. Oxford: James Currey.

Sherman, George. 1980. "What 'Green Desert'? The Ecology of Batak Grassland Farming." *Indonesia* 29:113–48.

Shipton, Parker. 1989. *Bitter Money: Cultural Economy and Some African Meanings of Forbidden Commodities*. Washington, D.C.: American Anthropological Association.

Shiva, Vandana. 1988. *Staying Alive: Women, Ecology, and Survival in India*. New Delhi: Kali for Women.

——. 1991. *The Violence of the Green Revolution: Third World Agriculture, Ecology, and Politics.* London: Zed.

——. 1993. *Monocultures of the Mind: Perspectives on Biodiversity and Biotechnology.* London: Zed.

Sinaga, Pangarimpunan. 1994a. *Hasil Rapat Ketua Nelayan Wilayah Suaka Margasatwa Danau Sentarum.* Bogor: Wetlands International-ODA.

——. 1994b. *Laporan Kegiatan dan Hasil Rapat Ketua Nelayan Wilayah Suaka Margasatwa Danau Sentarum.* KSDA-AWB (Konservasi Sumber Daya Alam [Natural Resources Conservation Units]-Asian Wetland Bureau).

——. 1994c. *Rapat Ketua Nelayan Wilayah Sungai Tengkidap di Nanga Sauk, Proyek Konservasi Suaka Margasatwa Danau Sentarum.* KSDA-AWB (Konservasi Sumber Daya Alam [Natural Resources Conservation Units]-Asian Wetland Bureau).

——. 1996. *Surat Pertanyataan para Ketua Nelayan dalam Wilayah Kecamatan Selimbau tentang Tidak Melakukan Penuaan Ikan dan Kesepakatan Adat Masyarakat yang Bermukim di Sekitar Suaka Margasataw Danau Sentarum.* KSDA-AWB (Konservasi Sumber Daya Alam [Natural Resources Conservation Units]-Asian Wetland Bureau).

Sinha, S., S. Gururani, and B. Greenberg. 1997. "The 'New Traditionalist' Discourse of Indian Environmentalism." *Journal of Peasant Studies* 24 (3): 65–99.

Sirait, Martua. 1998a. "Documentation Process on KDTI Internal Formulation of Ministry of Forestry Decree." Unpublished report. World Agroforestry Centre, Bogor.

——. 1998b. "The Krui Damar Agroforests." Unpublished report. World Agroforestry Centre, Bogor.

Sivalingam. G. 1993. *Malaysia's Agricultural Transformation.* Kuala Lumpur: Pelanduk Publications.

Sivaramakrishnan, K. 1999. *Modern Forests: Statemaking and Environmental Change in Colonial Eastern India.* Stanford, Calif.: Stanford University Press.

Skeat, Walter William, and Charles O. Blagden. [1906] 1966. *Pagan Races of the Malay Peninsula.* Vol. 1. London: Cass.

Skinner, A. M. 1878. "Geography of the Malay Peninsula, Part 1." *Journal of the Straits Branch of the Royal Asiatic Society* 1:52–62.

Slater, Candace. 1994. *Dance of the Dolphin: Transformation and Disenchantment in the Amazonian Imagination.* Chicago: University of Chicago Press.

——, ed. 2003. *In Search of the Rain Forest.* Durham, N.C.: Duke University Press.

Sletto, Bjørn Ingmunn. 2009. " 'We Drew What We Imagined': Participatory Mapping, Performance, and the Art of Landscape Making." *Current Anthropology* 50(4): 443–76.

Snodgrass, L. 1980. *Inequality and Economic Development.* Kuala Lumpur: Oxford University Press.

Sodhi, Navjot S., Greg Acciaioli, Maribeth Erb, and Alan Khee-Jin Tan, eds. 2008. *Biodiversity and Human Livelihoods in Protected Areas: Case Studies from the Malay Archipelago.* Cambridge: Cambridge University Press.

Sodikoff, Genese. 2009. "The Low-Wage Conservationist: Biodiversity and Perversities of Value in Madagascar." *American Anthropologist* 111 (4): 443–55.

Soepadmo, E. 1995. "Plant Diversity of the Malesian Tropical Rainforest and Its Phyto-geographical and Economic Significance." In *Ecology, Conservation, and Management of Southeast Asian Rainforests*, ed. R. Primack and R. Lovejoy, 19–40. New Haven, Conn.: Yale University Press.

Soja, Edward W. 1971. *The Political Organization of Space*. Washington, D.C.: American Association of Geographers.

———. 1989. *Postmodern Geographies: The Reassertion of Space in Critical Social Theory*. London: Verso.

Soong Ngin Kwi, G. Haridas, Yeoh Choon Seng, and Tan Peng Hua. 1988. *Soil Erosion and Conservation in Peninsular Malaysia*. Kuala Lumpur: Rubber Research Institute, Malaysia.

Soulé, Michael E., and Gary Lease, eds. 1995. *Reinventing Nature? Responses to Postmodern Deconstruction*. Washington, D.C.: Island.

Spence, Mark David. 1999. *Dispossessing the Wilderness: Indian Removal and the Making of the National Parks*. New York: Oxford University Press.

Spencer, J. E. 1966. *Shifting Cultivation in Southeast Asia*. Berkeley: University of California Press.

Speth, James Gustave. 2004. *Red Sky at Morning: America and the Crisis of the Global Environment*. New Haven, Conn.: Yale University Press.

Sponsel, Leslie. 2001a. "Do Anthropologists Need Religion, and Vice-Versa? Adventures and Dangers in Spiritual Ecology." In *New Directions in Anthropology and the Environment*, ed. Carol L. Crumley, 177–200. Walnut Creek, Calif.: AltaMira.

———. 2001b. "Is Indigenous Spiritual Ecology a New Fad? Reflections from the Historical and Spiritual Ecology of Hawai'i." In *Indigenous Traditions and Ecology: The Interbeing of Cosmology and Community*, ed. John A. Grim, 159–74. Cambridge, Mass.: Harvard University Center for the Study of World Religions.

———. 2005. "Trees—Sacred (Thailand)." In *Encyclopedia of Religion and Nature*, ed. Bron Taylor II, 1661–63. New York: Continuum.

Standing, G., ed. 1985. *Labour Circulation and the Labour Process*. London: Croom Helm.

Stearman, A. M. 1994. " 'Only Slaves Climb Trees': Revisiting the Myth of the Ecologically Noble Savage in Amazonia." *Human Nature* 5 (4): 339–57.

Stevens, S. 1997. "The Legacy of Yellowstone." In *Conservation through Cultural Survival: Indigenous Peoples and Protected Areas*, ed. Stevens, 13–32. Washington, D.C.: Island.

Stevens, W. E. 1968. "The Rare Large Mammals of Malaya." *Malayan Nature Journal* 22:10–25.

———. 1973. "The Conservation of Wild Life in West Malaysia." Unpublished report. Office of the Chief Game Warden. Federal Game Department, Ministry of Land and Mines, Seremban, Malaysia.

Stone, Glenn D. 2007. "Agricultural Deskilling and the Spread of Genetically Modified Cotton in Warangal." *Current Anthropology* 48 (1): 67–103.

Stonich, Susan, and Peter Vandergeest. 2001. "Violence, Environment, and Industrial Shrimp Farming." In *Violent Environments*, ed. Nancy L. Peluso and Michael Watts, 261–86. Ithaca, N.Y.: Cornell University Press.

Stott, Philip. 1991. "Recent Trends in the Ecology and Management of the World's Savanna Formations." *Progress in Physical Geography* 15 (1): 18–28.

Stutterheim, W. F. 1956. "Some Remarks on Pre-Hinduistic Burial Customs on Java." In *Studies in Indonesian Archaeology*, ed. W. F. Sutterheim, 65–90. The Hague: Martinus Nijhoff.

Sugishima, Takashi. 1994. "Double Descent, Alliance, and Botanical Metaphors among the Lionese of Central Flores." *Bijdragen* 150 (1): 146–70.

Sulistyawati, E. 2001. "An Agent-Based Simulation of Land-Use in a Swidden Agricultural Landscape of the Kantu' in Kalimantan, Indonesia." PhD diss., Australian National University.

Sulistyawati, E., I. R. Noble, and M. L. Roderick. 2005. "A Simulation Model to Study Land Use Strategies in Swidden Agriculture Systems." *Agricultural Systems* 85 (3): 271–88.

Sundaram, Jomo Kwame. 1988. *A Question of Class: Capital, the State, and Uneven Development in Malaya*. New York: Monthly Review.

Sundberg, Juanita. 2006. "Conservation, Globalization, and Democratization: Exploring the Contradictions in the Maya Biosphere Reserve, Guatemala." In *Globalization and New Geographies of Conservation*, ed. Karl S. Zimmerer, 259–76. Chicago: University of Chicago Press.

Sunderlin, W., I. A. P. Resosudarmo, E. Rianto, and A. Angelsen. 2000. *The Effect of Indonesia's Economic Crisis on Small Farmers and Natural Forest Cover in the Outer Islands*. Bogor: Center for International Forestry Research.

Susianto, Agus, Didik Purwadi, and John Pontius. n.d. "Kaligondang Sub-district: A Case History of an IPM Sub-district." Unpublished manuscript, Jakarta.

Sutlive, Vinson H., Jr. 1978. *The Iban of Sarawak: Chronicle of a Vanishing World*. Arlington Heights, Ill.: AHM.

——. 1992. *Tun Jugah of Sarawak: Colonialism and Iban Response*. Kuala Lumpur: Sarawak Literary Society and Penerbit Fajar Bahkti.

Suwito. 1996. "Transkripsi Penyuluhan Camat Pesisir Utara Krui Rencana Pengukuran Lahan untuk Areal Kelapa Sawit-PT Panji Padma Lestari" (Transcription of Pesisir Utara Sub-district head clarification speech regarding the planned land measurement for palm plantation area of PT Panji Padma Lestari). Unpublished manuscript, Lembaga Alam Tropika Indonesia, Bogor.

——. 1998. "Hasil Investigasi masalah HPT dan HL di Pesisir Krui." Internal notes.

Suyanto, B. Hariyadi, S. Budiyati, and Jaime Quizon. 1994. "Insecticide Use in Rice Farming on Java: A Preliminary Study Comparing the Behavior of SLPHT and Non-SLPHT Farmers." Agriculture Group Working Paper No. 20. Jakarta: AGWP-CPIS.

Tadem, Eduardo. 1992. "The Political Economy of Mindanao: An Overview. In *Mindanao: Land of Unfulfilled Promise*, ed. Mark Turner, Ron May, and Lulu Turner, 7–30. Quezon City: New Day Publishers.

Takane, T. 2007. "Customary Land Tenure, Inheritance Rules and Smallholder Farmers in Malawi." Discussion Paper No. 104. Chiba: Institute of Developing Economies, JETRO.

"Taman Negara: Introduction." 1971. *Malayan Nature Journal* 24:113–14.

Tarling, Nicholas. 1978. *Piracy and Politics in the Malay World: A Study of British Imperialism in Nineteenth-Century South-East Asia*. Nendeln, Liechenstein: Kraus Reprint.

Taussig, Michael T. 1980. *The Devil and Commodity Fetishism in South America*. Chapel Hill: University of North Carolina Press.

Teh Tiong Sa and Tunku Shamsul Bahrin. 1992. "Environmental Impacts of Land Development in Jengka Triangle, Pahang, Peninsular Malaysia." In *The View from Within: Geographical Essays on Malaysia and Southeast Asia*, ed. Voon Phin Keong and Tunku Shamsul Bahrin, 84–117. Kuala Lumpur: Department of Geography, University of Malaya.

Terborgh, John. 1999. *Requiem for Nature*. Washington, D.C.: Island.

Thompson, E. P. 1975. *Whigs and Hunters: The Origins of the Black Act*. New York: Pantheon.

Thompson M., M. Warburton, and T. Hatley. 1986. *Uncertainty on a Himalayan Scale: An Institutional Theory of Environmental Perception and a Strategic Framework for the Sustainable Development of the Himalaya*. London: Ethnographica.

Thongchai Winichakul. 1994. *Siam Mapped: A History of the Geo-Body of a Nation*. Honolulu: University of Hawai'i Press.

Thongmak, Seri, and David L. Hulse. 1993. "The Winds of Change: Karen People in Harmony with World Heritage." In *The Law of the Mother: Protecting Indigenous Peoples in Protected Areas*, ed. Elizabeth Kemf, 161–68. San Francisco: Sierra Club Books.

Tim Bantuan Teknis FAO. 1997. "Kecamatan PHT: Sebuah gambaran masyarakat PHT." Unpublished manuscript, Jakarta.

Topik, S., and W. G. Clarence-Smith. 2003. "Conclusion." In *The Global Coffee Economy in Africa, Asia, and Latin America, 1500–1989*, ed. Clarence-Smith and Topik, 385–410. Cambridge: Cambridge University Press.

Torquebiau, E. 1984. "Man Made Dipterocarp Forest in Sumatra." *Agroforestry Systems* 2 (2): 103–27.

Tregonning, K. 1965. *A History of Modern Sabah*. Singapore: University of Malaya Press.

Tsing, Anna L. 1993. *In the Realm of the Diamond Queen: Marginality in an Out-of-the-Way Place*. Princeton, N.J.: Princeton University Press.

———. 1999. "Becoming a Tribal Elder, and Other Green Development Fantasies." In *Transforming the Indonesian Uplands: Marginality, Power, and Production*, ed. Tania M. Li, 159–202. London: Berg.

———. 2003. "Agrarian Allegory and Global Futures." In *Nature in the Global South: Environmental Projects in South and Southeast Asia*, ed. Paul R. Greenough and Tsing, 124–69. Durham, N.C.: Duke University Press.

Tucker, Mary Evelyn. 2003. *Worldly Wonder: Religions Enter Their Ecological Phase.* Chicago: Open Court.

Tucker, Mary Evelyn, and John A. Grim, eds. 1994. *Worldviews and Ecology: Religion, Philosophy, and the Environment.* Maryknoll, N.Y.: Orbis Books.

Tunku Shamsul, Bahrin, and P. D. A. Perera. 1977. FELDA: *Twenty-One Years of Land Development.* Kuala Lumpur: Federal Land Development Authority.

Turnbull, M. 1972. *The Straits Settlements, 1826–67: Indian Presidency to Crown Colony.* Singapore: Oxford University Press.

United Nations. 1982. *Report of the Working Group on Indigenous Populations on Its First Session.* Geneva: United Nations

——. 1983. *Report of the Working Group on Indigenous Populations on Its First Session.* Geneva: United Nations.

US-AEP (United States–Asia Environmental Partnership). 1989. Country Assessment: Malaysia.

van Beek, Walter E. A., and Pieteke M. Banga. 1992. "The Dogon and Their Trees." In *Bush Base, Forest Farm: Culture, Environment, and Development,* ed. Elizabeth Croll and David Parkin, 57–75. London: Routledge.

Vandergeest, Peter. 1996. "Mapping Nature: Territorialization of Forest Rights in Thailand." *Society and Natural Resources* 9:159–75.

Vandergeest, Peter, and Nancy L. Peluso. 1995. "Territorialization and State Power in Thailand." *Theory and Society* 24:385–426.

Vandermeer, John, and Ivette Perfecto. 2005. *Breakfast of Biodiversit : The Political Ecology of Rain Forest Destruction.* 2nd ed. Oakland, Calif.: Food First Books.

Vasan, Sudha. 2006. *Living with Diversity: Forestry Institutions in the Western Himalaya.* Shimla, India: Indian Institute for Advanced Study.

Vayda, A. P. 1961. "Expansion and Warfare among Swidden Agriculturalists." *American Anthropologist* 63:346–58.

——. 2009. *Explaining Human Actions and Environmental Changes.* Lanham, Md.: AltaMira.

Velásquez Runk, Julie. 2009. "Social and River Networks for the Trees: Wounaan's Riverine Rhizomatic Cosmos and Arboreal Conservation." *American Anthropologist* 111 (4): 456–67.

Veth, P. J. 1854. *Borneo's Westafdeeling: Statistisch, historisch, vorafgegaan doore eene algemeene schets des ganschen eilands.* Nomen en Zoon: Zaltbommel.

Vincent, Jeffery R., and Yusuf Hadi. 1991. "Deforestation and Agricultural Expansion in Peninsular Malaysia." Discussion Paper No. 396. Cambridge Mass.: Harvard Institute for International Development.

——. 1993. "Malaysia." In *Sustainable Agriculture and the Environment in the Humid Tropics,* ed. National Research Council, 440–82. Washington, D.C.: National Academy Press.

Vincent, Jeffery R., and Rozali M. Ali, with Yii T. Chang. 1997. *Environment and Development in a Resource-Rich Economy: Malaysia under the New Economic Policy.* Cambridge Mass.: Harvard Institute for International Development.

Visser, Leontine E. 1989. *My Rice Field Is My Child: Social and Territorial Aspects of Swidden Cultivation in Sahu, Eastern Indonesia*. Trans. Rita DeCoursey. Dordrecht: Foris Publications.

Vollenhoven, Cornelis van. 1918–33. *Het Adatrecht van Nederlandsch-Indië*. Leiden, Netherlands: E. J. Brill.

Voon Phin Keong. 1976a. "Malay Reservations and Malay Land Ownership in Semenyih and Ulu Semenih Mukim." *Modern Asian Studies* 10 (4): 509–23.

———. 1976b. *Western Rubber Planting Enterprise in Southeast Asia, 1876–1921*. Kuala Lumpur: Penerbit Universiti Malaya.

Vosti, S. A., and J. Wittcover. 1996. "Slash-and-Burn Agriculture: Household Perspectives." *Agriculture, Ecosystems, and Environment* 58:23–38.

Wadley, Reed L. 1997. "Circular Labor Migration and Subsistence Agriculture: A Case of the Iban in West Kalimantan, Indonesia." PhD diss., Arizona State University.

———. 1999. "The History of Population Displacement and Forced Resettlement in and Around Danau Sentarum Wildlife Reserve, West Kalimantan, Indonesia: Implications for Co-Management." Paper presented at the "Displacement, Forced Settlement, and Conservation" conference, Oxford University, September 9–11.

———. 2001. "Community Co-operatives, Illegal Logging, and Regional Autonomy: Empowerment and Impoverishment in the Borderlands of West Kalimantan, Indonesia." Paper presented at the meeting titled "Resource Tenure, Forest Management, and Conflict Resolution: Perspectives from Borneo and New Guinea," Australian National University, Canberra, April 9–11.

———. 2003. "Lines in the Forest: Internal Territorialization and Local Accommodation in West Kalimantan, Indonesia (1865–1979)." *South East Asia Research* 11:91–112.

Wadley, Reed L., and Carol J. Pierce Colfer. 2004. "Sacred Forest, Hunting, and Conservation in West Kalimantan, Indonesia." *Human Ecology* 32 (3): 313–38.

Wadley, Reed L., Carol J. Pierce Colfer, and Ian G. Hood. 1996. "The Role of Sacred Groves in Hunting and Conservation among the Iban of West Kalimantan, Indonesia." Paper presented at the 95th Annual Meeting of the American Anthropological Association, San Francisco, November 20–24.

———. 1997. "Hunting Primates and Managing Forests: The Case of Iban Forest Farmers in Indonesian Borneo." *Human Ecology* 25 (2): 243–71.

Walker, R., and Ako Homma. 1996. "Land Use and Land Cover Dynamics in the Brazilian Amazon: An Overview." *Ecological Economics* 18 (1): 67–80.

Wallington, Tabatha J., Richard J. Hobbs, and Susan A. Moore. 2005. "Implications of Current Ecological Thinking for Biodiversity Conservation: A Review of the Salient Issues." *Ecology and Society* 10 (1): 15–30.

Wardhani, M. A. 1992. "Developments in IPM: The Indonesian Case." In *Integrated Pest Management in the Asia-Pacific Region*, ed. Peter A. C. Ooi et al., 27–35 Wallingford, U.K.: CAB International.

Weber, Max. 1958. *From Max Weber: Essays in Sociology*. Ed. and trans. H. H. Gerth and C. Wright Mills. New York: Oxford University Press.

Wells, D. R. 1988. "Birds." In *Key Environments: Malaysia*, ed. Earl of Cranbrook, 167–95. Oxford: Pergamon.

——. 1999. *Birds of the Thai-Malay Peninsula*. London: Academic.

Wessing, Robert. 1986. *The Soul of Ambiguity: The Tiger in South Asia*. Dekalb: Northern Illinois University.

West Coast Resident. 1958. "Letter from the West Coast Resident to the Chief Secretary, April 28, 1958." Ranau Native Court Office.

West Coast Resident Chisholm. 1957. "Letter from West Coast Resident Chisholm to Other Residents, November 26, 1957." Ranau District Office Records.

West, Paige 2005. "Translation, Value, and Space: Theorizing an Ethnographic and Engaged Environmental Anthropology." *American Anthropologist* 107 (4): 632–42.

——. 2006. *Conservation Is Our Government Now: The Politics of Ecology in Papua New Guinea*. Durham, N.C.: Duke University Press.

Wharton, C. H. 1968. "Man, Fire, and Wild Cattle in Southeast Asia." *Annual Proceedings of the Tall Timbers Fire Ecology Conference* 8:107–67.

Whatmore, Sarah. 2001. *Hybrid Geographies: Natures, Cultures, Spaces*. London: Sage.

Whitten, Max, and William H. Settle. 1998. "The Role of the Small-Scale Farmer in Preserving the Link between Biodiversity and Sustainable Agriculture." In *Frontiers in Biology: The Challenges of Biodiversity, Biotechnology and Sustainable Agriculture*, ed. C. H. Chou and Kwang-Tsao Shao, 187–207. Taipei: Academia Sinica. Proceedings of 26th AGM, International Union of Biological Sciences, Taipei, Taiwan, November 17–23, 1997.

Whitten, Tony, Derek Holmes, and Kathy MacKinnon. 2001. "Conservation Biology: A Displacement Behavior for Academia?" *Conservation Biology* 15 (1): 1–3.

Wickham, T. 1997a. *Review of the Activities and Challenges of the Danau Sentarum Wildlife Reserve Conservation Project, West Kalimantan, Indonesia*. Bogor: Wetlands International-ODA.

——. 1997b. *Two Years of Community-Based Participation in Wetlands Conservation: A Review of the Activities and Challenges of the Danau Sentarum Wildlife Reserve Conservation Project, West Kalimantan, Indonesia*. Bogor: Wetlands International/PHPA/ODA.

Widjnarti, Enis H. 1996. *Checklist of Freshwater Fishes of Danau Sentarum Wildlife Reserve and Adjacent Areas, Kapuas Hulu, West Kalimantan, of Danau Sentarum Wildlife Reserve and the Kapuas Lakes Area, Kalimantan Barat, Indonesia*. Bogor: Wetlands International/PHPA/ODA.

Wikkramatileke, R. 1972. "The Jengka Triangle, West Malaysia: A Regional Development Project." *Geographical Review* 62 (4): 479–90.

Wilkie, D. S., and J. T. Finn. 1988. "A Spatial Model of Land Use and Forest Regeneration in the Ituri Forest of Northeastern Zaire." *Ecological Modelling* 41:307–23.

Wilkinson, R. J. 1959. *A Malay-English Dictionary*. 2 vols. London: Macmillan.

Williams, Raymond. 1980. *Problems in Materialism and Culture: Selected Essays*. London: Verso.

Wilmer, Franke. 1993. *The Indigenous Voice in World Politics: Since Time Immemorial*. London: Sage.

Wilmsen, Edwin N. 1989. *Land Filled with Flies: A Political Economy of the Kalahari*. Chicago: University of Chicago Press.

Winarto, Yunita Triwardani. 1995. "State Intervention and Farmer Creativity: Integrated Pest Management among Rice Farmers in Subang, West Java." *Agriculture and Human Values* 12 (4): 47–57.

———. 1996. "Farmers' Perspectives on Integrated Pest Management." *Agricultural Research and Extension Network* 34:16–20.

———. 1997a. "Knowledge in the Making: Learning to Sustain Yields in the Green Revolution Era." Paper presented at the International Conference on Creativity and Innovation at the Grassroots for Sustainable Natural Resource Management, Ahmedabad, January 11–14.

———. 1997b. "Maintaining Seed Diversity during the Green Revolution Era." *Knowledge and Development Monitor* 5 (4): 3–6.

———. 1998. "Hama dan musuh alami, obat dan racun: Dinamika pengetahuan petani padi dalam pengendalian hama." *Antropologi Indonesia* 22 (55): 53–68.

———. 2004a. "The Evolutionary Changes in Rice Crop Farming: Integrated Pest Management in Indonesia, Cambodia, and Vietnam." *Southeast Asian Studies Journal* 42 (3): 241–72.

———. 2004b. *Seeds of Knowledge: The Beginning of Integrated Pest Management in Java.* New Haven, Conn.: Yale University Southeast Asia Studies.

———. 2005a. "Empowering Farmers, Improving Techniques? The Integrated Pest Management in Cambodia and Thailand." Asia Research Institute Working Paper No. 54. http://www.ari.nus.edu.sg/pub/wps.htm.

———. 2005b. "Food Security and Food Sovereignty: Fight for Right in Discourse and Practice." Paper presented at Fourth International Symposium of *Antropologi Indonesia*, "Indonesia in the Changing Global Context: Building Cooperation and Partnership?," University of Indonesia, Depok, July 12–16.

———. 2006. "Pengendalian hama terpadu setelah lima belas tahun berlalu: Adakah perubahan dan kemandirian?" *Jurnal Analisis Sosial* 11 (1): 27–56.

Winarto, Yunita Triwardani, and Imam Ardhianto. 2007. "Becoming Plant Breeders, Rediscovering Local Varieties: The Creativity of Farmers in Indramayu, Indonesia." *The HoneyBee Magazine* 18 (April–June): 1–11, http://www.sristi.org/.

Winarto, Yunita Triwardani, and Ezra Muhamad Choesin. 1998. "IPM in Progress: Striving towards Empowerment and Prosperity." Unpublished manuscript, Jakarta.

Winarto, Yunita Triwardani, Maidi, and Darmowiyoto. 1999. "Pembangunan Pertanian: Pemasungan Kebebasan Petani." *Antropologi Indonesia* 23 (59): 66–79.

Winarto, Yunita Triwardani, Ezra Muhamad Choeson, Fadli, Anastasia Sri Handayani Ningsih, and Susanto Darmono. 2000. "Satu Dasa Warsa Pengendalian Hama Terpadu: Berjuang Menggapai Kemandirian dan Kesejahteraan." Research report. Jakarta: Indonesian Food and Agriculture Organization Inter Country Program.

Wolf, Eric R. 1982. *Europe and the People without History.* Berkeley: University of California Press.

Wolters, O. W. 1967. *Early Indonesian Commerce: A Study of the Origins of Srivijaya.* Ithaca, N.Y.: Cornell University Press.

Woodside, Alexander. 1976. *Community and Revolution in Modern Vietnam*. Boston: Houghton Mifflin.

Woolley, G. C. 1953a. *Dusun Adat*. Native Affairs Bulletin No. 4. Jesselton: North Borneo Government Printer.

———. 1953b. *Dusun Adat*. Native Affairs Bulletin No. 5. Jesselton: North Borneo Government Printer.

———. 1953c. *Kwijau Adat*. Native Affairs Bulletin No. 6. Jesselton: North Borneo Government Printer. Jesselton: North Borneo Government Printer.

———. 1953d. *Murut Adat*. Native Affairs Bulletin No. 3. Jesselton: North Borneo Government Printer.

———. 1953e. *Tuaran Adat*. Native Affairs Bulletin No. 2. Jesselton: North Borneo Government Printer.

———. 1962a. *Dusun Customs in Putatan District*. Native Affairs Bulletin No. 7. Jesselton: North Borneo Government Printer.

———. 1962b. *The Timoguns*. Native Affairs Bulletin No. 1. Jesselton: North Borneo Government Printer.

Worster, Donald. 1979. *Dust Bowl: The Southern Plains in the 1930s*. New York: Oxford University Press.

Wright, Angus, and Wendy Wolford. 2003. *To Inherit the Earth: The Landless Movement and the Struggle for a New Brazil*. Oakland, Calif.: Food First Books.

WWFM (World Wild Life Fund Malaysia). 1983. *Proposals for a Conservation Strategy for Trengganu*. Petaling Jeya: WWF Malaysia.

———. 1986. *The Development of Taman Negara*. Petaling Jeya: Malaysia: WWF Malaysia.

———.1991. *A Conservation Strategy for Negeri Kelantan Darulnaim, 1991*. 3 vols. Petaling Jeya: WWF Malaysia.

Wyatt-Smith, J. 1995. *Manuel of Malayan Silviculture for Inland Forest*. Kuala Lumpur: Forest Research Institute, Malaysia.

Yahya Bin Mohamad. 1990. Perkembangan Ugama Kristian di Sabah Setelah Meredeka: Sorotan pada Zaman-Zaman Kerajaan Part Perikatan USNo-SCA Kerajjan Pari Berjaya dan Kerajaan Part PBS. Universiti Kebangsaan Malaysia, Bangi.

Yapa, Lakshman. 1993. "What Are Improved Seeds? An Epistemology of the Green Revolution." *Economic Geography* 69 (3): 254–73.

Yong, Frank S. K. 1992. "Environmental Impact of Tourism on Taman Negara National Park, Malaysia." *Malayan Nature Journal* 45 (1–4): 579–91.

Zaharah Haji Mahmud. 1992. "The Traditional Malay Ecumene of the Peninsula: A Historical-Geographical Essay on the Position of Wet Rice." In *The View from Within: Geographical Essays on Malaysia and Southeast Asia*, ed. Voon Phin Keong and Shamsul Bahrin Tengku, 297–321. Kuala Lumpur: University of Malaya.

Zahid Emby. 1990. "The Orang Asli Regrouping Scheme: Converting Swiddeners to Commercial Farmers." In *Margins and Minorities: The Peripheral Areas and Peoples of Malaysia*, ed. Victor T. King and Michael J. G. Parnwell, 94–109. Hull, U.K.: Hull University Press.

Zakaria, Yando. 1999. "Kembalikan Kedaulatan Ulayat Masyarakat *Adat*." *Wacana Jurnal Ilmu Sosial Transformatif* 2:124–50.

Zerner, Charles. 1994a. "Through a Green Lens: The Construction of Customary Environmental Law and Community in Indonesia's Maluku Island." *Law and Society Review* 28 (5): 1079–1123.

———. 1994b. "Tracking Sasi: The Transformation of a Central Moluccan Reef Management Institution in Indonesia." In *Collaborative and Community-Based Management of Coral Reefs: Lessons from Experience*, ed. A. T White et al., 19–32. West Hartford, Conn.: Kumarian.

Zimmerer, Karl S. 2000. "The Reworking of Conservation Geographies: Nonequilibrium Landscapes and Nature-Society Hybrids." *Annals of the American Association of Geographers* 90 (2): 356–69.

Zimmerer, Karl S., and Thomas J. Bassett, eds. 2003. *Political Ecology: An Integrative Approach to Geography and Environment-Development Studies.* New York: Guilford.

Zimmerer, Karl S., and K. Young, eds. 1998. *Nature Geography: New Lessons for Conservation in Developing Countries.* Madison: University of Wisconsin Press.

CONTRIBUTORS

UPIK DJALINS is currently working on her doctoral degree at Cornell University in the field of development sociology. Her research interrogates state formation in late colonial Indonesia by focusing on how knowledge of land rights was mutually constructed by both Indonesian and European scholars. Her interdisciplinary approach combines sociology, anthropology, history, and postcolonial studies, and her theoretical grounding is informed by the tensions between poststructuralist and historical materialist theories. She received her master of environmental sciences degree from the Yale School of Forestry and Environmental Studies.

AMITY A. DOOLITTLE received her PhD from the Yale School of Forestry and Environmental Studies. She currently is a lecturer and associate research scientist at the Yale School of Forestry and Environmental Studies. Her work uses an interdisciplinary approach combining perspectives from anthropology, political science, environmental history, and political ecology to explore property relations and conflicts over resources use. She is the author of *Property and Politics in Sabah, Malaysia (North Borneo): A Century of Native Struggles over Land Rights, 1881–1996* (2005).

MICHAEL R. DOVE earned his PhD in anthropology at Stanford University. He is the Margaret K. Musser Professor of Social Ecology and the director of the Tropical Resources Institute in the School of Forestry and Environmental Studies, a professor of anthropology, and the curator of anthropology in the Peabody Museum of Natural History at Yale University. His research focuses on the environmental relations of local communities in late-developing countries, especially in South and Southeast Asia. His most recent books are *Conserving Nature in Culture: Case Studies from Southeast Asia* (coedited with Percy E. Sajise and Amity A. Doolittle, 2005), *Environmental Anthropology: A Historical Reader* (coedited with Carol Carpenter, 2007), *Southeast Asian Grasslands: Understanding a Folk Landscape* (2008), and *The Banana Tree at the Gate: The History of Marginal Peoples and Global Markets in Borneo* (2011).

LEVITA DUHAYLUNGSOD, at the time the essay for his volume was written, was an associate professor in the Department of Agricultural Education and Rural Studies and the School of Environmental Science and Management, University of the Philippines, Los Baños. While doing periodic fieldwork as an anthropologist among the T'boli from 1999 to 2001, she linked up with local and national NGOs doing advocacy work for indigenous peoples in the Philippines. Her fieldwork on the T'boli resulted in a coauthored book and a number of popular and scholarly articles. She is currently a professorial lecturer in development studies at De La Salle University, Manila, and continues to visit T'boli communities.

EMILY E. HARWELL is an independent consultant with nearly two decades of experience researching the anthropology and political economy of natural-resource property and conflict. She earned her PhD in environmental anthropology at the Yale University School of Forestry and Environmental Studies and her bachelor of science in ecology from the University of the South. Her current research interests are the linkage between natural resources and violence, ethnic identity, and social networks in resource use, as well as postconflict peace-building efforts including resource management reform and the reintegration of ex-combatants. Some of her recent work includes "Whose Resources? Whose Common Good? Towards a New Paradigm of Environmental Justice and the National Interest in Indonesia" (coauthored with Owen Lynch and ELSAM, 2002).

JEYAMALAR KATHIRITHAMBY-WELLS received her PhD from the School of Oriental and African Studies, London University, and formerly held the chair of Asian history at the University of Malaya, Kuala Lumpur. She has in recent years expanded her earlier research interest in the Southeast Asian state and economy to include environmental and conservation history. Apart from publishing numerous articles and monographs, she contributed to the *Cambridge History of Southeast Asia* (1992), edited with John Villiers *The Southeast Asian Port and Polity* (1990), and more recently authored *Nature and Nation: Forests and Development in Peninsular Malaysia* (2005). She currently researches and teaches in Cambridge, U.K.

PERCY E. SAJISE is an Honorary Research Fellow of Bioversity International, an adjunct fellow of the Southeast Asian Regional Center for Graduate Study and Research in Agriculture (SEARCA) and an adjunct professor at the School of Environmental Science and Management (SESAM) at the University of the Philippines, Los Baños. He stepped down recently after serving for eight years as the regional director for Asia, Pacific, and Oceania of Biodiversity International. He has also worked for the Southeast Asia Ministers of Education Organization (SEAMEO) Regional Center for Graduate Study and Research in Agriculture (SEARCA) and spent fifteen years on the faculty of the University of the Philippines, Los Baños. A plant ecologist by training, he obtained his PhD at Cornell University. He became a member of the World Academy of Science and Arts in 2001.

ENDAH SULISTYAWATI earned her PhD from the Research School of Biological Sciences at the Australian National University, Canberra. She works as lecturer in the

School of Life Sciences and Technology, Institut Teknologi Bandung, in Bandung, Indonesia. She is currently in charge of coordinating the master's program in bio-management. She has research interest in plant ecology, landscape ecology, and human ecology. Her current research projects include the ecology of Mount Papandayan, carbon stock assessment in various ecosystem types, and development of remote sensing-based models for carbon stock estimation.

LYE TUCK-PO earned her PhD in anthropology at the University of Hawai'i, Manoa. She is currently a lecturer in the School of Social Sciences, Universiti Sains Malaysia, Penang, Malaysia. Previous work includes conservation consultancy, climate change activism, and various research fellowships in Japan and Australia. Her primary research focuses on the environmental knowledge and relations of local communities, especially in Southeast Asia. She has conducted fieldwork with Batek hunter-gatherers in Malaysia and Kuay and Khmer farmers and traders in Cambodia, with an emphasis on local perceptions of the forests and hydrological systems, respectively. She has most recently conducted fieldwork on dam-related resettlement for Penan communities in Sarawak. She is the author of "Changing Pathways: Forest Degradation and the Batek of Pahang, Malaysia" (2004) and has coedited (with W. de Jong and K. Abe) volumes on the political ecology of tropical forests (2003) and the social ecology of tropical forest migration (2005).

YUNITA T. WINARTO is a professor in the Department of Anthropology at Universitas Indonesia, where she currently holds the position of Academy Professor in Social Sciences and Humanities of the Academy Professorship Indonesia program. She is pursuing her fieldwork among Integrated Pest Management farmers in several locations in Java, focusing on farmer science and empowerment and the dialectics between scientific and local knowledge. She has published numerous articles as well as a book titled *Seeds of Knowledge: The Beginning of Integrated Pest Management in Java* (2004) and is now working on farmer responses to climate change.

Page numbers in italics indicate figures.

animism, 96

anthropology, 2, 6, 7, 107–8, 119 n. 52

Asia Indigenous Peoples' Pact, 220

bagi dua (split-in-two) system, 100, 115 nn. 22–23

bananas, 163–64, 228

Banjar kingdom, 107

Bantam, Sultanate of, 131

Barlow, H. S., 39, 41

Bartlett, Harley Harris, 9, 10

Bataks, 65, 102

Bateks, 12, 19, 37, 55–57, 59 n. 2; as "boundary" people, 55; dehumanized by Malaysian state, 18; founding of Taman Negara National Park and, 42, 43; government agencies and, 57–58; hunting-gathering way of life, 54, 56; Malay perceptions of, 51–53; "natural man" discourse and, 21; nature-culture dichotomy and, 22; as "Pangan Tanah," 40, 41; premodern image of, 16; role in maintaining forest, 46; wilderness ideal and, 24; wildlife and, 51. *See also* Orang Asli ("Sakai" people)

Bauer, P. T., 68, 70, 72

bee trees. *See* honey (bee) trees

belian tree, 192

Benjamin, Patricia A., 5

benzoin trees, 102

Berkes, Fikret, 216, 221

Berry, Sara, 19

biodiversity, 2, 5, 71; conservation of, 85; *damar* forest gardens and, 130, 138; decline in, 126; forests as reservoirs of, 232, 239; loss of, 80, 84–85, 88; parks' failure to protect, 38; sacred forests and, 7; in South Sumatra, 21

birds, 86, 87

Bloch, Maurice, 98

Boeck, Filip de, 106

Borneo, 91, 98, 107, 265; foreign trade history of, 213 n. 19; property relations in, 162; rain forest of, 96; rubber cultivation in, 94, 100, 111; swidden cultivation in, 99. *See also* Kalimantan (Indonesian Borneo); North Borneo; Sabah; Sarawak

Boschwezen (forested area), 133, 134, 139, 140

boundaries, 29, 180, 181; of community, 164, 168; decentralization of forest control and, 199–201; histories of, 192–96; *kecamatan* (subdistrict), 194–95; local practices in resource boundaries, 201–8; in Malay theories of personhood, 51–52; mapping technologies and, 210; of parks and reserves, 42, 60 n. 12, 124, 148 n. 14, 196–97; sovereignty over resources within, 183; state control and spatial boundaries, 160, 183; between villages and forested lands, 132, 134, 139. *See also* mapping; territory/territoriality

Bouquet, Mary, 108

Bourdieu, Pierre, 108, 119 n. 52

Brazil, 88, 127, 219, 274

British colonial regime, 94, 192

Brooke, Charles, 194

Brosius, J. Peter, 57

brown plant hopper (BPH), 279–80, 288, 292–93

Bumiputera (Sons of the Soil) policy, 76

Burkill, H. M., 39, 41

Burma, 104, 116 n. 32, 118 n. 48

Buttel, Frederick H., 8

Cameroon, 88

camphor (*Dryobalanops aromatica*), 62, 109

capitalism, 26, 75, 180–81, 191. *See also* cash crops; commodity production

cash crops, 25, 87; commodity-price fluctuations and, 268, 271–72; export crop production, 17; indigenous production of, 88; Malays and, 62, 63;

swidden systems complemented by, 102–4. *See also specific crops*

cassava (*Manihot esculenta*), 62, 242

Catholic Church, 97

cattle ranching, 227

cengal (*Balanocarpus heimii*), 71

Cernea, Michael, 235

Chapin, M., 5

Chayanov, A. V., 11, 247

China, 34 n. 10, 213 n. 19

Chinese, in Southeast Asia, 63, 64, 68, 94, 102

Chipko Indians, 7, 219

Christianity, 97, 191, 225–26

citizenship, 188, 190, 191, 223

civil society, 221, 300

class differences, 144, 147

Clay, Jason, 114 n. 13

Clements, F. E., 3

Clifford, Hugh, 65

climate change (global warming), 111, 126, 297–99

cloves (*Eugenia caryophyllus*), 138

cocoa, 80, 87

coconut, 80, 102, 229

coffee, 62, 64, 66, 102; *damar* forest gardens and, 138; forest cleared for, 88; Krui community and, 130

Colchester, Marcus, 220

Colfer, Carol J. Pierce, 203

Colombo Plan, 85

colonialism, 14, 53; forest peoples seen as noble savages, 55; indigenous peoples' relations with, 151–52; internal, 217, 225; marginalization of indigenous peoples and, 223; native property rights and, 152; plantation monoculture and, 63; postcolonial continuance of legacy of, 25; slavery and, 60 n. 15

Comite Tani Lampung (CTL), 132

commodity production, 16, 26, 75, 79, 233; oil-palm plantations and, 83; smallholders and, 93; subsistence

production and, 94. *See also* capitalism; cash crops; economy, global

communists, 196

communities, local, 123–24, 181; colonial state and, 152; differentiation between, 29; discourses of rights of, 27–30; diversity within households, 12; environmental degradation blamed on, 2; knowledge of ecosystem, 155; mapping of boundaries and, 211; natural resource management and, 10; state relations with, 21. See also *adat* (customary law); *desa* (administrative village); longhouses

computer simulation modeling, 31, 239–44, 273–74; cash sufficiency and, 264–65; commodity-price fluctuations incorporated into, 268–70; components of, 244–45; flexible agricultural strategy, 265–68; land access and fallowing practices, 257–61; land-use decision-making module, 247–52; lessons from simulations, 271–72; limitations of, 272–73; population dynamics module, 245–47; production evaluation module, 252; rice sufficiency, 261–63; simulation scenarios, 253; swidden system dynamics analyzed, 254–57; vegetation dynamic module, 252

Conklin, Beth A., 147 n. 3

Consensual Forest Land Use Agreement, 134

conservation, 1–7, 23, 40, 127, 216; backlash against, 21–22, 34 n. 17; commerce in conflict with, 44; communities in, 123; *damar* forest gardens and, 138; international claims on resources and, 196–99; "primitive environmentalism" and, 13; religion-driven, 8; in Sabah, 176; social justice and, 155; tourism and, 48–49; Western traditions of, 10

"Conservation in Malaysia" (Nor and Ng), 62

Consortium of Governments in Agricultural Research (CGIAR), 134

Convention on Biological Diversity (CBD), 221, 222

Coombes, Brad, 6

Coordinating Body of Indigenous Organizations of the Amazon Basin (COICA), 123

cosmologies, traditional, 9, 97–99, 107, 110

cover crops, 69, 82

Cramb, R. A., 247, 268

creation and destruction, cosmic cycle of, 97–102

crisis, discourse of, 4, 5

Cronon, William, 8–9, 44

Cultural Survival, 220

damar (*Shorea javanica*) forest gardens, 12–13, 22, 124, 130, 147, 148 n. 8; *adat* argument and, 27, 125, 145; under Dutch colonial regime, 133; history of cultivation of, 131, 148 n. 11; Indonesian government policy toward, 134; Kalpataru award and, 138–40; ministerial decree and, 140–43; villagers' conception of *adat* and, 135–38

Danau Sentarum Wildlife Reserve (DSWR), 29, 181, 183, 184; changes in reserve boundaries, 196–97; fishing practices in, 205, 206; location in West Kalimantan, *182*

Dayak groups, 11, 14, 90 n. 6; cosmology of, 97; Dusun language/ethnicity, 156, 158; Embaloh, 192; headhunting symbolism of, 118 n. 51; Islamicized, 113 n. 6; mast fruiting of trees and, 96; people–tree metaphors and, 107, 109–10; rubber cultivation and, 94, 99, 101, 111, 119 n. 57; in Second World War, 214 n. 26; smallholder rubber producers, 91; territorial boundaries of, 194; *tuba* fish poison used by, 206. *See also* Iban ethnic group; Kantu' ethnic group

"dead land" (*tanah mati*), 26, 91–94, 101, 103, 112

death, 92, 98; creation-destruction cycle and, 97, 114 n. 16; headhunting and, 118 n. 51; in household demographics, 239, 244, 245; life-death cycle, 92, 101, 107, 114 n. 18; Malay and Ibanic terms for, 101

deforestation, 7, 8, 64, 83, 108, 126

Deleuze, Gilles, 108

Department of Wildlife and National Parks [DWNP] (Malaysia), 37, 39, 41, 46–49, 54, 55, 58 n. 2

desa (administrative village), 29, 129, 133, 150 n. 28, 193; *kecamatan* boundaries and, 195; lottery of fishing spots and, 205

development, economic, 21, 88; national discourses of, 165; social justice and, 220; sustainable, 218, 220–22, 233, 234; Western-style, 2, 7

Directory of Food Crop Protection [DFCP] (Indonesia), 295–96

discourse, 10–11, 19–21

Djalins, Upik, 12–13, 14, 15, 18, 21; on competing resource-related discourses, 27–28; on local resource knowledge, 19–20; on nature–culture dichotomy, 22

Dodge, Nicholas, 60 n. 15

Dogon people (Mali), 113 n. 10, 114 n. 18

Doolittle, Amity, 12, 15, 16, 18; on central planning discourse, 21; on competing resource-related discourses, 27, 28–29

Douglas, Mary, 50

Dove, Michael R., 11, 13, 14, 16, 248, 251; on condemnation of swidden cultivation, 241; on evolving sustainability, 16; historical approach to resource-management regimes, 24; on Kantu' households, 243; on land-killing rubber, 26; on land rights in Indonesia, 128; on rice sufficiency, 261; on smallholder practices/values, 19

farmers (*cont.*)

and, 75, 276–77; Iban, 186; Integrated Pest Management (IPM) and, 32–33, 34 n. 15, 277, 281, 286–92; macroecology and, 292–94; Malay, 44, 50; national policy of Indonesian state and, 298–300; native title and, 170, 171; state "intensification" schemes and, 296–98; swidden system and, 99, 241, 242, 251, 268, 272. *See also* peasants; smallholders

Farmers' Association (Persuatuan Masyarakat Petani Repong Damar), 140, 143–44, 146, 150 n. 28

Federated Malay States (FMS), 67, 71, 72, 73. *See also* Malaya, British

fertilizers, 280, 282, 286, 287, 297, 301 n. 11

fishing, 12, 114 n. 13, 185, 189–90, 272; *jermal* nets, 189, 205–6, 207; lottery of fishing spots, 204–5; mapping of boundaries for, 200; poisons used in, 206–8; restrictions on, 198; rivers as common waterways and, 201; territorial claims and, 194

floods, 80, 184

Food and Agricultural Organization, 280

Ford Foundation, 134–35, 139, 149 n. 18

forest products, collection of, 179 n. 27, 203; ancestors and, 182; Krui community and, 130; land tenure and, 152; scarcity of resources, 190; "traditional" life and, 47

forests: Batek role in maintaining, 46; biodiversity of, 84, 232, 239; colonial views of, 45; decentralization of forest control, 199–201; differing social views of, 109; dipterocarp, 71, 82, 96, 130, 229; forest gardens, 124, 129, 130; *hutan simpan* (reserved forest), 161–62; Malay perceptions of, 50–53, 54; mast fruiting of trees, 96, 97–98; percentage of natural forest cover, 86; periodic disturbance in, 98; plantation rubber

and attrition of, 71–72; resource politics and, 217; sacred forests, 6–10, 62; state forests, 141, 195, 211; Western world in relation to, 108; wilderness ideal and, 44–46. *See also* primary forests; secondary forests; silviculture

Foucault, Michel, 93, 145. *See also* governmentality

Fox, James, 109

Francis, A., 160

frankincense (*Boswellia* spp.), 109

Frazer, Sir James George, 7

Freeman, D., 118 n. 49

Frossard, David, 285

fruit trees, 56, 68, 87, 106; death of, 116 n. 26; Govuton villagers and, 156; land tenure and, 152; native rights to, 159; people–tree metaphors and, 107. *See also* orchards

gaharu or aloes wood (*Aquilaria malaccensis*), 62

gambier (*Uncaria gambir* Roxb.), 63, 64, 66

game laws, 39, 41

Gema Palagung 2001 program, 296

gender, 2, 112, 219, 234, 284

General Association of the Rubber Planters of East Sumatra, 70

getah rambung, 65

Giddens, Anthony, 4

globalization, 93

global warming (climate change), 111, 126, 297–98

Golden Bough, The (Frazer), 7

governmentality, 20, 93, 94, 103

Graham, Laura R., 147 n. 3

Great Depression, 70

green revolution, 13, 16, 32–33, 34 n. 15; chemical companies and, 292; crop intensification schemes and, 296; discipline and surveillance of farmers and, 20; disease outbreaks and, 281; as "era of stupidity," 277; human capability neglected by, 298; IPM programs contrasted with, 288; "modern varieties" (MVS) of crops and, 276; rice productivity and, 276–77; smallholders and, 75. *See also* insecticides; "miracle seeds"

Grueso, Libia, 220

Guattari, Félix, 108

gum benjamin (*Styrax* spp.), 109

gutta-percha (*Palaquium* and *Payena* spp), 62, 65, 89 n. 1, 117 n. 36

Guyot, Dorothy, 81

habitat, 37, 55, 180; diversity of, 41; of endangered species, 197; forest, 84; Iban protected forest and, 203; loss of, 57, 85; protection of, 196

Haeckel, Ernst, 107

Hantu Hutan (Spirits of the Forest), 51

Harwell, Emily, 12, 14–15, 18, 21; on competing resource-related discourses, 27, 29–30; on dispute management, 18; on Iban and mapping efforts, 20; on tenurial regimes, 17

Hatley, T., 3

Hawkins, Gerald, 43

herbicides, 83

Hevea brasiliensis (Para rubber tree), 84, 94, 117 n. 36, 119 n. 57; "dead land" (*tanah mati*) and, 91, 112; introduction of, into Malaysia, 65; rice swiddens and, 11. *See also* rubber cultivation

Higaonon ethnic group (Philippines), 225

Hill, Stephanie, 6

Hislop, J. A., 56

Holmes, Derek, 23

Homma, Ako, 274

honey, 186, 187, 190

honey (bee) trees (*tapang*), 14, 97, 194, 202–3, 209, 214 n. 30

Hood, Ian G., 203

horticulture, 108

households, 31, 115 n. 19, 153, 181; *adat* community and, 143; as agents of land-use decisions, 239, 247–52; agricultural strategy and, 265–66; cash sufficiency and, 264–65; communal orchards and, 203; *damar* forest gardens and, 138; demography of, 239–40, 242–43, 254, 262, 271, 274; distribution of landholdings, 257–59, 271; diversification of resource use, 11, 12; domestic cycle, 11; fallow land and, 128, 259–61; in Govuton village, 156; interhousehold exchange, 99, 100; labor needs of, 272, 273–74; partitions of, 101, 243, 245, 247; rice sufficiency and, 97, 114 n. 14, 269, 270, 273; smallholder rubber cultivation and, 70, 93, 101, 107, 116 n. 27; social networking and, 188; swidden simulation model and, 252, 253

Hubback, T. R. (Theodore), 39–44, 46, 48, 59 n. 9, 60 n. 11

human rights, 2, 220

human sacrifice, 96, 110

hunting-gathering, 7, 37, 272; bureaucratic agencies and, 47–48; community access to land for, 175; forest viewed by, 109; Malay view of, 54; sedentarization of, 61 n. 23; wilderness ideal and, 56

hutan simpan (reserved forest), 161–62, *163*, 171

hybrid seeds, 296, 297

Iban ethnic group (Kalimantan), 12, 118 n. 49, 184, 196, 213 n. 12, 247; Dutch colonial policies toward, 15, 192, 194; fishing practices, 207, 213 n. 14; household demography, 240; mapping efforts and, 20, 209; rubber-price fluctuations and, 268; social networks and territorial resource claims of, 186–91;

Iban ethnic group (Kalimantan) (*cont.*)
spiritual significance of forest for, 203;
state relations with, 29; timber logging
concessions and, 199–200, 214 nn. 25–
26. *See also* Dayak groups
Ibanic speakers, 94, 101
identities, 181, 182, 201; cultural, 176, 225,
226; indigeneity and, 217; indigenous
peoples and politics of identity, 218,
219, 236; marginalization as historic
process and, 225; relational, 229; territory and local identity, 183
illipe nut (*engkabang* [*Shorea* spp.]), 97
Imperata cylindrica (L.) Beauv. (sword
grass), 64
incest taboos, 244, 246
India, 10, 213 n. 19
Indian Council of South America, 220
India rubber (*Ficus elastica*), 65, 89 n. 4
indigeneity, 2, 6, 23, 30, 146, 218–19; as
chameleon concept, 217; as collective
political identity, 219; *damar* forest
gardens and, 138; historical construction of, 15; politics of identity
and, 236; romantic views of past
golden age and, 216; strategic use of
discourse of, 21; territorial decolonization and, 220
indigenous peoples, 2, 123–24, 217, 235;
assimilated into conservation areas,
45; colonial state's relations with, 151;
culturally marginalized in the Philippines, 222–26; excluded from protected areas, 43; movements of, 219–
20; politics of identity and, 218, 219,
236; romanticized view of, 216; in social and political discourses, 218–19;
sustainable resource use and, 220–22;
transnational symbolic politics and,
126–27. *See also* aborigines; *specific
ethnic groups*
Indigenous Peoples Biodiversity Network (IPBN), 222

Indigenous Peoples Rights Act (IPRA,
Philippines), 224
Indochina, 110, 112
Indonesia, 11, 12, 16, 20; *adat* (customary
law) in, 124; Basic Agrarian Law, 126,
128; Bukit Barisan Selatan National
Park, 124, 129, 133, 140, 148 n. 13; Confrontation with Malaysia (1964), 196;
democracy in, 211; discourse on *adat*
community, 127–29; economic crisis
in, 296, 298; ethnic groups, 125, 127;
Integrated Pest Management (IPM) in,
278, 279–80; ministerial decree on
damar forest gardens, 140–43; National Agrarian Agency [Badan Pertanahan Nasional] (BPN), 141; New
Order regime, 129, 133, 191, 211; outer
islands, 99, 135, 240; plantation agriculture in, 63; rice cultivation, 17;
rubber cultivation by smallholders in,
90 n. 13, 112, 113 n. 6. *See also* Java/
Javanese; Kalimantan (Indonesian
Borneo); Sumatra
inheritance, 152, 163, 167, 168, 201, 258;
ancestors and, 182; household demography and, 240, 247, 257
insecticides, 83, 87, 301 n. 4. *See also*
green revolution
insects/plant pests, 277, 280, 281; brown
plant hopper (BPH), 279–80, 288, 292–
93; soybean seed borer, 286, 290, 293;
white rice borer (WRB), 283, 284, 287–
88, 290, 292, 296
Integrated Conservation and Development Programs (ICDP), 123
Integrated Pest Management (IPM), 16,
18, 277, 279–80; decentralization of
agricultural management and, 32; empowerment of farmers and, 281–82;
extension system and, 294–96; farming culture transformed by, 286–92;
freedom and dignity regained
through, 282–86; green revolution

and, 32–33, 34 n. 15, 288; Indonesian villages where applied, *278*, 279; in national food production policy of Indonesia, 296–300; peasant agency and, 29
Integrated Social Forestry (ISF) project (Philippines), 232, 234, 236 n. 3
International Center for Research in Agroforestry (ICRAF), 134, 140–41
International Monetary Fund (IMF), 199
International Rubber Restriction Scheme (IRRA), 71
International Working Group on Indigenous Affairs (IWGIA), 220
ironwood tree (*Eusideroxylon zwageri*), 98, 114 n. 17, 192
Islam and Muslims, 112, 191, 192, 198, 225, 229

Jabatan Hal-Ehwal Orang Asli (JHEOA, Malaysia), 47–48, 54, 55, 57–58
Jakun ethnic group, 43
Java/Javanese, 8, 65, 102; British interregnum, 131–32; coffee cultivation, 88; green revolution in, 32; Integrated Pest Management (IPM) in, *278*, 291; wet-rice cultivation in, 104. *See also* West Java
jelutung (*Dyera costulata*), 65, 117 n. 36
Jengka Triangle scheme (Malaysia), *77*, 81–82
jingah tree, 107
Jorgenson, 114 n. 12
"jungle rubber," 105, 116 n. 35

Kalimantan (Indonesian Borneo), 105, 108, 162, 164–65, 192. *See also* West Kalimantan
Kalpataru award, 138–40, 143, 147
Kantu' ethnic group, 15, 26, 247; "dead" rubber lands of, 91–94, 101, 110, 112 n. 2; environmental knowledge of, 243–44; household demography, 240; land accumulation of households, 257;

land rights among, 128; people–tree metaphors and, 106; pepper cultivation by, 273; rubber cultivation by, 94–95, 111–12; sexual prohibitions, 96, 100, 107; social structure of, 242–43; swidden cultivation by, 241, 242, 272, 273; swidden cycle and, 99; West Kalimantan territory of, *92*. *See also* Dayak groups; households; longhouses
Kapuas Lakes region (West Kalimantan), 15, 17, 22, 184, 196–99, 212 n. 3
Kathirithamby-Wells, Jeyamalar, 11, 13, 14, 15, 91; on central planning discourse, 21; on FELDA, 16–17; historical approach to resource-management regimes, 24, 25–26; on historical rise of estate sector, 19; on nature-culture dichotomy, 22
Kayapo Indians (Brazil), 127, 219
Kedah state (Malaysia), 72
Kelantan state (Malaysia), 74, 82–83
Keller, Eva, 92
kenagarian system, 133
Kerau Reserve (Malaysia), 85
King George V National Park. *See* Taman Negara National Park (Malaysia)
Kingsbury, Benedict, 219
Kingston, Jeffrey B., 132
kinship, 119 n. 52, 181, 193, 208; *adat* and, 125, 129; rights of access to land and, 164, 165; village control of territory and, 199
Kirsch, Stuart, 6
knowledge, 281, 299; *adat* community and, 146; circulation of knowledge among farmers, 290–91, 296; farmers as inheritors of, 145; Integrated Pest Management (IPM) and, 286–88; of pest insects' life cycle, 287–88; scientific and local knowledge linked, 280; state systems of, 13; Western, 285
knowledge, indigenous, 2, 10, 19–21, 216, 222, 285; hunting techniques, 45;

knowledge (*cont.*)
    landscape transformation and, 241;
    reconstructed and represented, 30–33;
    rubber cultivation and, 69
Krui region (South Sumatra), 12, 15, 27,
    28, 124, 125; *adat* argument and, 145;
    damar forests of, 22; Kalpataru award
    and, 138–40; making of community
    in, 129–31; on map of Sumatra, *130*;
    ministerial decree and, 141; revival of
    adat community and, 143
Kuala Yong settlement (Malaysia), 48

labor migration, 67, 70, 272
Lampung Province (Indonesia), 129, 131,
    135, 139, 144, 149 n. 18; bio-pesticides
    used in, 294; Integrated Pest Manage-
    ment (IPM) in, *278*, 284, 290; location
    on map, *130*. *See also* Krui region
    (South Sumatra)
Land Capability Council Report (1970,
    Malaysia), 82
Land Conservation Act (1960, Malaysia),
    82
land development schemes, 75–85, *77*, 89
Landless Workers' Movement, 219
landscapes, 33, 45; household dynamics
    and, 274; land-use decisions and, 239;
    macroecology and, 292–94; mapping
    and, 208; natural and political, 9;
    struggles over terminology and, 217;
    swidden cultivation model and, 245,
    246; transformation of, 8, 88, 241, 273
LATIN (Indonesian NGO), 135, 141, 144, 149
    n. 18, 150 n. 30
Latour, Bruno, 23
law, colonial/state, 28, 151; differences
    with customary law, 161, *162*; local
    control of community land and, 174;
    native titles to land, 170; rhetoric of
    "protecting" natives and, 160
law, customary. See *adat* (customary
    law)

Lenoir, Timothy, 23
Loans-to-Planters-Scheme (Malaysia), 65
logging, 81, 83, 86, 199; blockades on log-
    ging roads, 7; honey trees and, 202; in
    the Philippines, 217, 227; selective, 85;
    swidden cultivators in opposition to,
    105. *See also* timber
longhouses, 97, 114 n. 17, 156; communal
    orchards and, 203; "dead" rubber
    lands and, 101; rights of avail to fallow
    land, 128; social networking and, 188;
    tenurial regimes and, 116 n. 27; Tikul
    Batu, 242, 253, 275 n. 2
Lowe, Celia, 6
Lumad-Mindanao (coalition of indige-
    nous organizations, Philippines), 225–
    26
Lumadnong Alyansa Alang sa Demok-
    rasya (Indigenous Alliance for
    Democracy, Philippines), 225
Lye, Tuck-Po, 12, 14, 16, 19; on Bateks and
    Malaysian state, 18; on central plan-
    ning discourse, 21; historical approach
    to resource-management regimes, 24–
    25; on wilderness ideal in Malaysian
    park policy, 24

Mackie, C., 251
MacKinnon, Kathy, 23
Madagascar, 92, 98, 106
maize, 62, 87, 242
Malaya, British, 39–40, 65, 73, 76, 90 n. 9.
    See also Federated Malay States (FMS);
    native reserves (colonial Malaya);
    North Borneo
Malay language, 101
Malay Reservations Enactment (1913), 66
Malays, 12, 55; Batek autonomy and, 56;
    cash crops cultivated by, 62–63; cash
    economy and, 66–67; in Danau Sen-
    tarum Wildlife Reserve (DSWR), 184;
    Dutch colonial policies toward, 192,
    194, 212 n. 9; fishing practices, 207;

forests and forest people in perceptions of, 38, 50–53; founding of Taman Negara National Park and, 42, 44; honey harvest and, 186, 187, 190, 195, 202; Kantu' rubber cultivation and, 94; nationalist movement, 75–76; reservation land of, 71; river-valley axis of settlement, 73; rubber cultivation and, 65, 90 n. 6; "Sakai" aboriginals and, 40, 60 n. 15; shifting cultivation among, 113 n. 6; social networks and territorial resource claims of, 186–91; state relations with, 29; sultanates, 15, 192–95; timber resources and, 203–4; tourism and, 49. *See also* peasants

Malaysia: Confrontation with Indonesia (1964), 196; East Malaysia, 12; estate development in, 14, 16–17; Federal Land Consolidation and Rehabilitation Authority (FELCRA), 78, 83; Federal Land Development Authority (FELDA), 16–17, 76–79, 83; in global economy, 25; government policy toward Bateks, 18; National Forest Policy (NFP), 81; National Land Council, 78; National Rural Development Council, 76; rice peasants viewed by state, 19; smallholders in, 12, 112

Malaysia, Peninsular, 12, 37, 41; biodiversity in, 84, 88; land development schemes, 77, 89; Malay settlers' expansion into, 52; oil palm introduced into, 74; plantation monoculture in, 63–65; population density in, 62; rice cultivation in, 15; rubber cultivation in, 63, 86, 94, 111

Malaysian states: Johor, 64, 72, 74, 85; Negeri Sembilan, 64, 67, 74; Pahang, 67, 74, 85; Perak, 64, 65, 66, 74; Selangor, 64, 65, 66; Straits Settlements, 40; Terengganu, 83. *See also* Sabah; Sarawak

Malay Sketches (Swettenham), 45

Manobo ethnic group (Philippines), 225
mapping, 18, 20, 183, 209–10; countermapping tactics, 208–9; of Danau Sentarum Wildlife Reserve (DSWR) boundaries, 198–99; ethnicity and, 200–201. *See also* boundaries; territory/territoriality

Marcos, Ferdinand (Philippine president), 223, 225

marga clans, 27, 129, 139–40, 143–44, 149 n. 25; *adat* as colonial product and, 144; Dutch colonial regime and, 131–33; leaders and followers, 12; Marga Ngaras, 137; Marga Penengahan-Laay, 138; ministerial decree and, 141–42; revival of *adat* community and, 149 n. 27; tenurial rules and, 131

Marga Foundation (Yayasan Penyimbang Marga), 139, 140, 143, 144

marriage, 245, 272, 273; *adat* community and, 125; household demography and, 239–40; incest taboos and, 246; intervillage, 272; social networks and, 186

mast fruiting, 96, 97, 113 n. 8

Mathews, Andrew, 6

matrilineality, 112

Maxwell, George, 50–51

Mendeling people, 65

meranti (*Shorea* spp.), 71

Merina people (Madagascar), 98, 106, 115 n. 19, 118 n. 43

Minangkabau ethnic group (West Sumatra), 65

Mindanao, 15, 21, 225–26. *See also* Philippines

mining, 44, 64, 199, 227

"miracle seeds," 276, 296, 298

missionaries, Christian, 10

modernity, 2, 4, 23, 165

modernization, 10, 21

Moniaga, Sandra, 128

monocultures, 25, 26, 63–65, 84, 136, 277

moral economy, 115 n. 20, 164

Mosse, David, 3, 6

Mualang ethnic group (West Kalimantan), 247

Muslims. *See* Islam and Muslims

myrrh (*Commiphora* spp.), 109

national parks: in Africa, 45; in United
States, 44, 49. *See also* Taman Negara
National Park (Malaysia)

Native Americans, 44, 213 n. 18

Native Chief, 157, 158

Native Municipality Ordinance of the
Outer Islands (Inlaandsche Gemeente
Ordonantie Buitengewesten [IGOB],
Dutch East Indies), 132–33

native reserves (colonial Malaya), 15, 28,
155–56; access to land within, 154;
colonialism and formation of, 159–60,
171, 176; colonial law in Sabah and, 174;
communal access to land in, 168–69;
local autonomy and, 154; private enclosure of land in, 167, 168, 174; property rights and land disputes in, 161–
62; protection of traditional lands
and, 156–58; usufruct rights in, 170;
villagers' use of resources in, 171–73

native title, 167, 170–71, 177 n. 4. *See also*
property rights; tenurial regimes; usufruct rights

naturalism, ideology of, 5

natural resource management, 6–7, 30,
126; *adat* and, 14, 15, 147; centralized
planning and, 18, 276; by central state,
93; community-based, 10, 220, 222;
competing local discourses of, 28;
complexity of the local and, 11–13;
customary law in Sabah and, 155;
emergence in particular localities, 24;
historical development of, 14, 15; Indonesian state and, 134, 141, 204; nature-
culture dichotomy and, 22; postcolonial regimes and, 14; postequilibrium
vision of, 16; resource rights and, 180,

209; rubber-swidden complementarity
and, 110

nature-culture relation, 10, 91; Cartesian
divide between, 8; community dynamics and, 11; creation-destruction
cycle and, 98, 107; Dayak headhunting
symbolism and, 118 n. 51; indigenous
peoples and, 235; morality of exchange
between, 102; people-forest relations
and, 44; postmodern view of, 4; science and, 21–24; simulated model of
swidden cultivation and, 274

Negritos. *See* Semang (Negritos)

neo-evolutionism, 43, 47

Netherlands East Indies. *See* Dutch colonial regime

New Economic Policy (Malaysia), 67

Ng, Francis, 62

Ngaju ethnic group (Central Kalimantan), 97, 107

NGOs (nongovernmental organizations),
5, 34 n. 17, 129, 146, 149 n. 21, 150 n. 30;
adat and, 27; in Mindanao, 231; in
West Java, 127, 135. *See also* LATIN;
WATALA

noble savage, myth of, 45, 46, 55, 123

Nor, Salleh Mohd, 62

North America, 6, 24

North Borneo, 152, 158–61, 177 n. 3. *See
also* Sabah

nutmeg (*Myristica fragrans*), 63

Obeyesekere, Gananath, 92

Odum, Eugene P., 3

oil palm, 63, 74–75, 77; biodiversity loss
and, 86, 136; Dayak opposition to
plantations, 105; global claims about
benefits of, 111; Iban elites and, 199–
200; land development and, 76; megaplantations of, 79–80, 82, 90 n. 10; pest
control and, 87, 89; pollution and, 83;
price of, *74*

Ooi Jim Bee, 90 n. 6

Orang Asli ("Sakai" people), 40, 45, 46, 55, 178 n. 10; colonial and Malay perceptions of, 54; conservation role of, 56; Jakun ethnic group, 43; in Taman Negara National Park, 46–47; tigers and, 51. *See also* aborigines; Bateks; Semang (Negritos)

orchards, 87, 156, 187; colonial taxation and, 160; rubber and, 66, 68; tenurial regimes and, 191, 203. *See also* fruit trees

Orientalism, 10, 108

'Other,' symbolic, 54, 55

Pan-Islamic Party (pmip, Malaysia), 81

peasants: environmentally friendly practices of, 87; indigenous peoples distinguished from, 219; Integrated Pest Management (ipm) and, 20; land development schemes and, 79; Malay nationalists and, 75–76; moral economy of, 164; postcolonial Malaysian state view of, 19; rice (paddy) cultivation and, 25, 66, 89; rubber cultivation and, 63, 65, 88. *See also* farmers; Malays; smallholders

Peet, Richard, 233

Peluso, Nancy L., 183, 213 n. 15, 217

Pelzer, Karl, 102, 103, 115 n. 25, 116 n. 32, 117 n. 35

Peminggir *adat* group (South Sumatra), 129, 148 n. 7

Penan communities (Sarawak), 7, 127

Peninsula. *See* Malaysia, Peninsular

people–tree metaphors, 106–10, 117 nn. 40–42, 118 n. 43

pepper, 62, 63, 64, 66, 94, 102; *damar* forest gardens and, 138; demands on land by, 102, 116 nn. 30–31; Iban and, 188; Kantu' cultivation of, 242; Krui community and, 130

Permanent Forest Estate of Peninsular Malaysia, 86

Perz, Stephen G., 274

pesaka (hereditary) lands, 66

Pesisir Krui community (South Sumatra), 129–30, 133, 137, 148 n. 7, 149 n. 19, 149 n. 27

pesticides, 83, 277, 279, 280, 283, 289–90; botanical, 286, 297; decline in use of, 288; extension system and, 295; macroecology and, 292–94; as "medicines," 277, 283, 286, 287, 289, 294; organic, 282; as "poisons," 287, 289; unintended consequences of, 281, 287. *See also* Integrated Pest Management

pests. *See* brown plant hopper; soybean seed borer; white rice borer

Peters, Charles, 190, 202

Philippines, 30, 217; colonial land policies in, 14; cultural marginalization of indigenous peoples in, 222–26; farmer-scientists in, 285. *See also* Mindanao

Pickering, A., 23

Pinus merkusii (pine resin), 109

plantations, 63, 64, 66, 199, 211; biodiversity loss and, 84–85, 86, 130; environmental impact of megaplantations, 79–83; exploitation of human labor on, 110; extinction of mammal species and, 85; soil degeneration and, 64. *See also* estates

poaching, 42

politics, 5, 10, 15, 210; of cash cropping and swidden cultivation, 102–4; of gender, 112; of identity, 218, 219, 236; of indigeneity, 217; Malay perceptions of aborigines and, 52–53; megaplantations and, 89; political ecology, 9, 16; of redistribution of power and resources, 13; of translation and representation, 222

pollution, water, 80, 82

pond-field production, 25

population pressure, 15, 31

poverty, 5, 76, 179 n. 28
primary forests, 11, 77, 253; biodiversity and, 86; cleared for swidden cultivation, 245, 257, 259–60, 266, 268; decrease in simulation model, 256–57, 268, 269, 273; distribution of land-holdings and, 271; labor input for swiddens in, 243, 244; in Mindanao, 231; rubber cultivation and, 84; swidden plots in, 248, 250
property rights, 151, 152, 153; access to land, ethic of, 165, 174–75, 190, 191; changing social relations and, 176; community-held, 162–71, 173–77; consolidation of property, 166; land laws in Sabah, 157; native reserves and, 154; organized by colonial state, 160; property as a thing, 180–81; resource rights, 180, 209; social contract and, 181; transition from community to individual rights, 170–71, 173. *See also* native reserves (colonial Malaya); native title; tenurial regimes; usufruct rights

Raffles, Thomas Stamford, 131–32
rainy season, 68, 179 n. 27, 184, 293, 295, 300 n.3, 301 n. 8
Rambo, A. Terry, 45–46
Rappaport, Roy A., 6
rats, 85–86, 87
rattan (*Calamus* spp.), 130, 186, 187, 190, 204
reciprocity, 96–97, 100, 102, 233; moneylending and, 168; "moral economy of peasants" and, 164; property relations and, 208; *s'basa* of T'boli, 228
Redford, Kent, 5
religion, 7, 8, 50, 96
reproduction, natural and social, 24–27
Restrepo, Eduardo, 6
rhino, Sumatran, 85
rice (paddy) cultivation, 114 n. 14, 186, 242, 300 n. 3; boom-and-bust cycle, 25;

Dewi Sri (rice goddess), 114 n. 16; government policy toward, 71; green revolution and, 13, 16, 276–77; industrialization and, 15; Integrated Pest Management (IPM) and, 279–80; irrigation of, 130, 136; in Kapuas Lakes region, 185; land area devoted to, 80; loss of local knowledge and, 32; by Malays in West Kalimantan, 187; people-plant linkage and, 117 n. 40; pesticides and, 83; rice debt in Kantu' households, 245, 247–48; rice prices, 269, 270; rice sufficiency, 261–63, 269, 270, 273; rubber and, 11–12, 65–67, 95, 248, 249–50; sugar cane and, 116 n. 33; swidden, 247, 250, 263; System of Rice Intensification, 286. *See also* white rice borer
Ridley, H. N., 56, 64, 86
Rivers, W. H. R., 107
Robbins, Paul, 3, 6
Rocheleau, Dianne, 5
Rose, Carol, 180, 213 n. 17
Rosero, Carlos, 220
Rouquette, J., 202
rubber cultivation, 26, 62, 77; biodiversity loss and, 86; cash sufficiency and, 264–65; clean weeding of, 63, 69, 115 n. 24, 116 n. 35; composite agricultural strategies and, 31; disease of, 67, 68, 70; as easy cash crop, 64–65; estate production, 13; in flexible agricultural strategy, 265–66; by Iban, 188; India rubber (*Ficus elastica*), 65, 89 n. 4; International Rubber Restriction Scheme (IRRA), 71; introduction of, into Peninsular Malaysia, 63; "jungle rubber," 105, 116 n. 35; by Kantu', 94–95; lack of complementarity with swidden cultivation, 104–6, 110; land development and, 76; land "killed" by, 26, 91–94; marginalization of smallholders and, 72–75; megaplantations of, 79–80; pollution from rubber fac-

tories, 83; prices of, 71, 74, *74*, 95, 268–70; as resource right, 209; rice paddy-rubber symbiosis, 65–67; smallholder rice farmers and, 11–12; social/ecological exchange in, 99–106; Stevenson Restriction Scheme, 67; synthetic, 72; tapping methods, 56, 64. *See also Hevea brasiliensis* (Para rubber tree)

Rubber Research Institute, Malaya (RRIM), 69, 70, 87

Rural Industrial Smallholdings Development Authority (RISDA, Malaysia), 76

Ruwiastuti, Maria, 128–29

Sabah (East Malaysia), 15, 59 n. 2, 152, 154, 170, 178 n. 15; *adat* community and, 161–62; colonial law in, 174; customary laws interpreted in, 153; Lands and Surveys Department, 157, 158, 167; Native Court, 169, 179 n. 25; native reserve in, 12, 14, 155–61; Ranau District, 153, *154*, 157–58, 159; resource ownership in, 151; village property regimes in, 28, 162–71, 173–77. *See also* North Borneo

Sack, Robert David, 183

sacred/profane dichotomy, 8

sacred trees, 8, 62. *See also* people–tree metaphors

Said, Edward, 10

Sakai. *See* Orang Asli ("Sakai" people)

Sarawak, 59 n. 2, 115 n. 19, 188, 199, 242; British rajahs of, 194; Iban farmers and rubber-price fluctuations in, 268; Penan people of, 127

Sather, Clifford, 115 n. 19

Schebesta, Paul, 52

Schneider, Jane, 113 n. 11, 114 n. 12

Schultes, R. E., 110

science, 11, 23, 54, 89, 124, 298; apparently neutral language of, 19; farmers' science, 284–86; folk/local and, 38; genealogical tree motif and, 107–8; planta-

tion agriculture and, 88; shift to non-equilbrium views, 17

Scott, James C., 115 n. 20

secondary forests, 11, 64, 68; abandoned swiddens and, 252; converted to oil palm, 85; fallow land conversion and, 256–57; labor input for swiddens in, 243–44; in Mindanao, 231; rubber cultivation and, 84; swidden plots in, 247, 250, 251

seladang (Malayan gaur, *Bos gaurus hubbackki*), 85

self-determination, right to, 220

Semang (Negritos), 43, 51–52, 53, 55, 61 n. 20

settlement, integrated, 77

Seventh Malaysia Plan, 88

sharecropping, 100

Sheridan, Michael, 9

silviculture, 71, 91, 134, 142, 145, 186. *See also* forests

Singapore, 63

Singapore Botanical Gardens, 64, 86

Sivaramakrishnan, K., 93

Skeat, Walter William, 45

SK KDTI (*Surat Keputusan Menteri Kawasan Dengan Istimewa*, Indonesia), 140

Slater, Candace, 92–93

smallholders, 12, 34 n. 10; assistance withheld from, 69; cash crops and, 89; chemical yield stimulants and, 87; Chinese, 64; estates in competition with, 67, 68–71, 72–75; global economy and, 93; Malay, 66; rubber cultivation by, 88, 95, 111; rubber-price fluctuations and, 268; state relations with, 18

socialization (*sosialisasi*), 210

social networks, 160, 167, 181, 186–91, 201, 212 n. 8

soils, 62, 63; clean weeding and, 69; degeneration of, 64; intercropping as soils (*cont.*)

protection for, 87; logging clearance and, 105; organic fertilizers and, 282; protection of, 117 n. 35; rainfall on cleared land, 80. *See also* erosion, soil

Soja, Edward W., 183

sorghum, 87

Southeast Asia, 8, 13, 91, 217; ethnic minorities, 30; indigenous peoples' identity in, 219; natural resource management, 110; nature and culture in, 10; rubber introduced into, 11; "sacred trees" in, 62; sustainable resource use in, 13; traditional cosmologies in, 9

soybean seed borer, 286, 290, 293

Spanish colonialism, 14, 223

Steinberg, Philip E., 5

Stephens, Donald, 158

Stevens, W. E., 52, 85

Stevenson Restriction Scheme, 67

Stockdale, Frank, 73

Strong, James, 229, 234

Subanon ethnic group (Philippines), 225

subsistence crops, 62–63, 93, 107

sugar cane, 64, 116 n. 33

Suharto (Indonesian president), 210–11, 214 n. 24

Sulistyawati, Endah, 11, 12, 14, 15, 30–31

Sumatra, 12, 14, 21, 90 n. 13, 113 n. 6, 131; rubber cultivation in, 94, 111; tobacco and swiddens in, 102, 103–4

Survival International, 220

sustainability, 16, 214 n. 32; exchange and, 93; global economy and, 26; indigenous peoples and, 216, 220–22; Integrated Pest Management (IPM) and, 298; moral ecology and, 115 n. 20; periodic disturbance and, 98; of swidden cycle, 11, 12; swidden–rubber system and, 111–12

swamps, 11, 64, 85, 184, 212 n. 4

sweet potato (*Ipomoea batatas*), 62

Swettenham, Frank, 45, 51

swidden cultivation, 11, 104, 130, 155; bureaucratic policies in Malaysia and, 47; burning and, 102, 104, 228, 244, 251; cash crops in complementarity with, 102–4; composite agricultural strategies and, 25, 31; computer simulation model of, 241, 244–53, 272–73; creation-destruction cycle and, 99; *damar* forest gardens and, 131; "dead land" and, 91; as flexible agricultural strategy, 265–68; land-labor dynamic in, 240, 273–74; land-use decision making and, 247–52; opponents and supporters of, 240–41; pattern of agricultural expansion, 254; population dynamics and, 245–47, 254, 255, 256–57; ritualized exchange and, 100; slashing and felling tasks, 102, 104, 228, 244, 251; sustainability of, 11, 12; by T'boli ethnic group, 228; vegetation classification and, 252

System of Rice Intensification (Indonesia), 286

Taman Negara National Park (Malaysia), 14, 25, 37, 85, 91; Bateks in, 19, 24, 46–48, 55–58, 59 n. 2, 61 n. 18; colonial legacy and, 39, 53; constructed boundaries and, 22; founding of, 39–44; irreconcilable ideas of the wild and, 50, 53–55; *Master Plan* of, 46–47, 49; tourism in, 48–49, 54; wilderness ideal and, 24, 44–46; wild–tame distinction and, 50–53

tanah kampong (village land), 161–71, 163, 173–77

tanah mati. See "dead land" (*tanah mati*)

tapioca, 63, 64

taro (*Colocasia esculenta*), 163–64

taungya system (Burma), 104, 116 n. 32

taxation, 140, 148 n. 15; by Dutch colonial regime, 193; of native lands in North Borneo, 159, 160; on rubber, 63, 72, 73

T'boli ethnic group (Philippines), 15, 226; ancestral domain of, 226–28, *227*, 228–31, *230*, 234; continuities and transformations in life of, 233–35; economic and social structure, 228; forest protection and, 231–32; human livelihood interventions and, 232–33; indigeneity and, 30

Team Krui, 135, 143, 144, 146–47, 149 n. 18, 150 n. 30; *adat* argument and, 145; Kalpataru award and, 138–40; ministerial decree and, 140–41

tenurial regimes, 17, 29, 105–6, 115 n. 21; *adat* community and, 143, 148 n. 6, 161–62; distribution of landholdings and, 271; Dutch colonialism in Sumatra and, 132; land rights in Indonesia, 128; location of swiddens and, 250; in Mindanao, 228; in North Borneo (Sabah), 152; *pesaka* (hereditary) lands, 66; restriction of boundaries and, 190; state lands, 78, 157. *See also* "dead land" (*tanah mati*); native reserves (colonial Malaya); property rights; *tanah kampong* (village land)

territory/territoriality, 20, 29, 181; of *adat* community, 148 n. 6; Batek, 47, 55, 59 n. 2, 61 n. 20; communal use and, 128; ethnoterritorial movements of indigenous peoples, 220; Kantu', 92, 97; local identity and, 183–84; local resource practices and, 201; Malay–Iban differences and, 188; *marga*, 132; resource conflict and, 182; struggles over terminology and, 217; village control of, 195–96, 198–99. *See also* boundaries; mapping

Thailand, 7, 118 n. 48, 183

Third Malaysia Plan (TMP), 81

Thompson, M., 3

tigers, 39, 51, 106

timber, 60 n. 14, 148 n. 8, 186, 187, 190; from *damar* forest gardens, 140; local resource practices and, 203–4. *See also* logging

tin mines, 66

tobacco, 89 n. 3, 102, 242

toman fish, 189, 205, 206, 213 n. 14

Torrens system (Philippines), 223

tourism, 48–49, 165

Tree of Jesse, 107

Tree of Life, 97, 107

Tsing, Anna L., 57

Tun Mustapha (Sabah), 158

Unfederated Malay States (UFMS), 73

United Malay National Organization (UNMNO), 81

United Nations (UN): charter, 221; Convention on Biological Diversity, 87; United Nations Development Program (UNDP), 83; Working Group on Indigenous Populations (UNWGIP), 220; World Commission on Environment and Development (WCED), 220, 221

United States: colonial land policies in the Philippines, 14, 223, 225; National Park Service, 44, 49; synthetic rubber production, 72

Upo (Mindanao), village of, 227, 228–31, *230*, 233–35; forest protection and, 231–32; human livelihood interventions in, 232–33

usufruct rights, 157, 162, 163, 170. *See also* property rights

Vandergeest, Peter, 183

van der Schot, Wim E. M., 48

Van Royen ethnological report, 131

Visayan ethnic group (Philippines), 225, 227, 229

Vollenhoven, Cornelis van, 126

Vosti, S. A., 239

Wadley, Reed L., 203

Walker, R., 274

Warburton, M., 3

"waste lands," 132, 133, 160

WATALA (Indonesian NGO), 135, 144, 149 n. 18, 150 n. 30

Watts, Michael, 217, 233

weeds, 103, 117 n. 35, 240

Wessing, Robert, 51

West, Paige, 6

Western world, 6, 9, 13; capitalist idea of property, 191; Christian missionaries from, 10; development schemes inspired by, 7; scientific knowledge of, 276; tree-based model of reality and, 108–9

West Java, 32, 284, 292, 300 n. 3; IPM villages in, 278, 279; NGOs in, 127, 135; pest insect outbreaks in, 283, 292. *See also* Java/Javanese

West Kalimantan, 11, 20, 30, 110, 241; Catholic Church in, 97; conflicts over resource claims in, 14; Kantu' dead land in, 26, 91; Kantu' territory, 92; land rights among Kantu', 128; resource-use paradigms in, 12. *See also* Danau Sentarum Wildlife Reserve (DSWR)

Wharton, C. H., 56

white rice borer (WRB), 283, 284, 287–88, 290, 292, 296, 297

Whitten, Max, 23

wild/wilderness, 19, 24, 56; American ideals of protection of, 44; colonial experience and, 50; as Euro-American concept, 37; irreconcilable visions of, 53–55; tame/domesticated in contrast with, 22, 50–53

wildlife, 38, 71; agriculture and, 40; in Danau Sentarum Wildlife Reserve (DSWR), 185; Euro-American perceptions of, 54; habitat loss and, 85; in Iban protected forest, 203; indigenous people compared to, 45; planters' perceptions of, 39

Wild Life Commission Report (Hubback), 39–43

Winarto, Yunita Triwardani, 12, 13, 14; on central planning discourse, 21; on indigenous environmental knowledge, 30–31, 32; on integrated pest management, 16; on local resource knowledge, 19–20; on nature-culture dichotomy, 22

Witcover, J., 239

Wolf, Eric R., 102, 103–4

Wollenberg, Eva, 210

Woodside, Alexander, 110

Woolley, G. C., 152

World Bank, 74, 78, 123, 147 n. 2, 221

World Park Congress (Durban), 124

World War, First, 67, 71

World War, Second, 68, 72, 214 n. 26

Zaire, 106

Zakaria, Yando, 128, 148 n. 6

Zerner, Charles, 57, 128

Michael R. Dove is the Margaret K. Musser Professor of Social
Ecology in the School of Forestry and Environmental Studies
at Yale University, where he is also the director of the Tropical
Resources Institute and a professor of anthropology.

Percy E. Sajise is an honorary research fellow at Bioversity
International, where he was formerly the regional director
for Asia, Pacific, and Oceania.

Amity A. Doolittle is a lecturer and associate research scientist
at the School of Forestry and Environmental Studies at Yale
University.

Library of Congress Cataloging-in-Publication Data
Beyond the sacred forest : complicating conservation in
Southeast Asia / edited by Michael R. Dove, Percy E. Sajise,
and Amity A. Doolittle.
p. cm. — (New ecologies for the twenty-first century)
Includes bibliographical references and index.
ISBN 978-0-8223-4781-1 (cloth)
ISBN 978-0-8223-4796-5 (pbk.)
1. Nature conservation—Southeast Asia. 2. Biodiversity—
Southeast Asia. I. Dove, Michael, 1949– II. Sajise, Percy E.,
1943– III. Doolittle, Amity Appell. IV. Title. V. Series: New
ecologies for the twenty-first century.
QH77.S644B496   2011
333.720959—dc22   2010038076

www.ingramcontent.com/pod-product-compliance
Lightning Source LLC
Chambersburg PA
CBHW070603290326
41929CB00061B/2578